AF736474

Springer Series in Biomaterials Science and Engineering

Volume 18

Series Editor

Min Wang, Department of Mechanical Engineering,
The University of Hong Kong, Pokfulam Road, Hong Kong

The Springer Series in Biomaterials Science and Engineering addresses the manufacture, structure and properties, and applications of materials that are in contact with biological systems, temporarily or permanently. It deals with many aspects of modern biomaterials, from basic science to clinical applications, as well as host responses. It covers the whole spectrum of biomaterials–polymers, metals, glasses and ceramics, and composites/hybrids–and includes both biological materials (collagen, polysaccharides, biological apatites, etc.) and synthetic materials. The materials can be in different forms: single crystals, polycrystalline materials, particles, fibers/wires, coatings, non-porous materials, porous scaffolds, etc. New and developing areas of biomaterials, such as nano-biomaterials and diagnostic and therapeutic nanodevices, are also focuses in this series. Advanced analytical techniques that are applicable in R&D and theoretical methods and analyses for biomaterials are also important topics. Frontiers in nanomedicine, regenerative medicine and other rapidly advancing areas calling for great explorations are highly relevant.

The Springer Series in Biomaterials Science and Engineering aims to provide critical reviews of important subjects in the field, publish new discoveries and significant progresses that have been made in both biomaterials development and the advancement of principles, theories and designs, and report cutting-edge research and relevant technologies. The individual volumes in the series are thematic. The goal of each volume is to give readers a comprehensive overview of an area where new knowledge has been gained and insights made. Significant topics in the area are dealt with in good depth and future directions are predicted on the basis of current developments. As a collection, the series provides authoritative works to a wide audience in academia, the research community, and industry.

This book series is indexed by the EI Compendex and Scopus databases.

If you are interested in publishing your book in the series, please contact Dr. Mengchu Huang (Email: mengchu.huang@springer.com).

More information about this series at https://link.springer.com/bookseries/10955

Andy H. Choi · Besim Ben-Nissan
Editors

Innovative Bioceramics in Translational Medicine II

Surgical Applications

Springer

Editors
Andy H. Choi
Faculty of Science
University of Technology Sydney
Ultimo, NSW, Australia

Besim Ben-Nissan
Faculty of Science
University of Technology Sydney
Ultimo, NSW, Australia

ISSN 2195-0644 ISSN 2195-0652 (electronic)
Springer Series in Biomaterials Science and Engineering
ISBN 978-981-16-7438-9 ISBN 978-981-16-7439-6 (eBook)
https://doi.org/10.1007/978-981-16-7439-6

© The Editor(s) (if applicable) and The Author(s), under exclusive license to Springer Nature Singapore Pte Ltd. 2022
This work is subject to copyright. All rights are solely and exclusively licensed by the Publisher, whether the whole or part of the material is concerned, specifically the rights of translation, reprinting, reuse of illustrations, recitation, broadcasting, reproduction on microfilms or in any other physical way, and transmission or information storage and retrieval, electronic adaptation, computer software, or by similar or dissimilar methodology now known or hereafter developed.
The use of general descriptive names, registered names, trademarks, service marks, etc. in this publication does not imply, even in the absence of a specific statement, that such names are exempt from the relevant protective laws and regulations and therefore free for general use.
The publisher, the authors and the editors are safe to assume that the advice and information in this book are believed to be true and accurate at the date of publication. Neither the publisher nor the authors or the editors give a warranty, expressed or implied, with respect to the material contained herein or for any errors or omissions that may have been made. The publisher remains neutral with regard to jurisdictional claims in published maps and institutional affiliations.

This Springer imprint is published by the registered company Springer Nature Singapore Pte Ltd.
The registered company address is: 152 Beach Road, #21-01/04 Gateway East, Singapore 189721, Singapore

Preface

In order to meet the demands of modern-day medical needs, it is of paramount importance that biomaterials and bioceramics transcend their current limitations of simply augmenting or replacing bodily components to a more innovative role where they can interact with cells and tissues. Although the presence of material requirements is common during the design and development of medical devices, relevant clinical prerequisites should also be incorporated so that appropriate prosthetics and implantable components are produced. This resulted in the creation of a highly interdisciplinary field known as translational medicine.

The definition of translational medicine generally agreed upon in the scientific community is that it is the constructive translation of new and innovative technique and information through advancements in basic research performed by interdisciplinary research teams into novel methodologies for preventing, diagnosing, and treating diseases for the benefit of patients and the public at large. Translational medicine has three main pillars: benchside (in the laboratory), bedside (clinical trials), and the community.

Bioceramics employed in medicine and surgery plays a crucial role in expanding the performance and function of medical devices. The science and technology of bioceramics is truly interdisciplinary, and consequently, improved or innovative bioceramics can only be achieved through advancements in physical and biological sciences, engineering, and medicine. There have been increasing demands on medical devices that they not only extend life but also improve its quality. Of even greater importance are the exciting and potential opportunities associated with the production of patient-matched ceramic components containing complex shapes with three-dimensional (3D) printing technology.

Gaining a deeper understanding into the correlations between material properties and biological performance will be useful in the design of innovative bioceramics and in addressing issues of implant failure and related infection. The challenge remains in providing safe and efficacious bioceramics with the required properties and an acceptable biocompatibility level. As the field of innovative biomaterials finds increasing applications in cellular and tissue engineering, it will continue to be used in new ways as part of the most innovative therapeutic strategies.

Divided into 2 volumes, the books comprise of 23 chapters written by top-notch international surgeons and experts in the fields of orthopedics, maxillofacial surgery, orthodontics, spinal surgery, and biomaterials. It is envisaged that each chapter will provide an in-depth examination of the latest research and clinical advances in hard tissue reconstruction and regenerations and in the treatment of bone diseases such as osteoporosis.

The first volume, ***Fundamental Research***, covers the basic principles and techniques used in the manufacture of bioceramics and biocomposites for various biomedical applications including drug delivery, implantable bionics and the development of the cardiac pacemaker, and bone tissue engineering. Furthermore, self-healing materials have been attracting increasing interest in both engineering and medical applications during the past two decades. Self-healing hydrogels are particularly interesting because of their ability to repair structural damages and recover their original functions, specifically in tissue engineering.

The current emphasis of tissue engineering has changed by seizing the advantage of combining the utilization of living cells with 3D scaffolds to transport vital cells and other biological materials such as stem cells and peptides to the damaged site of the patient with the intention of promoting tissue healing and regeneration. Clinical applications of bioactive composite scaffolds containing bioceramics and biodegradable polymers have attracted much attention during the past three decades. These composite grafts can also provide antibacterial properties when combined with therapeutic metal ions such as silver and copper. Similarly, functionalizing metallic surfaces and bioceramics with antimicrobial peptides would enable the creation of scaffolds and implants that can provide a mechanism against bacterial infection, while at the same time, stimulate bone formation.

The second volume, ***Surgical Applications***, covers the translation of innovative techniques and novel applications of bioceramics and bioceramics-based composite from the laboratory to a clinical environment in areas such as wound management following orthopedic surgical incisions and the application of bioresorbable bone fixation devices and ceramic–polymer biocomposite bone grafts for the repair of damaged tissues in dentistry and orthopedics. The advancement in personalized surgery and the manufacture of patient-specific 3D-printed bioceramic scaffolds for bone regeneration in craniomaxillofacial and spinal surgery are also thoroughly examined in this volume. Furthermore, the incorporation of biogenic materials such as bone morphogenetic proteins as well as regenerative pharmacologic agents like dipyridamole will allow for the development of a new generation of smart bioceramics-based scaffolds that promotes osteoconductivity and more importantly osteoinductivity.

It has been a well-established fact that bone undergoes a continuous process of remodeling or regeneration in which the activities of osteoclasts and osteoblasts are combined. Osteoporosis arises if this relationship became unbalanced and the quantity of bone resorbed exceeds the amount of new bone formed resulting in a reduction in bone strength and an increase in fracture risk. Reducing the fracture risk thus became the primary focus in the treatment of osteoporosis. Monoclonal antibodies have been applied in recent years in the treatment of osteoporosis. In this volume, we also intended to give our readers a fundamental insight into the basic properties, the technology used in their development, and their clinical application in the treatment of osteoporosis.

Finally, I would like to express my deepest gratitude to all my contributing authors from Australia, France, India, Italy, Japan, Portugal, South Korea, Tanzania, Turkey, the United Kingdom, the United States, and Vietnam for their time and valuable contributions to this informative book during the challenging time of the COVID-19 pandemic. I would also like to thank my great family for their support throughout this endeavor. Also, I would like to give very special thanks to my mentor and co-editor Prof. Besim Ben-Nissan for his friendship, support, and advice for over two decades. Finally, I would like to acknowledge the people at Springer Publishing, especially Mano Priya Saravanan, Ramesh Premnath, and Dharaneeswaran Sundaramurthy, and Prof. Min Wang for their help and for making these two books possible.

Sydney, Australia Andy H. Choi

Contents

Editors and Contributors

About the Editors

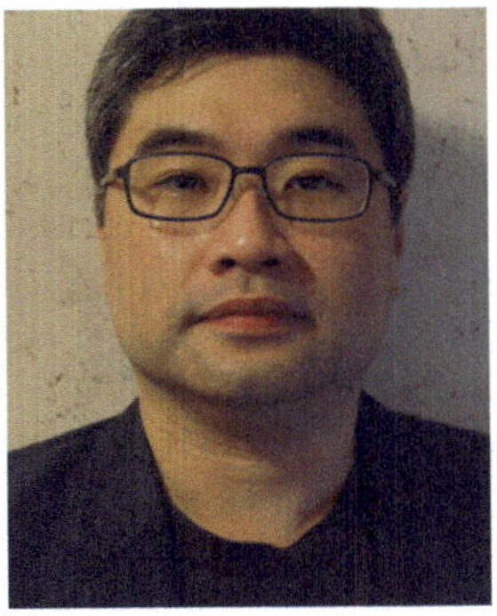

Dr. Andy H. Choi is an early career researcher who received his Ph.D. from the University of Technology Sydney (UTS) in Australia in 2004 on the use of computer modelling and simulation known as finite element analysis (FEA) to examine the biomechanical behavior of implants installed into a human mandible. After completing his Ph.D., he expanded his research focus from FEA to sol-gel synthesis of multifunctional calcium phosphate nano coatings and nano composite coatings for dental and biomedical applications.

In late 2010, Dr. Choi was successfully awarded the internationally competitive Endeavour Australia Cheung Kong Research Fellowship Award and undertook post-doctoral training at the Faculty of Dentistry of the University of Hong Kong focusing on the application of FEA in dentistry and the development of calcium phosphate nano-bioceramics.

He is currently serving as an associate editor for the Journal of the Australian Ceramic Society and as an editor for a number of dentistry-related journals. In addition, he is also serving as an editorial board member for several dentistry, nanotechnology, and orthopedics journals. To date, Dr. Choi has published over 50 publications including 5 books and 30 book chapters on calcium phosphate, nano-biomaterial coatings, sol-gel technology, marine structures, drug delivery, tissue engineering, and finite element analysis in nanomedicine and dentistry.

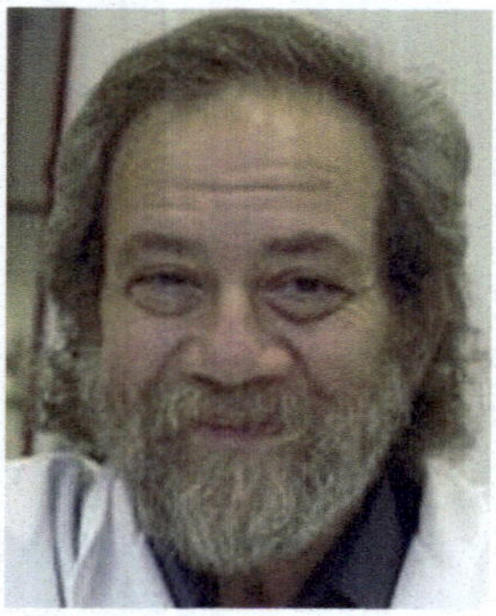

Prof. Besim Ben-Nissan has higher degrees in Metallurgical Engineering (ITU), Ceranic Engineering (University of New South Wales) and a Ph.D. in Mechanical and Industrial Engineering with Biomedical Engineering (University of New South Wales). Over the last four decades together with a large numbers of PhD students and post-doctoral fellows he has worked on production and analysis of various biomedical materials, implants, calcium phosphate ceramics, advanced ceramics (alumina, zirconia, silicon nitrides), sol-gel developed nanocoatings for enhanced bioactivity, corrosion and abrasion protections, optical and electronic ceramics and thermally insulating new generation composites.

He also has contributed to the areas of mechanical properties of sol-gel developed nanocoatings. In the biomedical field, he has involved with the development of materials for slow drug delivery, natural and marine material conversion, implant technology (bioactive materials including conversion of Australian corals to hydroxyapatite bone grafts), biomimetics (learning from nature and its application to regenerative medicine), bio-composites, investigative research on biomechanics and Finite Element Analysis (mandible, knee, hip joints, hip resurfacing), reliability and implant design (modular ceramic knee prosthesis, femoral head stresses). He was part of a research team which initiated the world's first reliable ceramic knee and hydroxyapatite sol gel derived nanocoatings. Since 1990 he has published over 260 papers in journals, five books and over 50 book chapters. He is one of the editors of the Journal of the Australian Ceramic Society and Editorial Board member of three international biomaterials journals. He was awarded "The Australasian Ceramic Society Award" for his contribution to "Ceramic education and research and development in Australia." He also received "Future Materials Award" for his contribution to the "Advanced nanocoated materials field".

He has collaborated with a number of international groups in Japan, USA, Thailand, Finland, Israel, France, UK, Germany and Turkey and held grants from the Australian Academy of Science and the Japan Society for Promotion of Science for collaborative work in the biomedical field in Europe, USA and Japan respectively. After serving as an academic for over 33 years he has retired, however still active and contributes to science by research in the biomedical field and supervising higher degree students.

Contributors

Howa Begam Centre for Healthcare Science and Technology, Indian Institute of Engineering Science and Technology, Shibpur, India

Besim Ben-Nissan Faculty of Science, School of Life Sciences, Biomaterials and Translational Medicine Group, University of Technology Sydney (UTS), Ultimo, Australia

Samuel H. Brill Tufts University, Medford, MA, USA

Giacomina Brunetti Department of Biosciences, Biotechnologies and Biopharmaceutics, University of Bari, Bari, Italy

Abhijit Chanda Department of Mechanical Engineering, Jadavpur University, Kolkata, India

Woojin Cho Department of Orthopaedic Surgery, Albert Einstein College of Medicine/Montefiore Medical Center, Bronx, NY, USA

Jee Ho Chong Tufts University, Medford, MA, USA

Paulo G. Coelho Department of Biomaterials, New York University College of Dentistry, New York, NY, USA;
Department of Mechanical and Aerospace Engineering, New York University Tandon School of Engineering, Brooklyn, NY, USA;
Hansjörg Wyss Department of Plastic Surgery, New York University Grossman School of Medicine, New York, NY, USA

Bruce N. Cronstein Department of Medicine, New York University Grossman School of Medicine, New York, NY, USA

Fatma Nur Depboylu Department of Bioengineering, Hacettepe University Institute of Science and Technology, Beytepe, Ankara, Turkey

Quang Ngoc Dong Department of Oral and Maxillofacial Surgery, Shimane University Faculty of Medicine, Izumo, Shimane, Japan;
National Hospital of Odonto-Stomatology, Hanoi, Vietnam

Sharbel Elhage Department of Plastic Surgery, Wake Forest University School of Medicine, Winston-Salem, NC, USA

David A. Essig Northwell Health at Long Island Jewish Medical Center, New Hyde Park, NY, USA;

Donald and Barbara Zucker School of Medicine at Hofstra/Northwell, Hempstead, NY, USA

Roberto L. Flores Hansjörg Wyss Department of Plastic Surgery, New York University Grossman School of Medicine, New York, NY, USA

Pietro Domenico Giorgi Orthopedics and Traumatology Unit, Emergency and Urgency Department, A.S.S.T. Grande Ospedale Metropolitano Niguarda, Milan, Italy

Maria Grano Department of Emergency and Organ Transplantation, Section of Human Anatomy and Histology, University of Bari, Bari, Italy

Kunio Ishikawa Faculty of Dental Science, Department of Biomaterials, Kyushu University, Fukuoka, Japan

Takahiro Kanno Department of Oral and Maxillofacial Surgery, Shimane University Faculty of Medicine, Izumo, Shimane, Japan

David H. Kim Department of Surgery, Division of Plastic and Reconstructive Surgery, Lewis Katz School of Medicine, Temple University, Philadelphia, PA, USA

Dongyoung Kim Rutgers New Jersey Medical School, Newark, NJ, USA

Feza Korkusuz Faculty of Medicine, Department of Sports Medicine, Hacettepe University, Sihhiye, Ankara, Turkey

Petek Korkusuz Faculty of Medicine, Department of Histology and Embryology, Hacettepe University, Sihhiye, Ankara, Turkey

Biswanath Kundu Bioceramic and Coating Division, CSIR-Central Glass & Ceramic Research Institute, Kolkata, India

Sanghyo Lee Department of Orthopaedic Surgery, Albert Einstein College of Medicine/Montefiore Medical Center, Bronx, NY, USA

Simona Legrenzi Orthopedics and Traumatology Unit, Emergency and Urgency Department, A.S.S.T. Grande Ospedale Metropolitano Niguarda, Milan, Italy

Sian Yik Lim Straub Clinic, Hawaii Pacific Health, Honolulu, HI, USA;
Pali Momi Medical Center, Hawaii Pacific Health, Honolulu, HI, USA

Innocent J. Macha Department of Mechanical and Industrial Engineering, University of Dar es Salaam, Dar es Salaam, Tanzania

Michito Maruta Department of Dental Engineering, Fukuoka Dental College, Fukuoka, Japan

James C. Melville Department of Oral & Maxillofacial Surgery, The University of Texas Health Science Center at Houston School of Dentistry, Houston, TX, USA

Matthew T. Morris Northwell Health at Long Island Jewish Medical Center, New Hyde Park, NY, USA

Prasenjit Mukherjee Department of Veterinary Clinical Complex, West Bengal University of Animal and Fishery Sciences, Mohanpur, Nadia, India

Samit Kumar Nandi Department of Veterinary Surgery and Radiology, West Bengal University of Animal and Fishery Sciences, Kolkata, India

Farhad B. Naini Kingston Hospital, Kingston upon Thames, UK;
St George's Hospital and Medical School, Tooting, London, UK

Vasudev Vivekanand Nayak Department of Biomaterials, New York University College of Dentistry, New York, NY, USA;
Department of Mechanical and Aerospace Engineering, New York University Tandon School of Engineering, Brooklyn, NY, USA

Francesco Puglia Orthopedics and Traumatology Unit, Emergency and Urgency Department, A.S.S.T. Grande Ospedale Metropolitano Niguarda, Milan, Italy;
University of Milan, Milan, Italy

Subhasis Roy Department of Veterinary Clinical Complex, West Bengal University of Animal and Fishery Sciences, Mohanpur, Nadia, India

Christopher M. Runyan Department of Plastic Surgery, Wake Forest University School of Medicine, Winston-Salem, NC, USA

Muhja Salah St George's Hospital, Tooting, London, UK

Giuseppe Rosario Schirò Orthopedics and Traumatology Unit, Emergency and Urgency Department, A.S.S.T. Grande Ospedale Metropolitano Niguarda, Milan, Italy

Lobat Tayebi Marquette University School of Dentistry, Milwaukee, WI, USA

Sara Todisco Department of Biosciences, Biotechnologies and Biopharmaceutics, University of Bari, Bari, Italy

Nick Tovar Department of Biomaterials, New York University College of Dentistry, New York, NY, USA;
Department of Oral and Maxillofacial Surgery, NYU Langone Medical Center and Bellevue Hospital Center, New York, NY, USA

Kanji Tsuru Department of Dental Engineering, Fukuoka Dental College, Fukuoka, Japan

Lukasz Witek Department of Biomaterials, New York University College of Dentistry, New York, NY, USA;
Department of Biomedical Engineering, New York University Tandon School of Engineering, Brooklyn, NY, USA

Evren Yasa Department of Mechanical Engineering, Eskişehir Osmangazi University, Eskisehir, Turkey

Simon Young Department of Oral & Maxillofacial Surgery, The University of Texas Health Science Center at Houston School of Dentistry, Houston, TX, USA

Chapter 1
Past and Future of Wound Dressing in Soft and Hard Tissue Surgery

Innocent J. Macha and Besim Ben-Nissan

Abstract Wound management following surgery is challenging because of high rates of wound infections due to the rise of antibiotic-resistant bacteria and increased risk of allergic reactions. On the other hand, several factors play significant roles in wound healing such as surgical techniques used, aging, oxygenation and pre-existing medical conditions such as diabetes. The application of wound dressings plays critical roles in wound healing and infectious prevention. The choice of proper wound dressing depends on the type of a wound. This chapter will comprehensively discuss advances in surgical wound dressings for soft and hard tissues with a focus on proper preparations techniques and characterizations. Different post-operative wound healing monitoring procedures will also be covered.

Keywords Wound healing · Wound dressing · Infection control · Dressing materials · Alginate · Hydrogel · Hydrocolloid

1.1 Introduction

Among the key elements of surgical wound care, appropriate dressings play critical roles in wound healing as well as prevention of potential risk of surgical site infection and associated complications such as wound dehiscence. It has been reported that about 27 million surgical procedures are performed in the US each year with up to 5% resulting in surgical site infection [1]. The burden of surgical site infection in EU/EEA is estimated at 543,149 cases annually [2]. In developing countries, few data are available on the healthcare associated infection and there is a need to improve surveillance and surgical site infection control practices [3].

I. J. Macha (✉)
Department of Mechanical and Industrial Engineering, University of Dar es Salaam, P.O BOX 35131, Dar es Salaam, Tanzania
e-mail: imacha@udsm.ac.tz

B. Ben-Nissan
Faculty of Science, School of Life Sciences, Biomaterials and Translational Medicine Group, University of Technology Sydney (UTS), Ultimo, Australia

© The Author(s), under exclusive license to Springer Nature Singapore Pte Ltd. 2022

A. H. Choi and B. Ben-Nissan (eds.), *Innovative Bioceramics in Translational Medicine II*, Springer Series in Biomaterials Science and Engineering 18,
https://doi.org/10.1007/978-981-16-7439-6_1

Dressing selection and dressing change protocol in surgical wounds is not always straightforward but necessitate taking into account patient's circumstances and personal preferences and education in psychosocial terms as well as wound's healing capacity without disturbing the healing process. It has been suggested that the ideal dressing does not cause trauma or disturb the wound bed, enable easy and pain-free removal and should create an undisturbed healing environment. "Undisturbed wound healing", a relatively new concept is gaining much attention in clinical set-up for acute and chronic wound management [4]. Undisturbed wound healing is connected to the choice of wound dressing materials which decrease frequency of dressing changes, provide moist wound healing and prevention of wound adherence. Principally, wound healing can be improved if the dressing remains in-situ especially during the initial stages of healing such as epithelialization while preventing peri-wound skin damage, bacterial infections and avoiding frequent dress removal which is associated with an alleviated pain and delay healing. Dressing change is advisable to base on clinical need rather routine.

This chapter will comprehensively discuss advances in surgical wound dressings for soft and hard tissues with a focus on proper preparations techniques and characterizations. Different post-operative wound healing monitoring procedures will also be covered.

1.2 Wound Healing

The American College of Surgeons and Surgical Infection Society suggest that surgical wounds heal follow one of the three mechanisms: primary, secondary or tertiary intention [5] with four phases namely hemostasis, inflammation, proliferation, and remodeling regardless of the aetiology of the wound. These four phases are separated but often occur concurrently. However, the mode of recovery of bone and soft tissues wounds differ. Primary intention healing refers to the healing or wound closure where by tissues are replaced in their original anatomic position similarly to their structure they had before injury and without any tissue loss. This healing is quick and if happens within eight hours after initial incision will prevent wound exposure to substantial contaminants and minimal scarring.

Secondary intention healing occurs in wounds that are not linear and may take longer time to heal. These are the wounds with high risk for dehiscence due to poor patient's conditions, exposed to substantial contamination, or subjected to extensive tension as in articulating joints [6]. Wounds that are not covered with epithelialized tissue due to intentional surgical procedures such as extraction sockets or accidental wounds where there is loss of tissues, fall in this category.

The term tertiary intention referred to delayed wound healing from either primary or secondary intention due to bacterial infections or excessive contamination. In that case, wounds would require debridement followed by monitoring to ensure tissue regeneration and wound closure. Also, traumatic injuries such as crush injuries which causes vascular damage and alteration in tissue perfusion undergo tertiary intention wound healing.

1.2.1 Phases of Acute Wound Healing

The four phases of wound healing mentioned above are continuous and overlapping phases. Naturally, the body will activate healing processes by providing various components of extracellular matrix (ECM) in response to different micro-environment conditions. These phases comprising hemostasis phase, inflammation phase, proliferation phase and remodeling phase and Fig. 1.1 shows different components of ECM at each healing phase. Each phase must occur in the appropriate time for normal wound healing but multitude of factors associated with patient's conditions may impair the wound healing in the post-operative period. It is therefore, important for surgeons and other health workers involved in wound management to have in-depth understanding of important factors affecting wound healing for successful wound management [8]. Aging has been reported to affect different phases of wound

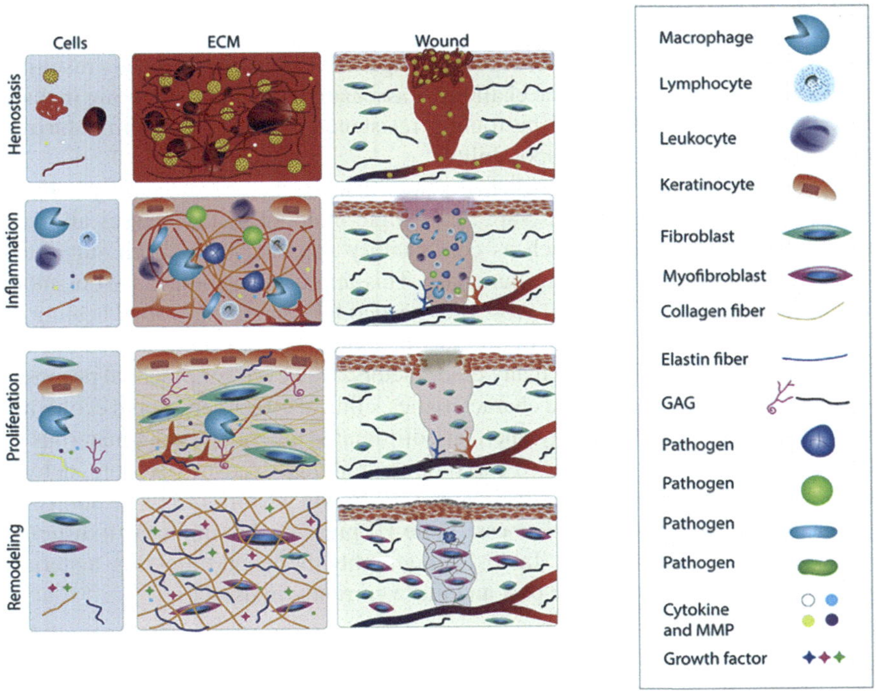

Fig. 1.1 Wound heling phases showing ECM and the main cell types (Adapted from [7] with permission)

healing and the course of chronic ulcer formation. Lifestyle and systemic diseases such as diabetes, cancer, kidney failure and high blood pressure are reported to significantly delay the healing process. Apart from patient's lifestyle which could be changed, controlling underlined diseases in support of wound healing could be challenging. In a healing process, as previously indicated, wound dressings play critical roles in wound healing and choosing the right type of dressings and application techniques for a particular wound need skilled and experienced personnel with a deep understanding of wound healing dynamics.

1.3 Wound Dressings

The continuing advancement of science and technology has changed the way wounds are being dressed. Since the first use of wound dressing developed from natural materials in 18th Century, significant modifications have been done in the last 50 years with the first modern wound dressing development in middle 1981 [9].

The modern wound dressings were more active with the ability to maintain a moist environment at the wound bed and absorbing exudate. Nowadays there are smart dressings with electronic sensors integrated to detect and give a wireless communication of the data in real time on the healing progress [10]. These smart dressings offer a major benefit of facilitation of monitoring a healing process remotely resulting in reduction of hospital cost. The challenges associated with smart dressing include biocompatibility, moisture and biofouling resistance, availability of miniaturized components, calibration and disposability.

The ideal wound dressing has characteristics discussed comprehensively in these reviews [11, 12]. However, there are no a single material that possesses all ideal criteria. Different strategies such as combinatory approach have been suggested in the design of wound dressing materials. This approach combines two or more different materials with different properties needed for supporting wound healing. It should be iterated that the choice of dressing materials depends on wound characteristics, patients' preferences and many other factors. However, for practical purposes only three factors such as wound characteristics, clinical effectiveness and economic factors are considered. Currently, there are different synthetic dressing materials in clinical use that can be categorized based on their ability to adsorb exudate, deliver drugs such as antibiotic, anti-inflammatory or growth factors [13, 14], ability to promote granulation and the activeness. Table 1.1 presents the summary of different wound dressing categories and Table 1.2 highlights the wound dressing materials that were previously used in the past and those currently in use.

Table 1.1 Wound dressing materials and their categories

Category/main function	Dressing materials	Suitability	Change frequency
Exudate absorbent	Hydrophobic polyurethane or silicone foam, alginate [15–17]	Moderate to highly exudating wounds	1–4 days
Delivery vehicle	Anticeptic (Silver, Iodine, Chlorhexidine, Polyhexamethyl-biguanide (PHMB), Honey, Acetic acid and Potassium permanganate) [18–20]	Prevent and treat wound infection	3 days
Promoting granulation	Hydrocolloid [21]	Granulated wounds with mild to moderate exudate	3–5 days
Maintain moist environment	Polyurethane films and semipermeable membrane, hydrogels (propylene glycol) [17, 22]	Superficial wounds, wounds with light exudate, wounds on elbows, heels or flat surface	1–3 times per week
Activeness	Poly (hydroxyethyl methacrylate) (pHEMA) and silicone [23]	Delivery and extraction of solutes via externally controlled convective mass transfer	2–3 days
Pressure	Compression bandages, stockings [21]	Venous laceration	3–6 months

1.4 Post-operative Wound Management

Proper management of post-operative wounds provides a supportive and protective environment for wound healing and surgical site infection respectively. Dressing being one of the major components of wound management that promote healing by providing a moist environment and protecting the wound from potential danger. In surgical wounds, the specific function of the dressing is to absorb blood or exudate fluid in the immediate postoperative phase that might lead to maceration of the wound. Each wound needs a specific dressing regime to accommodate wound characteristics and patient's preferences. Dressings applied on a surgical wound should ideally be left in place undisturbed unless to have the wound reviewed or when wound become stained by discharge or clinical signs of infection. It has been suggested that it is not necessary to dress a closed surgical wound after the removal of the initial. Typically, initial surgical dressings are to remain undisturbed for 48–72 h and some stay in place for up to seven days. However, some patients may prefer to have their wounds dressed.

Table 1.2 Past and present wound dressing materials, properties and usage

Materials	Properties and function	Usage
Animal-based (e.g.: collagen, jellyfish collagen, sponge)	Hydrophilic, can be porous, high absorption ability	Acute wounds, skin regeneration
Herbal origin (e.g.: cotton, bamboo viscose)	Cellulose, and/or viscose, double layered, and sorption ability	Burn wounds, chronic wounds and ulcerations
Synthetic origin (e.g.: Polyurethane, PU, PLA, N-isopropylacrylamide)	Inexpensive well distributed porous structures and high flexibility and strength	Easily produced thin films, they are suitable for relatively shallow wounds, and with low to moderate drainage, and high absorbance
Alginate	Highly absorbent, might need a secondary dressing	Infected and noninfected wounds, but not for dried ones, but can be made to aid blood coagulation
Hydrogel	Many different compositions, transparent, maintains a moist wound environment, facilitate autolytic debridement	For wounds with low to moderate exudate, burns, ulcers, surgical wounds, and skin repairs
Hydrocolloid	Ability to inhibit bacteria growth, appropriate for wounds with low to moderate drainage, encourage autolytic debridement, and excellent adhesion property	Traumatic injuries, leg ulcers, acute and chronic wounds, pressure sores, minor burns, inappropriate for infected wounds and diabetic foot ulceration
Medicated	Mainly for prevention of further infections, facilitate removal of necrotic tissues, and most importantly promotes tissue regeneration	Mainly for infected wounds

Most of the open surgical wounds that are not linear on involve surgical procedures such as extraction of sockets heal by secondary intention and they need appropriate dressings [24]. The risk of opting gauze-based dressing causes excessive pain on removal because of dress adherence on the wound and covered by the healing tissue. Studies have revealed that the use of hydrofibre dressing in surgical wounds tend to decrease pain during dress change, reduce length of stay, enhance healing and improve patient confidence. The recommended timeframe for staple or suture removal is one to two weeks unless for cosmetic concerns in the area such as face or eye they may require earlier removal to prevent scarring (three to five days). For wounds located at palms or soles normally take 2–3 weeks for dress change. Other wound closure supporting materials such as steri-strips or tissue/skin adhesives should be left in place until the fall off by themselves, usually five to ten days.

1.4.1 *Post-operative Wound Complications*

Surgical wounds suffer from two common complications such as surgical site infection and wound dehiscence. These normally occur when a healing process is stalled at the inflammatory healing phase. The severity of these complications ranges from mild cases needing wound care management and administration of antibiotics to serious cases with secondary surgical procedures and a high mortality rate [25]. The main contributing factors such as surgical technique and degree of contamination have been regarded as strong indicators for surgical site infection and wound dehiscence. Complication risk factors can be grouped based on patient, surgical procedures and post-operative factors. However, some scholars have refuted surgical technique contribution to pathogenesis of the complications due to mainly the advancement of surgical tools and appropriate procedures and training. Despite the advances in medical supplies and devices for surgery, postoperative wound complications still remain a serious challenge. Surgical site infection alone comprising of about 15% of all health care associated infections, reportedly with more than 500,000 yearly [26].

Preventive measures for wound complications begin before surgical procedures. This includes thorough assessment of the risk factors associated with the patient (pre-existing underlined medical conditions that may limit healing by preventing the delivery of oxygen and necessary nutrients to the healing tissues), intra-operative, and post-operation, using of good surgical techniques and meticulous hemostasis, proper technology choice to enhance healing and close monitoring strategies in the post-operative period. Pre- and intra-operative risk factors and mitigations discussion is beyond the scope of this chapter and only post-operative issues will be presented.

Different factors contributing to post-operative surgical site infection include saturated or leaking wound dressing, physical disruptions of staple, suture or glue that holds the tissue together before re-epithelialized which allow migration of bacterial to the wound. The Centre for Disease Control and Prevention's (CDC) guideline for prevention of surgical site infection describes the specific criteria required for an infection to be considered a surgical site infection [27]. It is indicated that surgical site infection can occur within 30 days postoperatively or a year later for an implanted device. An infection is considered surgical site infection if the tissues are affected and show histological signs including purulent drainage or abscess.

1.4.2 *Dressing Selection on Surgical Wounds*

Post-surgical incision care being an important part of wound healing, must consider all aspects of wound care in order to reduce the risk of infection and associated wound complications. Dressing selection, dressing protocol and dressing change time constitute to the ideal care of surgical wounds. It has been suggested that the concept of undisturbed wound healing should be given considerable attention in surgical wound care. While there are specific indicators which should trigger immediate dressing change such as potential dehiscence, excessive bleeding, dressing saturation and suspected local infection, a longer wear time should be considered more suitable for the healing. Apart from considering dressing and wound characteristics in wound dressing selection, different requirements for individual patients should also be considered. Table 1 shows suitability of different wound dressing materials based on the wound characteristics. The ideal dressing, on the other hand should possess specific properties required in managing post-surgical wounds [28]. Enough evidence should be needed to support the use of any specific type of dressing post-operatively for wounds healing by different mechanisms; primary, secondary or tertiary intentions. Acute surgical wounds sometimes are left open to heal by secondary intention but require moist environment in which the wound dressings should be able to keep moist environment and prevent bacteria from entering wound bed.

1.4.2.1 Dressing Characteristics

Ability to handle exudate: Exudate is produced as a normal part of a healing process to keep the wound bed wet. The wound fluid acts a medium for cells, nutrients and growth factors to migrate to the wound and help the healing process. For the normal wound healing, exudate production decreases over time but in chronic wounds exudate is produced over the prolonged period of time in higher amount than normal. Exudate contains high level of substance that are detrimental to cell-supporting extracellular matrix which also lead to wound pain, delayed healing, enlargement of the wound, skin maceration and local wound infection. Dressing that can minimize the effects of excessive exudate to the wound healing would be an important parameter to consider in choosing the dressing materials for surgical wounds.

Ability to stretch: This is a key important parameter for dressing materials for skin protection and patient comfort and mobility. Flexible and conforming dressing materials increases the levels of physical activity while recovery. With flexible material pain management is improved through wound contact layer and reduce the risk of blistering or irritation. The advancement in technology provides the possibility to tailor the properties of flexible materials for specialized wound management applications with an improved reliability and efficiency. Most of the dressing materials such as films, foams, adhesives, mesh and textile could be modified to have an improved flexibility and conforming traits.

Waterproof: Waterproof dressing offer protective barrier to the damaged tissues against contaminants. Patients wearing waterproof dressing can feel more secure and protected even in challenging environment. It increases the level of physical activities such as swimming without using an addition product on the wound. Most of waterproof dressing have transparent layers that can allow monitoring of wound without removing the dressing [29].

Transparency: Transparent dressing which are referred to as transparent films are impermeable to liquid, water and bacteria allowing one-way passage of moisture and gases. They are normally ultra-thin, flexible and waterproof, offering optimal protection for the wound. Transparency allows visualization of the wound bed make it easy to follow the wound healing progress without dressing removal. These dressings are suitable for the treatment of frictional wounds, closed surgical incision sites, skin graft and donor sites, catheter sites and areas of friction. They cannot be used for wounds with excessive exudate and usually changed three times a week [15].

1.4.2.2 Surgical Wounds

In this sub-section discussion is limited to selected common surgical wounds seen in clinical and community settings but we are aware that there are many different types of surgical wounds that may be encountered.

Total Joint Arthroplasty

It has been reported elsewhere that the ideal dressing for orthopedic surgical incisions should be absorbent, able to act as complete barrier, transparent, non-adherent, able keep the mound moist and require minimum changes. However, there is no single dressing material that encompasses all of the above parameters [15]. It envisaged that the combinatory approach in the development of dressing materials will result into a product with all important parameters. Hydrofiber and hydrocolloid dressings can handle excess exudate but keep the wound moist, have high absorptive capacity and permeability. They have low blistering rates and require minimum change thus reduce the risk of surgical site infection.

Caesarian (C) Section

C-section surgery is a surgical birthing method attributed with more risks of post-surgery complications such as premature rupture, wound infections, pelvic peritonitis than traditional vaginal births. Similar abdominal surgery to C-section include hysterectomy, appendectomy, hernia surgery and laparotomy. Proper dressing material should be able to handle exudate from the wound, provide optimal healing conditions, protect the surgical area and prevent stitches catching clothes. The skin edges normally seal within one to two days after the operation but varies from a person to another. After two days wound can usually be left open, however, some people prefer their wounds be covered.

Skin Graft and Donor Site

Skin graft is surgical procedures undertaken to remove a section of a skin of varied thickness from one part of the body (donor site) such as upper thigh and placed on a site of the injury (recipient site). Donor site wound care would require application of wound dressing capable of supporting moist environment [30]. Since the healing at donor site is by epithelialization, a transparent dressing or fine mesh gauze are preferred. Skin grafts are normally sutured, stapled or glued at the recipient site, treated like surgical wound and covered with an appropriate dressing. The choice of dressing remains with the surgeon responsible. Post-surgical risks are bacterial infection, graft contracture and hematoma. The healing takes up to six weeks and up to two years for scars to mature and inflammation to cease [31].

1.5 Future Directions

It has been becoming more clearer in the translational medicine that by incorporation of the therapeutic agents into the wound dressings, it is possible to repair the wounds more quickly and more efficiently. Among the therapeutic agents, antibiotics can be used to prevent wound infections, growth factors to revitalize damaged tissues and supplements, such as vitamins and minerals. In the past, plain gauze and thin paraffin-impregnated gauze were utilized as drug carriers. Nowadays, hydrocolloids, hydrogels, alginates, polylactic acid and polyurethane films/foams are the materials used to deliver therapeutic agents. One factor that limits the wound healing process is an infection and early detection.

Knowing the currently encountered problems, wound dressing researchers in addition to new material development are directing their efforts in next-generation wound dressings with the abilities of diagnosis during early stages, real-time monitoring, and on-demand therapy. Number of investigators, combined bioelectronics, and a smart flexible electronics-integrated wound dressing with a double-layer structure, the upper layer of which is polydimethylsiloxane-encapsulated flexible electronics integrated with a temperature sensor and ultraviolet (UV) light-emitting diodes, and the lower layer of which is a UV-responsive antibacterial hydrogel [32]. It was reported that this dressing is expected to provide early infection diagnosis via real-time wound-temperature monitoring by the integrated sensor and on-demand infection treatment by the release of antibiotics from the hydrogel by in situ UV irradiation. Their animal trials showed that this integrated system possesses a good flexibility, excellent compatibility, and high monitoring sensitivity and durability. It is envisaged that this technology can be extended to patches using current nanotechnology.

1.6 Conclusion Remarks

The need for proper wound dressing materials to appropriately manage surgical wounds is gradually increasing as a major aspect to prevent post-operative complications, a growing area of concern for patients and surgeons. This is an opportunity for the scientific community and biomedical industries to put more effort in development of advanced wound dressing materials with ideal characteristics. A single material, even with chemical and surface modifications cannot possess all necessary traits of an ideal dressing, to support healing of surgical wounds. With the advances in technology, we believe the combinatory approach in dressing development could be employed to address the problem. With the large number of wound dressing products available in clinical settings, the appropriate choice remains largely to the surgeons and health-care professionals. The best practice for surgical wound dressing choice recommended in this chapter when combined with the well evidence-based interventions will equip healthcare professionals with ability to make the needed decisions for the proper wound management.

References

1. Infectious Disease Advisor (2017) Surgical site infections.Hospital infection control. https://www.infectiousdiseaseadvisor.com/home/decision-support-in-medicine/hospital-infection-control/surgical-site-infections/. Accessed 16 May 2021
2. European Centre for Disease Prevention and Control (ECDC) (2021) Facts about surgical site infections. https://www.ecdc.europa.eu/en/surgical-site-infections/facts. Accessed 16 May 2021
3. Allegranzi B, Bagheri Nejad S, Combescure C et al (2011) Burden of endemic health-care-associated infection in developing countries: systematic review and meta-analysis. Lancet 377:228–241
4. Stephen-Haynes J (2015) The benefits of undisturbed healing using ALLEVYN Life™. Available via Wounds International. https://www.woundsinternational.com/resources/details/the-benefits-of-undisturbed-healing-using-allevyn-lifetm. Accessed 16 May 2021
5. Ban KA, Minei JP, Laronga C et al (2017) American college of surgeons and surgical infection society: surgical site infection guidelines, 2016 update. J Am Coll Surg 224:59–74
6. Young A, McNaught CE (2011) The physiology of wound healing. Surgery 29:475–479
7. Kordestani SS (2019) Wound healing process. Atlas of wound healing. Elsevier, Amsterdam, pp 11–22
8. Tahir M, Chaudhry EA, Zimri FK et al (2020) Negative pressure wound therapy versus conventional dressing for open fractures in lower extremity trauma. Bone Joint J 102-B:912–917
9. Shah JB (2011) The history of wound care. J Am Col Certif Wound Spec 3:65–66
10. O'Callaghan S, Galvin P, O'Mahony C et al (2020) "Smart" wound dressings for advanced wound care: a review. J Wound Care 29:394–406
11. Boateng JS, Matthews KH, Stevens HN et al (2008) Wound healing dressings and drug delivery systems: a review. J Pharm Sci 97:2892–2923
12. Borda LJ, Macquhae FE, Kirsner RS (2016) Wound dressings: a comprehensive review. Curr Derm Rep 5:287–297
13. Macha IJ, Sufi S (2020) Novel slow drug release bioceramic composite materials for wound dressing applications: potential of natural materials. SN Appl Sci. https://doi.org/10.1007/s42452-020-1977-z

14. Macha IJ, Muna MM, Magere JL (2018) In vitro study and characterization of cotton fabric PLA composite as a slow antibiotic delivery device for biomedical applications. J Drug Deliv Sci Technol 43:172–177
15. Criscitelli T (2012) Fast facts for wound care nursing: practical wound management in a nutshell. AORN J 96:563
16. Jones V, Grey JE, Harding KG (2006) Wound dressings. BMJ 332:777–780
17. Lionelli GT, Lawrence WT (2003) Wound dressings. Surg Clin North Am 83:617–638
18. Wounds International (2008) Wound infection in clinical practice: a WUWHS international consensus. https://www.woundsinternational.com/resources/details/wound-infection-clinical-practice-wuwhs-international-consensus. Accessed 16 May 2021
19. McDonnell G, Russell AD (1999). Antiseptics and disinfectants: activity, action, and resistance. Clin Microbiol Rev 12:147–179. Erratum in: Clin Microbiol Rev 2001 Jan; 14(1):227
20. Brennan SS, Leaper DJ (1985) The effect of antiseptics on the healing wound: a study using the rabbit ear chamber. Br J Surg 72:780–782
21. Dabiri G, Damstetter E, Phillips T (2016) Choosing a wound dressing based on common wound characteristics. Adv Wound Care 5:32–41
22. Eisenbud D, Hunter H, Kessler L et al (2003) Hydrogel wound dressings: where do we stand in 2003? Ostomy Wound Manage 49:52–57
23. Cabodi M, Cross VL, Qu Z et al (2007) An active wound dressing for controlled convective mass transfer with the wound bed. J Biomed Mater Res B Appl Biomater 82:210–222
24. Andersen BM (2019) Prevention of postoperative wound infections. Prevention and control of infections in hospitals. Springer, Cham, pp 377–437
25. Dhaigude BD, Shree S, Shah P et al (2018) Post-operative wound complications following emergency and elective abdominal surgeries. Int Surg J. https://doi.org/10.18203/2349-2902.isj20175901
26. National Center for Health Statistics (US) (2016) Health, United States, 2016: With chartbook on long-term trends in health. Report No.: 2017-1232
27. Centers for Disease Control and Prevention (2017) Guideline for prevention of surgical site infection. https://www.cdc.gov/infectioncontrol/guidelines/ssi/index.html. Accessed 16 May 2021
28. Bennett-Marsden M (2010) How to select a wound dressing. Pharm J 2:363–366
29. Wounds UK (2016) Medical adhesive-related skin injuries (MARSI): made easy. https://www.wounds-uk.com/resources/details/medical-adhesive-related-skin-injuries-marsi-made-easy. Accessed 16 May 2021
30. Chowdhry M, Chen AF (2015) Wound dressings for primary and revision total joint arthroplasty. Ann Transl Med 3:268
31. Beldon P (2007) What you need to know about skin grafts and donor site wounds. Wound Essentials 2:149–155
32. Pang Q, Lou D, Li S et al (2020) Smart flexible electronics-integrated wound dressing for real-time monitoring and on-demand treatment of infected wounds. Adv Sci 7:1902673

Innocent J. Macha has extensive experience in research and teaching for more than nine (9) years in the area of biomaterials synthesis and characterization, drug delivery devices, cell culture and bacteria biofilm. Dr. Macha is also one of the Associate Editors of the Journal of The Australian Ceramic Society and has published more than 26 articles and 10 book chapters.

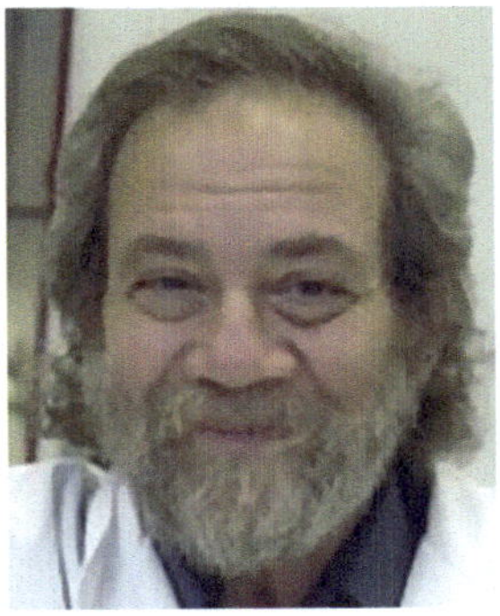

Besim Ben-Nissan has higher degrees in Metallurgical Engineering (ITU), Ceranic Engineering (University of New South Wales) and a Ph.D. in Mechanical and Industrial Engineering with Biomedical Engineering (University of New South Wales). Over the last four decades together with a large numbers of Ph.D. students and post-doctoral fellows he has worked on production and analysis of various biomedical materials, implants, calcium phosphate ceramics, advanced ceramics (alumina, zirconia, silicon nitrides), sol-gel developed nanocoatings for enhanced bioactivity, corrosion and abrasion protections, optical and electronic ceramics and thermally insulating new generation composites.

He also has contributed to the areas of mechanical properties of sol-gel developed nanocoatings. In the biomedical field, he has involved with the development of materials for slow drug delivery, natural and marine material conversion, implant technology (bioactive materials including conversion of Australian corals to hydroxyapatite bone grafts), biomimetics (learning from nature and its application to regenerative medicine), biocomposites, investigative research on biomechanics and Finite Element Analysis (mandible, knee, hip joints, hip resurfacing), reliability and implant design (modular ceramic knee prosthesis, femoral head stresses). He was part of a research team which initiated the world's first reliable ceramic knee and hydroxyapatite sol–gel derived nanocoatings. Since 1990 he has published over 260 papers in journals, five books and over 50 book chapters. He is one of the editors of the Journal of the Australian Ceramic Society and Editorial Board member of three international biomaterials journals. He was awarded "The Australasian Ceramic Society Award" for his contribution to "Ceramic education and research and development in Australia." He also received "Future Materials Award" for his contribution to the "Advanced nanocoated materials field".

He has collaborated with a number of international groups in Japan, USA, Thailand, Finland, Israel, France, UK, Germany and Turkey and held grants from the Australian Academy of Science and the Japan Society for Promotion of Science for collaborative work in the biomedical field in Europe, USA and

Japan respectively. After serving as an academic for over 33 years he has retired, however still active and contributes to science by research in the biomedical field and supervising higher degree students.

Chapter 2
3D Printing and Bioprinting of Biomaterials and Bioceramic Scaffolds: Clinical Outcomes and Implications in Bone Tissue Engineering and Maxillofacial Reconstructive Surgery

Muhja Salah, Farhad B. Naini, and Lobat Tayebi

Abstract Three-dimensional (3D) printing is a type of additive manufacturing that works by the application of material inks layer by layer using data from computer-aided design (CAD) to help to place the ink in a predefined place, thus producing a highly accurate product even with complex geometry. The goal in using 3D bioprinting is to develop a biological scaffold that resembles the desired tissue to be replaced, including the cells and the growth factors, in a specific spatial relationship. The developments in bone tissue engineering (BTE) and 3D bioprinting are revolutionizing osseous craniofacial reconstructive surgery. This chapter aims to describe 3D bioprinting of biomaterial and bioceramic scaffolds for bone tissue engineering and maxillofacial reconstructive surgery.

Keywords Additive manufacturing · Layer by layer · Bioprinting · Biological scaffold · Bioactive glass · Calcium phosphate · Hydroxyapatite · Mesenchymal stem cells · Induced pluripotent stem cells · Exosome · Biomimetics · Self-assembly

---

M. Salah (✉)
St George's Hospital, Tooting, London, UK

F. B. Naini
Kingston Hospital, Kingston upon Thames, UK

St George's Hospital and Medical School, Tooting, London, UK

L. Tayebi
Marquette University School of Dentistry, Milwaukee, WI, USA

© The Author(s), under exclusive license to Springer Nature Singapore Pte Ltd. 2022

A. H. Choi and B. Ben-Nissan (eds.), *Innovative Bioceramics in Translational Medicine II*, Springer Series in Biomaterials Science and Engineering 18,
https://doi.org/10.1007/978-981-16-7439-6_2

2.1 Introduction

Three-dimensional (3D) printing is a type of additive manufacturing that was first invented in 1984 for engineering and industrial purposes. It works by the application of material inks layer by layer using data from computer-aided design (CAD) to help to place the ink in a predefined place, thus producing a highly accurate product even with complex geometry [1, 2]. The technology found its way to the health sector through dentistry when additive manufacturing was used to print a solid block of dental implants, crowns, and bridges from a biocompatible and bioinert material that does not elicit an immune reaction [3].

Scientists were overly ambitious realizing the precision of the end-product when 3D printing was used. They decided to unleash the power of 3D printing and use it for medicinal purposes to bioprint tissues. The first bioprinting attempt was undertaken early in 1988, using an inkjet printer depositing cell drops on-demand approach. The goal in using 3D bioprinting is to develop a biological scaffold that resembles the desired tissue to be replaced, including the cells and the growth factors, in a specific spatial relationship. It is a customizable, patient-specific solution meeting the patient's need at a macro level (i.e., shape and size), and on a micro level resembling patients' tissue structure and architecture [4, 5]. The development in bone tissue engineering and 3D bioprinting also aims to solve the crisis in the shortage in organs needed for transplantation [6].

Tissue loss in the craniofacial region can occur due to a craniofacial genetic deformity, trauma, or surgical excision as a treatment of tissue malignancy [7]. Facial disfigurement has a severe negative impact on individuals, both socially and psychologically, and requires rapid, precise, and aesthetic rebuilding producing a functional, harmonious, and symmetrical face [8]. Osseous craniofacial reconstruction traditionally employs a graft harvested from the iliac crest or the ribs, which serve as the bridge needed to direct the 3D bone growth (osseoconduction), as well as inducing the differentiation and the recruiting of osteoblasts (osseoinduction) into the injured area to promote bone healing [9]. However, placing a graft is not without risk; autogenous bone grafting carries the risk of morbidity (pain in the donor site, neuralgia, blood loss, or infection), while the allogenic bone graft is associated with the possibility of transmitting infection or eliciting an immune reaction [10]. Moreover, facial reconstruction using a bone graft does not always provide aesthetic results due to the anatomical complexity of bone, soft tissue, and the hollow cavities in the face. 3D bioprinting, on the other hand, may provide a more precise alternative that fits the defects, reducing the need to count on the surgeon's ability to harvest or carve the graft to fit the surface.

2.2 Bone Tissue Engineering

Bone tissue engineering has received much attention in the last few decades, and it showed tremendous progress due to the improved understanding of bone biology, along with the advances in the biomaterials. It focuses on:

(a) Developing biomaterials that can provide the same physical and biological properties as natural bone [11].
(b) Producing scaffolds from these biomaterials, having the same architecture and topography that ensure nutrient and oxygen passage, micro-vessels, and nerve ingrowth, as well as regulating the stem cell differentiation down the osteogenic fate [12, 13].
(c) Incorporating mesenchymal stem cells (MSC) that are directed toward differentiating into osteogenic cell lineage [14].
(d) Incorporating bone growth factors; bone morphogenic proteins (BMP), insulin-like growth factor-2 (IGF-1), vascular endothelial growth factors (VEGF), and others that enhance osteogenesis [15].

2.3 Biomaterials in Bone Tissue Engineering

Bone is composed of 60–70% inorganic phase in the form of hydroxyapatite ($Ca_{10}(PO_4)_6(OH)_2$), while the organic phase is mostly formed of collagen type I with some other proteins and growth factors. The simplicity of the natural bone composition enabled the progress in bone tissue engineering. Biomaterials used to fabricate scaffolds should be biocompatible, biodegradable to be replaced by the newly generated bone, and bioactive to enhance bone regeneration, having physical strength and mechanical properties, which enable it to support the load the natural bone is supporting [16]. Examples of biomaterials used in bone tissue engineering include demineralized bone matrix as well as a number of bioceramics and bioglasses.

2.3.1 Demineralized Bone Matrix

These are allografts treated with chemical acid to demineralize as well as removing the inorganic component of the graft, leaving the matrix proteins, mainly collagen I and bone growth factor BMP, and are then treated with radiation to decrease the possibility of eliciting an immune reaction [17]. Demineralized bone matrix has been used for decades in clinical applications, and has shown tremendous success being osteoconductive and osseoinductive, but because the end-product is in a powder form, making it is difficult to handle during surgery, which consequently has limited its use [18]. Solutions implemented to ease the manipulation of the powder were based on using the powder mixed with a viscous carrier to enable it to condense and pack into bony defects [17].

Wagner et al. reported using demineralized bone matrix for mandibular reconstruction by wrapping it in an acellular dermal matrix to confine the demineralized bone matrix paste and placing it over a bent plate [19]. The patients were followed up for five years and showed evidence of bone healing. In a recent study, Driscoll et al. used demineralized bone matrix mixed with hydroxyapatite crystals in different ratios in a 3D printer to print scaffolds for spinal repair, and it was tested in rat models [20]. The preclinical studies showed successful fusion, with the developed biomaterial being a hybrid encompassing the osseoinductive properties of the demineralized bone matrix carrying the bone growth factor along with the osteoconductive properties of the hydroxyapatite.

2.3.2 Bioceramics and Bioglasses

These are inorganic oxides including hydroxyapatite, calcium phosphate, tricalcium phosphate (TCP), and calcium silicate. They are considered bioactive as they bond to bone and elicit osteogenesis [21].

2.3.2.1 Hydroxyapatite

This makes the bulk of the bone composition, thus, it has been studied extensively as a bone substitute because its composition resembles natural bone. Both calcium and phosphate ions present in hydroxyapatite promote bone regeneration. Calcium ions stimulate osteoblasts by activating ERK1/2 pathways, which protect them from apoptosis, as well as having a central role in bone maturation by deposition in immature bone [22]. Phosphate ions activate the IGF-1 pathway in osteoblasts, which is implicated in cell survival, growth, and protein synthesis [23]. Besides, it is osseoinductive and osseoconductive, making it ideal to be a synthetic bone substitute. But, it is organized in a highly arranged crystalline microstructure that hinders it from degrading, and it also inherits a low compressive and tensile strength making it brittle when loaded [24]. To reduce the brittleness of hydroxyapatite, Mukherjee et al. investigated the effect of adding carbon nanotubes (CNT) to hydroxyapatite and found that it increased the fracture toughness of the scaffold. They tested the scaffold on animal models and found that the addition of CNT was biologically safe with no toxicity shown in either the liver or kidney, but with enhanced bone regeneration on the implanted site [25]. However, the data was stated to be preliminary and incomplete to proceed onto clinical studies.

In relations to the 3D printing of hydroxyapatite, Seitz et al. were able to use hydroxyapatite powder sprayed with a polymeric binder dissolved in water to ensure ink flow to produce a porous scaffold with fully interconnected channels sixteen years ago, which are further compacted after printing in a 1250 °C furnace to remove and achieve binder pyrolysis [26]. Six years ago, Shao et al. proposed the use of 3D-gel printing (3DGP) instead of regular 3D printing with the use of hydroxyethylmethacrylate (HEMA) gelation system to produce a flowable slurry. Their new system has the advantage of being appropriate to a wide range of materials from metal to ceramics, keeping the cost low while achieving high printing efficiency, producing complex shapes due to the flow of the slurry. They used it with stainless steel, zirconia, and hydroxyapatite, and tested the fabricated scaffolds and their mechanical properties [27–29]. Most studies on the fabrication of hydroxyapatite scaffolds were carried out using biomaterials only, without embedding cells within the scaffold and the seeding of cells occurred after fabrication, thus it should not be confused with bioinks, which incorporate both biomaterials and cells [30].

2.3.2.2 Tricalcium Phosphate

In 1920, Albee was the first to report that rhombohedral β-form, β-tricalcium phosphate (β-TCP), enhances osteogenesis [31]. Tricalcium phosphate is composed of calcium and phosphate ions just like hydroxyapatite, which renders it to have the same effect on osteoblasts resulting in bone regeneration. Gao et al. showed the in vivo osteogenic potential of the tricalcium phosphate granules placed in a titanium porous scaffold and implanted in a femur defect on animal models [32].

In contrast to hydroxyapatite, tricalcium phosphate has a crystalline structure that is not highly organized, which makes it more susceptible to resorption and degradation, which is an ideal property for a scaffold material [33]. Ishikawa et al. compared the mechanical properties and recorded the histological findings of the newly generated bone when tricalcium phosphate and hydroxyapatite were used [34]. The study confirmed tricalcium phosphate has higher solubility than hydroxyapatite, which explains why more bone is found around the implanted tricalcium phosphate than the hydroxyapatite.

Degradation of the tricalcium phosphate is a desired property when constructing a scaffold, but degradation of the material should be coordinated with the speed of the osteogenesis process. To adjust tricalcium phosphate degradation, it has been doped with mineral oxides like magnesium oxide (MgO) and strontium oxide (SrO), which affect the crystalline orientation of the tricalcium phosphate and make it less soluble and alter both the mechanical and biological properties of the tricalcium phosphate [35]. Banerjee et al. confirmed slower degradation of the implanted MgO/SrO-doped β-TCP than pure β-TCP on animal models [36]. They also showed that the doped implant had more cell attachment, which increases cell differentiation and proliferation. Analysis of osteocalcin and type I collagen inside the implants indicated faster osteogenesis and remodeling. Recently, Gu et al. used Mg-doped tricalcium phosphate and 3D-printed an interconnected-pores scaffold with mechanical properties close to bone [37]. They further seeded the scaffold with MSCs derived from bone marrow and umbilical cord and showed that both osteogenesis and angiogenesis were enhanced. In an animal model, Kim et al. transplanted 3D-printed scaffolds from a composite of tricalcium phosphate and polycaprolactone polymer and used it to repair the maxilla in a dog after resecting a tumor with success [38].

In 1986, another strategy was proposed to adjust the solubility of tricalcium phosphate in physiological conditions by combining hydroxyapatite with tricalcium phosphate in different ratios to achieve the best physical and mechanical properties for the desired load-bearing application referred to as biphasic calcium phosphate [39]. Increasing the hydroxyapatite content in the biphasic calcium phosphate leads to a more stable material, while increasing the tricalcium phosphate results in a material that is more soluble, thus, it can be easily tailored. Liu et al. 3D-printed scaffolds using biphasic calcium phosphate and examined the in vivo behavior using rabbit calvarial defects which showed an increase in osteogenesis and high bone density [40].

2.3.2.3 Calcium Phosphate Cement

This cement was accidentally invented in the 1980s by the American Dental Association Health Foundation Paffenbarger Research Centre (ADAHF-PRC) who were trying to develop a cement to treat and remineralize early dental caries. A mixture of tetra calcium phosphate, dicalcium phosphate anhydrous, and dicalcium phosphate dihydrate with water was found to rapidly produce hydroxyapatite. A decade after that, the FDA approved calcium phosphate cement for clinical use, and since then a tremendous number of studies have been conducted [41]. The cement was found to promote osteogenesis, being osteoconductive, and most importantly it is injectable, which makes it easier for clinical use. Injecting the material into the site of surgery will allow it to mould into the shape of the deformity without the need for further drilling at the surgical site to match the size and shape of the scaffold. Yu et al. reported the success of calcium phosphate cement in bone regeneration when they performed an in vivo study in which injectable calcium phosphate cement was implanted into a femoral condyle defects of rabbits [42]. Lin et al. cultured three types of cells: induced pluripotent stem cells (iPSC), human umbilical vein endothelial cells (HUVECs), and pericytes; into scaffolds made from calcium phosphate cement and implanted them into cranial defects created on rats [43]. They found that the tri-culture group had elevated angiogenic and osteogenic markers, and mineralization.

2.3.2.4 Bioactive Glass

These are silicate-based ceramics composed of silicon dioxide, calcium oxide, phosphorus oxide, potassium oxide, magnesium oxide, and boric oxide. The composition and percentage of these oxides vary, but the key component, silicate, always constitutes 45–52% of its weight [44]. Bioglasses possess the capability to form a strong chemical bond with the bone tissue that is created through the polycondensation of a silicone-rich layer on the surface of the bioactive glass due to ion exchange between ions in the physiological fluid and leaching of ions from the surface of the bioglass [44]. Moreover, the electronegative silicone-rich layer on the surface is considered osseoinductive as it adsorbs protein that in turn attracts macrophages and MSCs [44].

As a bone substitute material, bioactive glass proved its worthiness in an in vivo animal study carried out by Moimas et al. where bioactive glass implanted into tibial defects were found to be completely resorbed in six months and be replaced by bone tissue [45]. It was shown that the composition of the bioactive glass affects the union chemical reaction and stimulation of cells to promote osteogenesis. Bioactive glass went through optimization of its formula, and 45S4 was invented composed of 45% SiO_2, 24.5% Na_2O, 24.5% CaO, 6% P_2O_5 (wt%), characterized by a high amount of Na_2O and CaO, which make the surface of the material very bioactive [46]. Scaffolds made from 45S5 were found by Detsch et al. to drive umbilical cord-derived MSCs down the osteogenic differentiation pathway [47].

Recently, the development of the sol–gel method, adding ammonia to the sol phase to transform it into a gel and then freeze-dry it, produced 58S bioactive glass composing of 60 mol.% SiO_2, 36 mol.% CaO and 4 mol.% P_2O_5. 58S bioactive glass has the benefit of achieving a homogeneous biomaterial compared to the melting method used originally where phosphate becomes volatile at high temperature [48]. Wheeler et al. compared in vivo bone regeneration capacity between 45S4 and 58S scaffolds after implantation within critical-sized distal femoral cancellous bone defects in a rabbit model and the results showed that the 58S degraded much quicker but was able to form bone earlier than 45S4 at 4 weeks, which is normalized at 12 weeks [49].

The 3D printing of scaffolds composed of bioactive glass have been investigated in a number of in vitro and in vivo studies El-Rashidy et al. comprehensively reviewed the in vivo studies undertaken on the regeneration of bone with 3D-printed bioactive glass scaffolds [50]. Recently, Kolan et al. compared the osteogenic potential of bioactive glass scaffolds made by 3D printing with and without the use of BMP after implantation into cranial defects in rats [51]. Their study concluded that the addition of BMP to the scaffold greatly enhanced bone regeneration.

2.4 Cells in Bone Tissue Engineering

3D bioprinting includes both biomaterial and cells in the bioink to fabricate a scaffold. Ideal biomaterials for bone substitutes should stimulate the seeded stem cells to differentiate into osteoblasts responsible for the bone regeneration. Gao et al. proposed different molecular mechanisms by which biomaterials interact with stem cells to promote osteogenesis [14]. The exact process is not known, but they postulated that phosphorus, magnesium, and strontium ions released from the biomaterial activate the BMP pathway and increase the concentration of the calcitonin gene-related peptide. The following section describes the various types of cells used in bone tissue engineering.

2.4.1 Mesenchymal Stem Cells (MSCs)

These are used for their pluripotency, ability to differentiate into osteoblasts, and immune modulative effect [52]. They can also be derived from a variety of sources ranging from bone marrow, umbilical cord, placenta, dental pulp, adipose tissue, and other sources [52]. Injecting MSCs derived from adipose tissue along mandibular fracture lines were found to enhance osseointegration and bone quality, as well as promoted bone healing as observed in the study by Castillo-Cardiel et al. [53]. In 2016, a study by Chamieh et al. discovered that implanting collagen scaffolds seeded with dental pulp-derived MSCs into calvarial defects in rats resulted in accelerated bone regeneration compared to rats having a collagen scaffold with no seeded cells, demonstrated by evaluating the variations in bone density and through histological examination [54]. Fahimipour et al. reported in a recent article the utilization of the bioprinter to bioprint collagen matrix to mimic the extracellular matrix of natural bone with the MSC and BMP [55]. The matrix has the benefit of confining BMP as well as preventing it from escaping the scaffold, which is known to cause ectopic bone formation or osteomas [56]. The 3D printing was used again to 3D print a scaffold that represents the mineralized part of the bone, which is then used to support the MSC-BMP collagen matrix. It was found that using this method enhanced MSCs seeding, and proliferation while the availability of BMP enhanced the osteogenic potential of the MSCs [55]. A recent report by Dong et al. showed that the presence of osteoclasts is crucial for bone regeneration as well as osteoblasts [57]. In their study, a proteomic analysis was performed, and mass spectrometry was used to identify proteins secreted in extracellular matrix. The analysis showed the presence of more than 608 protein presents, among which two proteins are known to be secreted by pre-osteoclasts, CXCL12 and IGFBP5 proteins, both are responsible for MSC cells' migration and osteogenic differentiation, respectively [57]. They confirmed their hypothesis by implanting scaffolds made from decalcified bone matrix seeded with co-cultured MSCs and pre-osteoclasts into femur defects in rats showing significant enhancements in bone regeneration compared to implanting scaffolds seeded with MSCs only.

2.4.2 Induced Pluripotent Stem Cells (iPSCs)

These are another exciting source of cells that can differentiate into any cell type, mimicking embryonic stem cells. However, with the ethical dilemma that has risen by extracting embryonic stem cells, which results in the destruction of human embryos, motivated scientists to look for other sources of cells that have the same pluripotency [58]. To circumvent this issue, iPSCs were produced by Takahashi et al. in 2007 by transducing four factors: Oct3/4, Sox2, Klf4, and c-Myc, present in embryonic stem cells, in fibroblast turning them into cells mimicking pluripotency [59]. Xie et al. investigated the osteogenic differentiation of iPSCs seeded on a scaffold made from a composite of hydroxyapatite-chitosan-collagen and found the proliferation of iPSCs into osteoblasts and an increase in bone protein secretion [60]. Moreover, they implanted these scaffolds into cranial defects of animal models, and compared the density of bone with the scaffold seeded with iPSC and other seeded with MSCs and found that the iPSCs scaffold has nearly double the bone density than when MSCs were used alone.

The osteogenic differentiation of iPSCs was studied by a number of research groups [61]. A study by Kao et al. discovered that resveratrol has a supporting effect on the osteogenic differentiation of iPSCs [62]. Later, a study by Ji et al. examined the osteogenic differentiation of human iPSCs regulated by nano-hydroxyapatite/chitosan/gelatine 3D scaffolds with nano-hydroxyapatite in different ratios [63]. Investigation was also carried out to reprogram iPSCs to functional osteoblasts using only the small molecule exogenous adenosine [64]. However, iPSCs still carry the potential of tumorigenicity and teratoma formation, which still limits its use clinically, and further investigation should be conducted to optimize its use and safety [65].

2.4.3 Exosome

Recently, increasing interest was diverted into cell-free therapies after the discovery that MSCs cause tissue regeneration due to its paracrine effect. This approach carries the benefit of avoiding tumorigenicity, resistance to apoptosis, triggering an immune response, and genetic instability, which are all present in MSCs utilization [66]. It will also permit the repeated injections or administration of the therapy without the fear of accumulation of cells in non-targeted tissue, especially the lungs [67].

The cell-free approach uses the exosomes, which are membrane-bound vesicles, produced by endosomes in the cell containing a specific cargo either: micro-RNA, messenger-RNA, proteins, or other biomolecules, and get excreted outside the cell to be communicated into another cell [68]. Exosomes are produced by most cell types as a way of communication and crosstalk between cells. Exosomes from MSCs regulate the paracrine effect that enhances the regeneration of tissues [69]. Several studies have been conducted and showed the potential of using exosomes for bone regeneration. Lu et al. extracted exosomes from adipose-derived MSCs and used a TNF-α pre-conditioned medium, which was found to positively promote osteogenesis and bone repair [70]. Zhao et al. proposed that exosomes extracted from bone marrow-derived MSCs and co-cultured with osteoblasts, result in the activation of the MAPK pathway on the osteoblasts, which is important for the cell cycle and growth, and results in their proliferation, thus promoting bone regeneration [71].

More importantly, Diomede et al. demonstrated the ability of an implanted 3D-printed scaffold to heal calvarial defect in rats that is composed of a polymer poly-lactic acid (PLA), seeded with exosomes and gingiva-derived MSCs [72]. Furthermore, Zhang et al. also worked extensively on exosomes and in one of their study, they showed that a scaffold made with tricalcium phosphate combined with exosomes derived from iPSCs healed calvarial defects on rats via activating the PI3K/Akt signalling pathway [73]. In a later study, they used exosomes derived from umbilical cord-derived MSCs combined in hydrogel and transplanted at the femoral fracture site in the animal model. They found that implanted exosomes promoted angiogenesis, which in turn enhanced fracture healing [74]. Although the results of these studies are promising, still, a consensus on exosome extraction and purification has not been achieved which is important in translational medicine.

2.5 3D Bioprinting Approaches

In the process of bioprinting, deposition of both the biomaterial and the cells occur simultaneously. 3D bioprinting is achieved by one or a combination of the following strategies.

2.5.1 Biomimicry

This is a straightforward approach using the bioprinter to replicate the original architecture of the tissue, thereby providing the right environmental factors that guide cells to differentiate into the right type of cells. This approach of bioprinting is extremely dependent on the material ink used to construct the scaffold. A scaffold is the parallel of the extracellular matrix, that should be able to provide the chemical and physiological cues important for cell viability, differentiation, and expansion [75]. Scaffold biomaterials should be biocompatible, permeable to nutrients, having adequate stiffness to withstand loading and deformation while at the same time, able to undergo degradation at the same pace that allows the growth of new bone tissue and eventually replaces the scaffold [76]. All these requirements are crucial in choosing the most ideal scaffold bioink and they are also the primary factors that determines the success of the printed scaffold. After bioprinting, a bioreactor is used to regulate environmental factors such as the oxygen, temperature, nutrient diffusion, and the gravitational force needed for cell infiltration to the depth of the printed scaffold [77].

2.5.2 Self-assembly Approach

This is a scaffold-free approach that eliminates the need for scaffold biomaterials and mitigating the difficulties faced using the scaffold. The approach adopts the same embryological development process which utilizes interaction and signals between adjacent cells and their extracellular matrix to self-organize into the tissue intended for engineering [78]. High-density initial cell seeding ensures cell–cell interaction, resulting in cell producing their own extracellular matrix and forming cell aggregates in the form of spheres or sheets, and carries the advantage of efficiency to produce tissues faster than using scaffold bioink. Various methods are used to form these cell aggregates from magnetic levitation, hanging drop, hydrogel microwell, and others, each with its pros and cons [79]. Spheroids and sheets are then used in a 3D printer to form the engineered tissue. The advantage of using this approach is the ability to use different types of cells as well as regulating their ratios. This allowed for the co-culturing endothelial cells with MSCs, which promotes angiogenesis in the final construct, while the MSCs differentiate into the desired cell type [80]. Yamasaki et al. created a scaffold-free construct from adipose tissue-derived MSCs by using the needle array 3D printing method and implanted them into femoral defects of pigs which showed enhanced osteochondral regeneration [81]. Recently, Heo et al. described a method to 3D print spheroid aggregates, made from human umbilical vein endothelial cells (HUVECs) and MSCs and called it the aspiration-assisted bioprinting (AAB) technique, in which they showed that it allowed for better and more precise positioning of the spheroids to produce scaffold-free bone tissue [82].

2.5.3 3D Bioprinting in Bone Tissue Engineering and Craniofacial Reconstruction

3D bioprinting is offering an exciting future for bone tissue engineering and craniofacial reconstruction, but the technology is still in its early stages. Few studies were carried out or are currently in progress that shows promising results. In 2014, Goh et al. implanted a polycaprolactone scaffold fabricated by 3D printing in sockets of newly extracted teeth to preserve the height of maxillary and mandibular ridges [83]. In the same year, a Chinese team, led by Zhang who worked extensively in BTE, published the results of their clinical trial on 23 female patients reconstructing the mandibular angle after ostectomy [84]. They demonstrated that using 3D bioprinting titanium scaffolds, led to greater bone regeneration, shorter operation time, and better aesthetic results. In 2015, Sumida et al. published the results of their clinical trial of implanting 3D printed scaffolds for maxillary and mandibular ridges in 13 patients without randomization and reported favorable outcomes [85]. 3D printing is also used by neurosurgeons for the correction of calvarial defects after resecting brain tumours. Kilstrom et al. reported in 2019 the results of using 3D printing to fabricate calcium phosphate-titanium reinforced scaffolds implanted on the skull of 52 patients with the intention to promote bone regeneration and osteointegration [86].

A search in the clinical trial government website (www.clinicaltrial.gov) in March 2021 revealed the presence of 342 clinical trials with different statuses, when searching MSCs and bone regeneration, of which 6 trials are concerned with bone regeneration in the craniomaxillofacial region, listed in Table 2.1.

At the same time, only 4 studies are concerned with using 3D printing for the correction of bone defects in the craniomaxillofacial region, listed in Table 2.2.

Table 2.1 Applications of MSCs and bone regeneration in the craniomaxillofacial region

Title	Status	Study results	Conditions	Interventions	Locations
Efficacy in alveolar bone regeneration with autologous MSCs and biomaterial in comparison to autologous bone grafting (NCT04297813)	Recruiting	No results available	Alveolar bone atrophy	• Combination product: advanced medicinal therapy (MSC combined with biomaterial) • Procedure: autologous bone graft	• Syddansk Universitet SDU (University Hospital of Southern Denmark), Odense, Denmark • Assistance Publique—Hôpitaux De Paris, Créteil, France • CHU Nantes, Centre de Soins Dentaires, Nantes, France • University of Bergen, Institute of Clinical Dentistry, Bergen, Hordaland, Norway • Universidad Complutense De Madrid, Madrid, Calle Fernando De Castro Rodriguez, Spain • Universitat Internacional De Catalunya, Barcelona, Spain
Bone tissue engineering with dental pulp stem cells for alveolar cleft repair (NCT03766217)	Completed	No results available	Cleft lip and palate	• Combination product: Mesenchymal stem cells associated with biomaterials • Combination Product: Iliac crest autogenous bone graft	Hospital Sírio-Libanes, São Paulo, Brazil
Reconstruction of jaw bone using mesenchymal stem cells (NCT02751125)	Completed	No results available	Bone atrophy	Drug: BCP with autologous mesenchymal stem cells (MSC)	Institute of Clinical Dentistry, University of Bergen, Bergen, Hordaland, Norway

(continued)

Table 2.1 (continued)

Title	Status	Study results	Conditions	Interventions	Locations
Autologous alveolar bone marrow mesenchymal stem cells for the reconstruction of infrabony periodontal defects (NCT02449005)	Completed	No results available	Chronic periodontitis	• Biological: BM-MSCs/fibrin glue/collagen fleece • Other: fibrin glue/collagen fleece • Procedure: open flap debridement	Dental School, Aristotle University, Thessaloniki, Greece
Use of mesenchymal stem cells for alveolar bone tissue engineering for Cleft Lip and Palate Patients (NCT01932164)	Completed	Has results	Cleft lip and palate	• Procedure: maxillary alveolar graft by tissue engineering • Procedure: Bone tissue engineering using mesenchymal stem cells	Hospital Sírio Libanês, São Paulo, Brazil
Treatment Of maxillary bone cysts with autologous bone mesenchymal stem cells (MSV-H) (NCT01389661)	Completed	No results available	Maxillary cyst Bone loss of substance	• Biological: MSV treatment	• Río Hortega University Hospital, Valladolid, Valladid, Spain • Bionand, Parque Tecnológico de Andalucía, Universidad de Málaga, Malaga, Spain • Instituto de Biologia y Genetica Molecular, Valladolid, Spain

Table 2.2 Applications of 3D printing for the correction of bone defects in craniomaxillofacial region

Title	Status	Study results	Conditions	Interventions	Locations
Efficiency of 3D-printed implant versus autograft for orbital reconstruction (TOR-3D) (NCT03608280)	Not yet recruiting	No results available	Significant bone defect in the orbit	• Procedure: bone autograft • Procedure: orbital reconstruction by 3D-printed porous titanium implant	
Craniofacial applications of 3D printing (NCT03292679)	Unknown status	No results available	Facial fracture	Procedure: 3D template	
Three-dimensional printing of patient-specific titanium Plates in Jaw Surgery: A Pilot Study (3DJP16) (NCT03057223)	Recruiting	No results available	Mandibular neoplasms Maxillary neoplasms Dentofacial deformities Maxillofacial injuries	Device: 3D-printed patient-specific titanium plates	The Prince Philip Dental Hospital, Hong Kong, Hong Kong
Personalized titanium Plates vs CAD/CAM surgical splints in maxillary repositioning of orthognathic Surgery (NCT02914431)	Completed	No results available	Malocclusion abnormalities, jaw	Device: 3D printing Personalized Titanium Plate	Ninth People's Hospital, Shanghai JiaoTong University School of Medicine, Shanghai, Shanghai, China

2.6 Concluding Remarks

Developing bone tissue engineering is important, as the need for bone implants increases due to increasing population, increasing facial injuries, orthognathic surgeries, tumors, and craniofacial deformities. Translation of this technology would be the only solution to treat large defects and non-union fractures and when technology is combined with 3D printing, it allows potentially more aesthetic facial reconstruction and reduced surgery time. However, the technology needs further investigation to optimize the biomaterial to ensure both optimal osteogenesis and angiogenesis to enable vascularization of the scaffolds. Biomaterials used should also provide the mechanical properties needed for the implanted site, as bone engineered to be implanted in a load-bearing bone should be different from scaffolds created for non-load bearing bone. Enhancing 3D printing technology enables it to provide scaffolds exactly mimicking the natural bone with the highest resolution. Also, the ease and availability of the biomaterial, 3D printer, and expertise in hospital settings should be discussed to allow its translation directly to patients.

References

1. Khorram Niaki M, Nonino F (2017) Additive manufacturing management: a review and future research agenda. Int J Prod Res 55:1419–1439
2. Wohlers T, Gornet T (2014) History of additive manufacturing. http://www.wohlersassociates.com/history2016.pdf. Accessed 12 Apr 2021
3. Javaid M, Haleem A (2019) Current status and applications of additive manufacturing in dentistry: a literature-based review. J Oral Biol Craniofac Res 9:179–185
4. Murphy SV, De Coppi P, Atala A (2020) Opportunities and challenges of translational 3D bioprinting. Nat Biomed Eng 4:370–380
5. Mandrycky C, Wang Z, Kim K et al (2016) 3D bioprinting for engineering complex tissues. Biotechnol Adv 34:422–434
6. Abouna GM (2008) Organ shortage crisis: problems and possible solutions. Transplant Proc 40:34–38
7. Zhang W, Yelick PC (2018) Craniofacial tissue engineering. Cold Spring Harb Perspect Med 8:a025775
8. Singh VP, Moss TP (2015) Psychological impact of visible differences in patients with congenital craniofacial anomalies. Prog Orthod 16:5
9. Elsalanty ME, Genecov DG (2009) Bone grafts in craniofacial surgery. Craniomaxillofac Trauma Reconstr 2:125–134
10. Sohn HS, Oh JK (2019) Review of bone graft and bone substitutes with an emphasis on fracture surgeries. Biomater Res 23:9
11. Wang W, Yeung KWK (2017) Bone grafts and biomaterials substitutes for bone defect repair: a review. Bioact Mater 2:224–247
12. Cun X, Hosta-Rigau L (2020) Topography: a biophysical approach to direct the fate of mesenchymal stem cells in tissue engineering applications. Nanomaterials 10:2070
13. Chen X, Fan H, Deng X et al (2018) Scaffold structural microenvironmental cues to guide tissue regeneration in bone tissue applications. Nanomaterials 8:960
14. Gao C, Peng S, Feng P et al (2017) Bone biomaterials and interactions with stem cells. Bone Res 5:17059
15. De Witte TM, Fratila-Apachitei LE, Zadpoor AA et al (2018) Bone tissue engineering via growth factor delivery: from scaffolds to complex matrices. Regen Biomater 5:197–211
16. Hutmacher DW (2000) Scaffolds in tissue engineering bone and cartilage. Biomaterials 21:2529–2543
17. Zhang H, Yang L, Yang XG et al (2019) Demineralized bone matrix carriers and their clinical applications: an overview. Orthop Surg 11:725–737
18. Gruskin E, Doll BA, Futrell FW et al (2012) Demineralized bone matrix in bone repair: history and use. Adv Drug Deliv Rev 64:1063–1077
19. Wagner JD, Wagner JP, Wagner SA (2017) Mandibular defect reconstruction utilizing preoperative 3D modeling, plate pre-bending and a novel triad graft: a case report. Stomatological Dis Sci 1:97–102
20. Driscoll JA, Lubbe R, Jakus AE et al (2020) 3D-printed ceramic-demineralized bone matrix hyperelastic bone composite scaffolds for spinal fusion. Tissue Eng Part A 26:157–166
21. Ribas RG, Schatkoski VM, do Amaral Montanheiro TL et al (2019) Current advances in bone tissue engineering concerning ceramic and bioglass scaffolds: a review. Ceram Int 45:21051–21061

22. Danciu TE, Adam RM, Naruse K et al (2003) Calcium regulates the PI3K-Akt pathway in stretched osteoblasts. FEBS Lett 536:193–197
23. Guntur AR, Rosen CJ (2013) IGF-1 regulation of key signaling pathways in bone. Bonekey Rep 2:437
24. Mondal S, Pal U (2019) 3D hydroxyapatite scaffold for bone regeneration and local drug delivery applications. J Drug Deliv Sci Technol 53:101131
25. Mukherjee S, Nandi SK, Kundu B et al (2016) Enhanced bone regeneration with carbon nanotube reinforced hydroxyapatite in animal model. J Mech Behav Biomed Mater 60:243–255
26. Seitz H, Rieder W, Irsen S et al (2005) Three-dimensional printing of porous ceramic scaffolds for bone tissue engineering. J Biomed Mater Res B Appl Biomater 74:782–788
27. Shao H, He J, Lin T et al (2019) 3D gel-printing of hydroxyapatite scaffold for bone tissue engineering. Ceram Int 45:1163–1170
28. Shao H, Zhao D, Lin T et al (2017) 3D gel-printing of zirconia ceramic parts. Ceram Int 43:13938–13942
29. Ren X, Shao H, Lin T et al (2016) 3D gel-printing—an additive manufacturing method for producing complex shape parts. Mater Des 101:80–87
30. Groll J, Burdick JA, Cho DW et al (2018) A definition of bioinks and their distinction from biomaterial inks. Biofabrication 11:013001
31. Albee FH (1920) Studies in bone growth: triple calcium phosphate as a stimulus to osteogenesis. Ann Surg 71:32–39
32. Gao P, Zhang H, Liu Y et al (2016) Beta-tricalcium phosphate granules improve osteogenesis in vitro and establish innovative osteo-regenerators for bone tissue engineering in vivo. Sci Rep 6:23367
33. Bohner M, Santoni BLG, Döbelin N (2020) β-tricalcium phosphate for bone substitution: synthesis and properties. Acta Biomater 113:23–41
34. Ishikawa K, Miyamoto Y, Tsuchiya A et al (2018) Physical and histological comparison of hydroxyapatite, carbonate apatite, and β-tricalcium phosphate bone substitutes. Materials 11:1993
35. Gallo M, Le Gars SB, Douillard T et al (2019) Effect of grain orientation and magnesium doping on β-tricalcium phosphate resorption behavior. Acta Biomater 89:391–402
36. Banerjee SS, Tarafder S, Davies NM et al (2010) Understanding the influence of MgO and SrO binary doping on the mechanical and biological properties of beta-TCP ceramics. Acta Biomater 6:4167–4174
37. Gu Y, Zhang J, Zhang X et al (2019) Three-dimensional printed Mg-doped β-TCP bone tissue engineering scaffolds: effects of magnesium ion concentration on osteogenesis and angiogenesis in vitro. Tissue Eng Regen Med 16:415–429
38. Kim SE, Shim KM, Jang K et al (2018) Three-dimensional printing-based reconstruction of a maxillary bone defect in a dog following tumor removal. In Vivo 32:63–70
39. Daculsi G (1998) Biphasic calcium phosphate concept applied to artificial bone, implant coating and injectable bone substitute. Biomaterials 19:1473–1478
40. Liu F, Liu Y, Li X et al (2019) Osteogenesis of 3D printed macro-pore size biphasic calcium phosphate scaffold in rabbit calvaria. J Biomater Appl 33:1168–1177
41. Chow LC, Takagi S (2001) A natural bone cement-a laboratory novelty led to the development of revolutionary new biomaterials. J Res Natl Inst Stand Technol 106:1029–1033
42. Yu L, Li Y, Zhao K et al (2013) A novel injectable calcium phosphate cement-bioactive glass composite for bone regeneration. PLoS One 8:e62570
43. Zhang C, Hu K, Liu X et al (2017) Novel hiPSC-based tri-culture for pre-vascularization of calcium phosphate scaffold to enhance bone and vessel formation. Mater Sci Eng C Mater Biol Appl 79:296–304
44. Välimäki VV, Aro HT (2006) Molecular basis for action of bioactive glasses as bone graft substitute. Scand J Surg 95:95–102
45. Moimas L, Biasotto M, Di Lenarda R et al (2006) Rabbit pilot study on the resorbability of three-dimensional bioactive glass fibre scaffolds. Acta Biomater 2:191–199

46. Fiume E, Barberi J, Verné E et al (2018) Bioactive glasses: from parent 45S5 composition to scaffold-assisted tissue-healing therapies. J Funct Biomater 9:24
47. Detsch R, Alles S, Hum J et al (2015) Osteogenic differentiation of umbilical cord and adipose derived stem cells onto highly porous 45S5 Bioglass®-based scaffolds. J Biomed Mater Res A 103:1029–1037
48. Vuong BX, Hiep DT (2019) Bioactive glass 58S prepared using an innovation sol-gel process. Process Appl Ceram 13:98–103
49. Wheeler DL, Eschbach EJ, Hoellrich RG et al (2000) Assessment of resorbable bioactive material for grafting of critical-size cancellous defects. J Orthop Res 18:140–148
50. El-Rashidy AA, Roether JA, Harhaus L et al (2017) Regenerating bone with bioactive glass scaffolds: a review of in vivo studies in bone defect models. Acta Biomater 62:1–28
51. Kolan KCR, Huang YW, Semon JA et al (2020) 3D-printed biomimetic bioactive glass scaffolds for bone regeneration in rat calvarial defects. Int J Bioprint 6:274
52. Liu C, Zhang H, Tang X et al (2018) Mesenchymal stem cells promote the osteogenesis in collagen-induced arthritic mice through the inhibition of TNF-α. Stem Cells Int 2018:4069032
53. Castillo-Cardiel G, López-Echaury AC, Saucedo-Ortiz JA et al (2017) Bone regeneration in mandibular fractures after the application of autologous mesenchymal stem cells, a randomized clinical trial. Dent Traumatol 33:38–44
54. Chamieh F, Collignon AM, Coyac BR et al (2016) Accelerated craniofacial bone regeneration through dense collagen gel scaffolds seeded with dental pulp stem cells. Sci Rep 6:38814
55. Fahimipour F, Dashtimoghadam E, Mahdi Hasani-Sadrabadi M (2019) Enhancing cell seeding and osteogenesis of MSCs on 3D printed scaffolds through injectable BMP2 immobilized ECM-Mimetic gel. Dent Mater 35:990–1006
56. Blackwood KA, Bock N, Dargaville TR et al (2012) Scaffolds for growth factor delivery as applied to bone tissue engineering. Int J Polym Sci. https://doi.org/10.1155/2012/174942
57. Dong R, Bai Y, Dai J et al (2020) Engineered scaffolds based on mesenchymal stem cells/preosteoclasts extracellular matrix promote bone regeneration. J Tissue Eng 11:2041731420926918
58. Lo B, Parham L (2009) Ethical issues in stem cell research. Endocr Rev 30:204–213
59. Takahashi K, Tanabe K, Ohnuki M et al (2007) Induction of pluripotent stem cells from adult human fibroblasts by defined factors. Cell 131:861–872
60. Xie J, Peng C, Zhao Q et al (2016) Osteogenic differentiation and bone regeneration of iPSC-MSCs supported by a biomimetic nanofibrous scaffold. Acta Biomater 29:365–379
61. Csobonyeiova M, Polak S, Zamborsky R et al (2017) iPS cell technologies and their prospect for bone regeneration and disease modeling: a mini review. J Adv Res 8:321–327
62. Kao CL, Tai LK, Chiou SH et al (2010) Resveratrol promotes osteogenic differentiation and protects against dexamethasone damage in murine induced pluripotent stem cells. Stem Cells Dev 19:247–258
63. Ji J, Tong X, Huang X et al (2016) Patient-derived human induced pluripotent stem cells from gingival fibroblasts composited with defined nanohydroxyapatite/chitosan/gelatin porous scaffolds as potential bone graft substitutes. Stem Cells Transl Med 5:95–105
64. Kang H, Shih YR, Nakasaki M et al (2016) Small molecule-driven direct conversion of human pluripotent stem cells into functional osteoblasts. Sci Adv 2:e1600691
65. Gorecka J, Kostiuk V, Fereydooni A et al (2019) The potential and limitations of induced pluripotent stem cells to achieve wound healing. Stem Cell Res Ther 10:87
66. Barkholt L, Flory E, Jekerle V et al (2013) Risk of tumorigenicity in mesenchymal stromal cell-based therapies–bridging scientific observations and regulatory viewpoints. Cytotherapy 15:753–759
67. Boltze J, Arnold A, Walczak P et al (2015) The dark side of the force—constraints and complications of cell therapies for stroke. Front Neurol 6:155
68. de la Torre GC, Goreham RV, Bech Serra JJ et al (2018) "Exosomics"—a review of biophysics, biology and biochemistry of exosomes with a focus on human breast milk. Front Genet 9:92
69. Nikfarjam S, Rezaie J, Zolbanin NM et al (2020) Mesenchymal stem cell derived-exosomes: a modern approach in translational medicine. J Transl Med 18:449

70. Lu Z, Chen Y, Dunstan C et al (2017) Priming adipose stem cells with tumor necrosis factor-alpha preconditioning potentiates their exosome efficacy for bone regeneration. Tissue Eng Part A 23:1212–1220
71. Zhao P, Xiao L, Peng J et al (2018) Exosomes derived from bone marrow mesenchymal stem cells improve osteoporosis through promoting osteoblast proliferation via MAPK pathway. Eur Rev Med Pharmacol Sci 22:3962–3970
72. Diomede F, Gugliandolo A, Cardelli P et al (2018) Three-dimensional printed PLA scaffold and human gingival stem cell-derived extracellular vesicles: a new tool for bone defect repair. Stem Cell Res Ther 9:104
73. Zhang J, Liu X, Li H et al (2016) Exosomes/tricalcium phosphate combination scaffolds can enhance bone regeneration by activating the PI3K/Akt signaling pathway. Stem Cell Res Ther 7:136
74. Zhang Y, Hao Z, Wang P et al (2019) Exosomes from human umbilical cord mesenchymal stem cells enhance fracture healing through HIF-1α-mediated promotion of angiogenesis in a rat model of stabilized fracture. Cell Prolif 52:e12570.
75. Kim TG, Shin H, Lim DW (2012) Biomimetic scaffolds for tissue engineering. Adv Funct Mater 22:2446–2468
76. Velasco MA, Narváez-Tovar CA, Garzón-Alvarado DA (2015) Design, materials, and mechanobiology of biodegradable scaffolds for bone tissue engineering. Biomed Res Int 2015:729076.
77. Nikolova MP, Chavali MS (2019) Recent advances in biomaterials for 3D scaffolds: a review. Bioact Mater 4:271–292
78. Bishop ES, Mostafa S, Pakvasa M et al (2017) 3-D bioprinting technologies in tissue engineering and regenerative medicine: current and future trends. Genes Dis 4:185–195
79. Khoshnood N, Zamanian A (2020) A comprehensive review on scaffold-free bioinks for bioprinting. Bioprinting 19:e00088
80. Ovsianikov A, Khademhosseini A, Mironov V (2018) The synergy of scaffold-based and scaffold-free tissue engineering strategies. Trends Biotechnol 36:348–357
81. Yamasaki A, Kunitomi Y, Murata D et al (2919) Osteochondral regeneration using constructs of mesenchymal stem cells made by bio three-dimensional printing in mini-pigs. J Orthop Res 37:1398–1408
82. Heo DN, Ayan B, Dey M et al (2020) 3D bioprinting of co-cultured osteogenic spheroids for bone tissue fabrication. bioRxiv. https://doi.org/10.1101/2020.06.16.155143
83. Goh BT, Teh LY, Tan DB et al (2015) Novel 3D polycaprolactone scaffold for ridge preservation—a pilot randomised controlled clinical trial. Clin Oral Implants Res 26:271–277
84. Shen C, Zhang Y, Li Q et al (2014) Application of three-dimensional printing technique in artificial bone fabrication for bone defect after mandibular angle ostectomy. Zhongguo Xiu Fu Chong Jian Wai Ke Za Zhi 28:300–303
85. Sumida T, Otawa N, Kamata YU et al (2015) Custom-made titanium devices as membranes for bone augmentation in implant treatment: clinical application and the comparison with conventional titanium mesh. J Craniomaxillofac Surg 43:2183–2188
86. Kihlström Burenstam Linder L, Birgersson U, Lundgren K et al (2019) Patient-specific titanium-reinforced calcium phosphate implant for the repair and healing of complex cranial defects. World Neurosurg 122:e399–e407

Muhja Salah is a specialist orthodontist graduated from University of Khartoum in the Republic of the Sudan. Dr. Salah was also awarded a Master of Science degree with merit from Aston University in the UK on stem cell and regenerative medicine.

Farhad B. Naini is Consultant Orthodontist at Kingston and St George's University Hospitals in London, UK. He is Clinical Lead for Maxillofacial Surgery and Dentistry at Kingston Hospital NHS Foundation Trust, and Director of Research in dentistry at St George's Hospital. Dr. Naini is also a Fellow and examiner for the Royal College of Surgeons of England, Fellow of the Royal College of Surgeons of Edinburgh, Fellow and Odontology Section Council member of the Royal Society of Medicine and a Fellow of the Higher Education Academy. He has over 120 publications in peer-reviewed journals, and over 45 book chapters. He is co-editor of the textbook 'Orthodontics: Principles and Practice', co-editor of the textbook 'Preadjusted Edgewise Fixed Orthodontic Appliances: Principles and Practice', co-editor of the 61 chapter, 95-author tome 'Orthognathic Surgery: Principles, Planning and Practice', and author of the reference textbook, 'Facial Aesthetics: Concepts and Clinical Diagnosis'.

Lobat Tayebi Dr. Tayebi is currently an Associate Professor and Director of Research at Marquette University School of Dentistry. After her graduation from University of California-Davis in 2011, Dr. Tayebi joined Oklahoma State University as an Assistant Professor of Materials Science and Engineering. She moved to Marquette University School of Dentistry in 2014. She has published over 280 peer-reviewed articles and book chapters and her research interests include biomaterials, 3D printing, regenerative medicine and tissue engineering and personalized medicine.

Chapter 3
Bioresorbable Bone Fixation Devices for Oral and Maxillofacial Surgery

Quang Ngoc Dong and Takahiro Kanno

Abstract Osteosynthetic devices play a critical role in the field of maxillofacial surgery and can be applied in various operative procedures from facial reconstruction and trauma to corrective dentofacial jaw surgery. Titanium material has been utilized to make bone fixation hardware for a very long time and has survived the test of time with reliable characteristics. However, as a permanent foreign body, titanium devices also bring some morbidities which lead to secondary operation for removal. As a result, bioresorbable alternatives are developed to surpass the weakness of non-resorbable materials. Several types of bioresorbable material have been made commercially available and are widely used in the world. Recently, innovative bioactive/osteoconductive composites of polymers and bioceramics as well as bioresorbable metal with favorable features have attracted more attention from the science community and may potentially become the new standard of osteosynthetic materials in Oral and Maxillofacial surgery.

Keywords Osteosynthesis materials · Hydroxyapatite · Fracture fixation · Magnesium · Polyglycolic acid · Poly(lactic acid) · Resorbable material · Maxillofacial surgery

3.1 Introduction

Facial fractures are commonly encountered by oral and maxillofacial surgeons. These fractures are usually treated with open reduction and internal fixation to facilitate bone healing. Bone fixation is also required to immobilize bone fragments after surgeries involving osteotomy, such as orthognathic surgeries and neoplasm ablation. Therefore, bone fixation devices are very important for oral and maxillofacial surgeries.

Q. N. Dong · T. Kanno (✉)
Department of Oral and Maxillofacial Surgery, Shimane University Faculty of Medicine, Izumo, Shimane, Japan
e-mail: tkanno@med.shimane-u.ac.jp

Q. N. Dong
National Hospital of Odonto-Stomatology, Hanoi, Vietnam

© The Author(s), under exclusive license to Springer Nature Singapore Pte Ltd. 2022

A. H. Choi and B. Ben-Nissan (eds.), *Innovative Bioceramics in Translational Medicine II*, Springer Series in Biomaterials Science and Engineering 18,
https://doi.org/10.1007/978-981-16-7439-6_3

There are three stages of bone healing: inflammation, proliferation, and remodeling. Bone and soft tissue healing have similar stages, except that calcification is only seen in bone healing. After bleeding from an injury stops, a blood clot forms between the bone fragments. Inflammatory and pluripotent mesenchymal cells migrate to the interfragmentary blood clot, remove necrotic tissue, and produce a callus. The callus, consisting of fibrous tissue, cartilage, and immature bone, connects the bone fragments. If the fractured bone is immobilized, vascularized, and free of infection, the callus will gradually ossify during the remodeling stage to complete bone healing. This healing process is termed indirect bone healing, but surgeons prefer direct bone healing. In direct bone healing, bone fragments are surgically reduced into either their original positions or any other suitable position and stabilized with fixation devices. Lamellar bone, without callus, forms between bone fragments. This healing process is termed healing by primary intention and has the best outcomes. If the bone fragments are inadequately immobilized, the interfragmentary gap may increase and delayed union or non-union may occur [1]. Therefore, the importance of bone fixation devices cannot be overemphasized.

Titanium is the standard material used for bone plates and screws in multiple surgical specialties, including oral and maxillofacial surgery. Compared to other conventional materials, titanium has many favorable properties such as biocompatibility, osseointegration, high mechanical strength, ease of handling, radiopacity, and minimal scatter on computed tomography scans that makes it the ideal choice in stabilizing bone fragments [2–4]. However, titanium is non-bioresorbable and remains in the body permanently. A second surgery may sometimes be required to remove the titanium hardware because of complications, such as stress-shielding, migration, thermal irritability, and infection [5]. Surgical removal of titanium increases the cost and complications associated with general anesthesia and surgery. Because temporary bone fixation is required in most cases, bioresorbable materials may be used for bone fixation in maxillofacial surgeries.

The structure and function of the facial bones are unique: the hollow paranasal sinuses and orbits communicate with the external environment. Defects in the facial bones are not uncommon, and mastication places a relatively large load on the bones. Therefore, the ideal bioresorbable material for bone fixation should have adequate strength, minimal foreign body reactions, and the ability to immobilize, support, and regenerate bones.

Various bioresorbable bone plates and screws with different characteristics have been developed. Earlier generations of bioresorbable devices had limited clinical utility because of insufficient stiffness, rapid loss of strength, long resorption time, and adverse reactions [6–8]. However, later generations of bioresorbable bone plates overcame many of these limitations. These later generations possess greater mechanical strength and demonstrate equivalent outcomes to titanium [9]. Recently, biomaterials consisting of bioceramics and polymers/co-polymers have proven to be appropriate alternatives for maxillofacial surgeries because of their favorable physical features, bioactivity, and shorter resorption times [10, 11]. Additionally, bioresorbable metals, such as magnesium (Mg) alloys, have been evaluated as alternatives to titanium and polymer-based materials, but their use in oral and maxillofacial surgeries remains limited.

In this chapter, we will describe bioresorbable bone fixation devices including later-generation devices with innovative features, used in maxillofacial surgeries.

3.2 Polymer-Based Osteosynthesis Materials

Polymer-based fixation devices have been extensively studied, and many devices are commercially available. Table 3.1 summarizes the bioresorbable implants approved by the Japanese government for use in maxillofacial surgeries.

The most commonly used commercially available bioresorbable osteosynthesis materials contain polyhydroxy acids, which are homopolymers or co-polymers of poly-L-lactic acid (PLLA), poly-D-lactic acid (PDLA), polyglycolic acid (PGA), or polydioxanone sulfate [12]. Homopolymers such as PLLA, with high crystallinity and hydrophobicity, have long resorption times. Conversely, the use of co-polymers accelerates degradation and reduces the resorption time. Bioceramics, such as unsintered and uncalcined hydroxyapatite (u-HAp), with bioactive and osteoconductive properties have recently been used as osteosynthesis materials. For the reason that the above-mentioned materials stimulate bone regeneration, surgeons have many options during maxillofacial reconstruction, especially of the mid-face and orbits.

3.2.1 Polyglycolic Acid

Polyglycolic Acid (PGA) was the first bioresorbable polymer to be used in clinical practice. However, PGA has limited utility for osteosynthesis because of its rapid degradation and loss of strength. The material loses its mechanical strength within 4–7 weeks after implantation into the human body, which is insufficient for bone healing [13]. Furthermore, rapid PGA degradation releases a large quantity of glycolic acid, which lowers the pH and leads to complications such as sterile sinus formation, osteolysis, and intraosseous fluid accumulation [14]. Therefore, PGA is used to manufacture biodegradable sutures instead of bone fixation devices.

Table 3.1 Commercially available resorbable and bioresorbable plate systems for maxillofacial osteosynthesis approved by the Japanese government (PLLA: poly-L-lactic acid; PDLLA: poly-D-L-lactic acid; PGA: polyglycolic acid; u-HAp: unsintered hydroxyapatite; wt.: weight)

Plate system	Plate/screw composition	Plate thickness (mm)	Screw diameter (mm)	Indication	Manufacturer	Biodegradation period
GRAND FIX®	PLLA (100%)	1.0/1.5	2.2	Mid-face/mandible	GUNZE, Kyoto, Japan	>3 years
GRAND FIX® (flat type)	PLLA (100%)	0.95	2.2	Mid-face	GUNZE, Kyoto, Japan	>3 years
SonicWeld Rx®	PDLLA (100%)	0.8/1.0 (0.3/0.6)	1.6/2.1	Mid-face	KLS Martin GmbH & Co., Tuttlingen, Germany	12–30 months
LactoSorb®	PLLA (82%) + PGA (15%)	0.9/1.4 (0.5)	1.5/2.0	Mid-face	Biomet Inc.,m Jacksonville, FL, USA	12–18 months
RapidSorb®	PLLA (85%) + PGA (15%)	0.8/1.2 (0.5)	1.5/2.0	Mid-face	DePuy Synthes CMF, West Chester, PA, USA	12 months
FIXORB-MX®	PLLA (100%)	1.5	2.0	Mid-face/mandible	TEIJIN Medical Corp., Osaka, Japan	5.5 years
SuperFIXORB-MX® (OSTEOTRANS MX)®	**Plate:** PLLA (60 wt.%) + u-HAp (40 wt.%) **Screw:** PLLA (70 wt.%) + u-HAp (30 wt.%)	1.0/1.4 90.3/0.5)	2.0	Mid-face/mandible	TEIJIN Medical Corp., Osaka, Japan	5.5 years
SuperFIXORB-EX®	PLLA/PGA (90 wt.%) + u-HAp (10 wt.%)			Mid-face/mandible	TEIJIN Medical Corp., Osaka, Japan	

3.2.2 *Poly(Lactic Acid): Poly-L-Lactic Acid and Poly-D-Lactic Acid*

Poly(lactic acid) (PLA), a common bioresorbable polymer, is used as an osteosynthesis material, and has a high molecular weight and high degree of crystallinity. PLA has two stereoisomeric forms: poly-l-lactic acid (PLLA) and poly-d-lactic acid (PDLA), differentiated by the basis of the optically active carbon atom in the lactic acid [15]. PLLA has been used since 1990, and is considered the "first generation" of bioresorbable bone materials [13, 16, 17] (Fig. 3.1). As a result of its crystallinity and hydrophobicity, PLLA is resistant to hydrolysis and bioresorption in vitro; complete loss of strength takes more than 2 years [13, 17]. The resorption time in vivo for PLLA is more than 3.5 years [18, 19]. Two PLLA devices: FixsorbMX® (TEIJIN Medical Corp., Osaka, Japan) and GRAND FIX® (GUNZE, Kyoto, Japan), have been utilized in maxillofacial surgery [6, 9]. Problems associated with these devices include insufficient strength of materials, foreign body reactions, and a late degradation tissue response [9, 18, 19]. Conversely, PDLA has a lower degree of crystallinity and is less resistant to hydrolysis. PDLA also shows slower degradation, which makes it highly biocompatible and useful for extensive facial osteosynthetic surgeries, especially of the mid-face and mandible. Although crystalline particles offer a resistance to degradation, they may elicit an inflammatory response [6, 15, 18].

Despite advances in the polymer technology of maxillofacial osteosynthesis systems, there has been no significant improvement in the strength of materials.

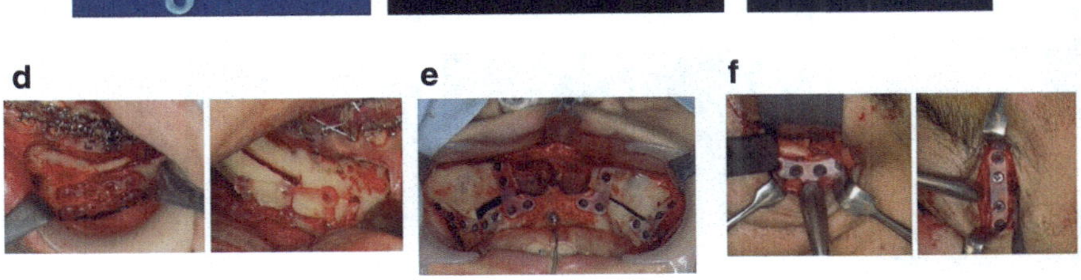

Fig. 3.1 Maxillofacial osteosynthesis systems using first-generation bioresorbable materials [12]. **a** GRAND FIX® system; **b** FIXORB-MX system; **c** GRAND FIX® flat-type system; **d** three-dimensional and single type bioresorbable plates used during bilateral sagittal split ramus osteotomy for mandibular setback using the GRAND FIX® system in orthognathic surgery; **e** thin/flat-type bioresorbable plate (GRAND FIX®, flat type) used for internal fixation of the Le Fort I maxillary osteotomy in orthognathic surgery and **f** used for left infraorbital rim and frontozygomatic suture in a complex zygomatico-maxillary fracture

Increased thickness of bioresorbable plate systems, which is required to maintain strength, leads to complications such as palpability [9]. Several years are needed for the complete resorption of PLLA plate systems; therefore, the thickness of osteosynthesis plates is an important consideration. For this reason, the GRAND FIX flat-type plate (Gunze) was modified to obtain a novel, commercially available thin and flat bioresorbable plate. This system is ideal for mid-facial osteosynthesis because of its durability and strength, although it has limited approval for use in Japan. The plate is stiff and, because of its width, has mechanical strength similar to that of other PLLA plates. This novel system reduces palpability at easily palpable facial sites, such as the periorbital rim and frontozygomatic sutures [9].

3.2.3 Co-polymers of Polyglycolic Acid, Poly-L-Lactic Acid, and Poly-D-Lactic Acid

Co-polymers of PGA, PLLA, and PDLA are categorized as rapidly bioresorbable second-generation osteosynthetic materials, and are preferred over PGA and PLLA alone [6, 13] (Fig. 3.2). Their properties can be controlled by modifying the ratio of glycolide to lactide. The rates of hydration and hydrolysis increase when crystalline PGA is co-polymerized with PLLA [6]. Therefore, different mechanical properties and degradation rates can be achieved by co-polymerization of various derivatives of α-hydroxy acids. The degradation time of the co-polymers depends on the ratio of

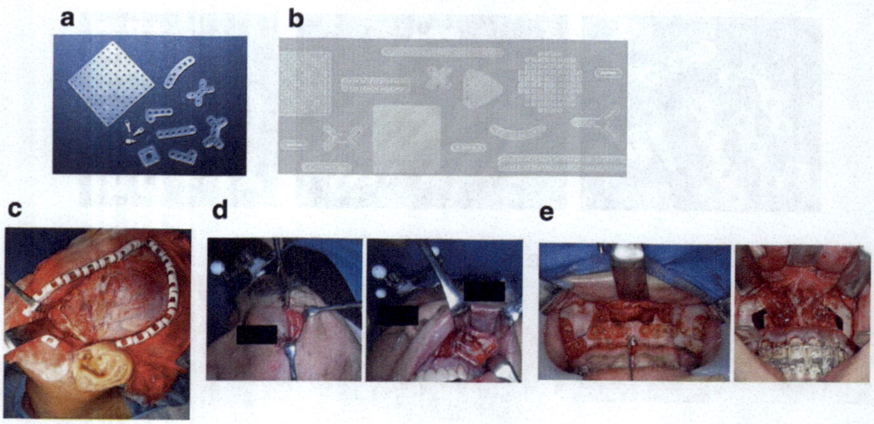

Fig. 3.2 Maxillofacial osteosynthesis systems using second-generation rapidly bioresorbable materials [12]. **a** LactoSorb® system; **b** RapidSorb® system; **c** Three-dimensional bioresorbable plate for Le Fort II/III mid-face fractures using the LactoSorb® system; **d** Bioresorbable plate (RapidSorb®) for internal fixation of a complex left zygomatico-maxillary fracture; **e** double-L-shape bioresorbable plate (LactoSorb®) for internal fixation of Le Fort I advancement and anterior maxillary osteotomy in orthognathic surgery

constituent monomers [13]. In general, co-polymers with higher glycolide content degrade more rapidly. For example, the degradation time for an 85:15 PDLA:PGA co-polymer is 5 months. As an exception to this rule, 50:50 PGA:PLLA co-polymers degrade most rapidly [6, 20].

Co-polymers of L- and D-lactides, such as SR-P(L/DL)LA 70/30 (composed of 70% PLLA and 30% PDLA), lose strength in vitro within 48 weeks of implantation. Co-polymers of L-lactide and glycolide (PLGA) are used extensively because of their favorable physicochemical properties. LactoSorb® (Biomet Inc., Jacksonville, Florida, USA) is a copolymer of PLLA (82%) and PGA (18%) [20–23]. RapidSorb® manufactured by DePuy Synthes CMF (West Chester, PA, USA) is composed of the same polymers and almost in the same ratio, as LactoSorb (85:15). RapidSorb and LactoSorb are ideal for mid-face and maxillary osteosynthesis, but have limited approval for use in Japan [24]. These co-polymers are designed to provide adequate strength for 6–8 weeks and have a resorption time of 12–18 months, which is ideal for mid-facial osteosynthesis [13, 20, 22, 25]. A retrospective study by Sukegawa et al. found the PLLA/PGA co-polymer plate system was appropriate for maxillofacial osteosynthesis, with only minor postoperative complications. These co-polymers are metabolized via the citric acid cycle, and excreted by the lungs as carbon dioxide and water [23, 24, 26]. Due to their amorphous structure, copolymers do not release crystalline particles, and their slow degradation makes them highly biocompatible [6, 13].

3.2.4 Unsintered Hydroxyapatite and PLLA Bioactive/Bioresorbable Material

Recently, hydroxyapatite has been added to PLLA because of its osteoconductive properties [27, 28]. SuperFIXORB-MX® (TEIJIN Medical Corp.), also known as OSTEOTRANS MX® outside of Japan, is composed of fine particles of unsintered hydroxyapatite, carbonate ions, and PLLA [3, 29]. Composites of unsintered hydroxyapatite/PLLA, as third-generation bioactive/bioresorbable osteosynthetic materials, are processed by machining and milling to create miniscrews and miniplates containing 30 and 40% weight fractions of unsintered hydroxyapatite particles (raw hydroxyapatite; neither calcined nor sintered), respectively [3, 27–30] (Fig. 3.3). Due to the fact that they are osteoconductive and biodegradable, the unsintered hydroxyapatite/PLLA nano-composites are completely replaced by bone [27, 28, 30, 31].

Bioactive/bioresorbable osteosynthetic devices maintain a bending strength equal to that of human cortical bone for 25 weeks in vivo [27, 28, 30, 31]. After PLLA is implanted, hydrolysis by body fluids and biodegradation begin. The molecular weight of PLLA decreases, and the unsintered hydroxyapatite fraction increases for almost 2 years [27, 28]. The composite is free from PLLA after 4 years, and the majority of unsintered hydroxyapatite particles will be replaced by bone after

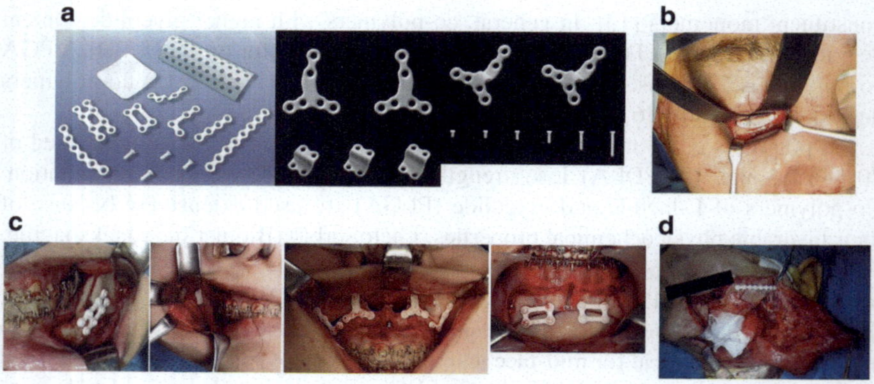

Fig. 3.3 Maxillofacial osteosynthesis systems using third-generation bioactive/osteoconductive and bioresorbable materials [12]. **a** SuperFIXORB-MX® (OSTEOTRANS MS®) system; **b** bioresorbable sheet and tack fixation for right naso-orbito-ethmoidal (midfacial) fracture reconstruction using the SuperFIXORB-MX® (OSTEOTRANS MS®) system and infraorbital rim fixation using the RapidSorb® system; **c** three-dimensional and pre-formed bioresorbable plate applied during bilateral sagittal split ramus osteotomy, intraoral vertical ramus osteotomy, Le Fort I osteotomy and genioplasty using the SuperFIXORB-MX® (OSTEOTRANS MS®) system for orthognathic surgery. **d** Use of the SuperFIXORB-MX® (OSTEOTRANS MS®) plate system for zygoma and zygomatic arch replacement and fixation with modified Crockett's method following total maxillectomy of a maxillary squamous cell carcinoma

5.5 years [27, 30]. The prolonged duration of resorption may cause complications, such as palpable discomfort over the plate beneath thin facial skin (e.g., periorbital region), and the discomfort may worsen over time because of an increase in plate volume due to the deposition of fibrous tissue [27, 30]. In comparison with other older bioresorbable polymers, unsintered hydroxyapatite/PLLA osteoconductive composites provide greater stability and are approved for use in osteosynthetic surgeries, especially those involving the mid-face and mandible [27, 30, 31].

The unsintered hydroxyapatite/PLLA composite has higher mechanical strength compared to PLLA devices, including bending strength, bending modulus, shear strength, and impact strength. A recent in vitro study involving biomechanical loading of a three-dimensional mandibular model of bilateral sagittal split ramus osteotomy (BSSRO) demonstrated high efficacy and stability of unsintered hydroxyapatite/PLLA plates [32]. Unsintered hydroxyapatite/PLLA plates possess osteoconductivity and bond directly to bone. Recent studies have evaluated the effects of this plate system on bioactive osteoconductive bone healing through scanning capacitance microscopy of incompletely exposed plates removed from maxillofacial regions [5, 29–34]. These studies found deposits of bone-like regenerative tissue on the surface of removed plates and screws that were in contact with the bone [3, 33], suggesting that the plates bonded with the bone and accelerated its healing.

To study the bioactive and osteoconductive effects of unsintered hydroxyapatite/PLLA on facial bones, Dong et al. [11] and Ngo et al. [10] covered a critical-sized defect in rat mandible with a thin sheet of unsintered hydroxyapatite/PLLA, and observed a large quantity of new bone inside the defect using histopathology and micro-computed tomography (Fig. 3.4). The new bone was attached firmly to the parent bone and unsintered hydroxyapatite/PLLA material, which may be a result of the osteoconductive properties of unsintered hydroxyapatite. Shikinami and Okuno [35] observed a calcium phosphate layer surrounding the unsintered hydroxyapatite/PLLA material after 3 to 6 days, which completely covered the entire surface within 7 days of immersion in simulated body fluid at 37 °C. Kokubo [36] suggested that the interaction between unsintered hydroxyapatite and the biological environment results in the formation of a calcium phosphate surface layer and serum protein adsorption. This surface layer modifies the structure of the adsorbed serum proteins (i.e., fibronectin). The bonds within the molecules are then exposed to osteoblasts, and their progenitors attach to the material surface. Cellular differentiation, bone matrix formation, and mineralization are regulated by intracellular signaling mechanisms, which stimulate the formation and bonding of bone. This phenomenon is responsible for the osteoconductive properties of unsintered hydroxyapatite/PLLA.

Osteoconductive bioactivity is advantageous for maxillofacial osteosynthesis and early functional improvements after fracture or osteotomy (Figs. 3.5 and 3.6).

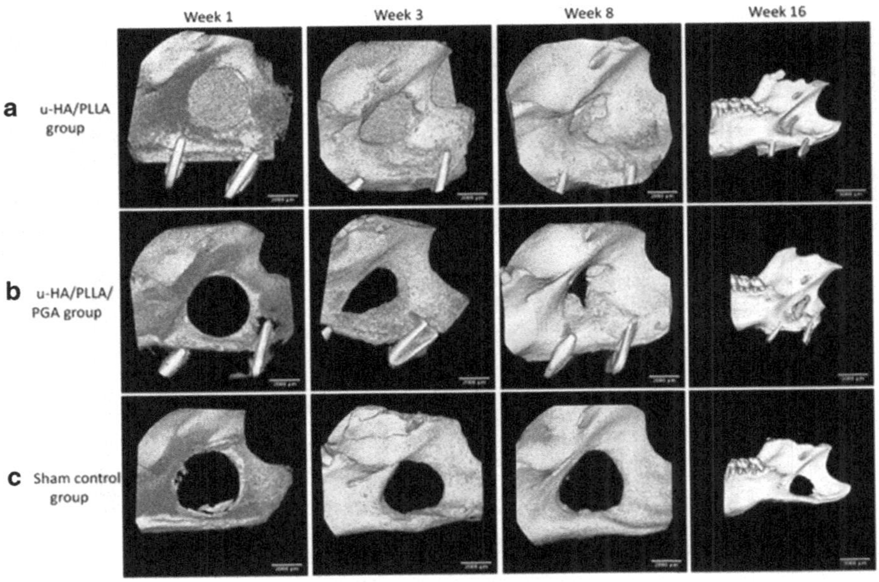

Fig. 3.4 Three-dimensional views of the unsintered hydroxyapatite/poly-L-lactic acid (PLLA) group **a** [10]; **b** three-dimensional views of the unsintered hydroxyapatite/PLLA group, showing an increase in newly formed bone over time; **c** three-dimensional views of the sham control group without new bone. For an adequate view, images at week 16 are shown at a lower magnification than those at other weeks. The unsintered hydroxyapatite/PLLA/polyglycolic acid material was not visible due to its low radiopacity. Scale bars: 2000 μm (white), 5000 μm (yellow)

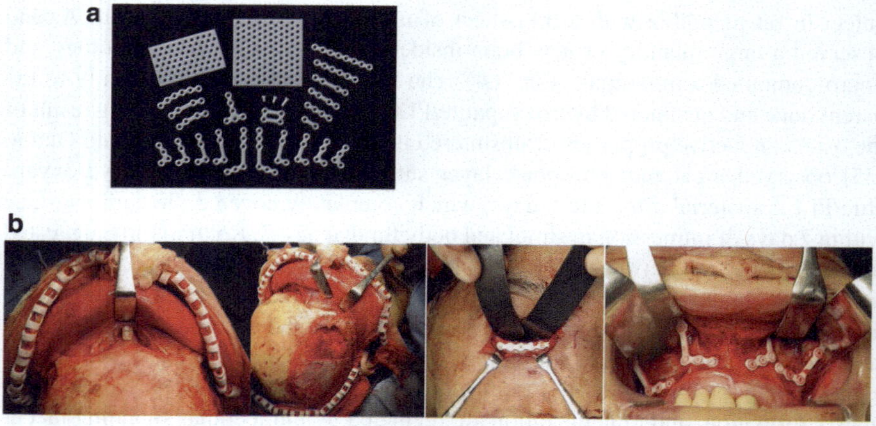

Fig. 3.5 Midfacial osteosynthesis systems using fourth-generation bioactive/osteoconductive and rapid bioresorbable materials. **a** SuperFIXORB-EX® system; **b** fourth-generation bioactive/osteoconductive and rapid bioresorbable for Le Fort III/II/I severe midfacial fracture open reduction and internal osteosynthesis, and the nasal cantilever technique using a calvaria bone grafting

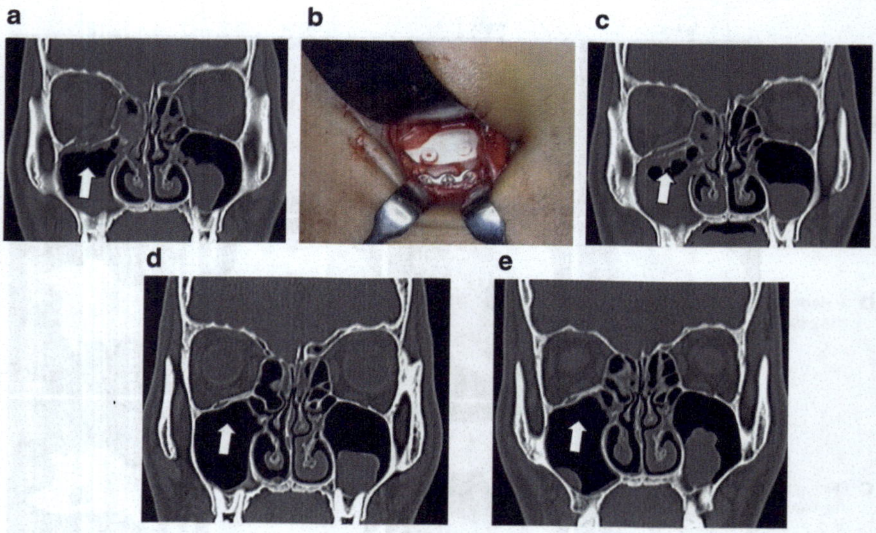

Fig. 3.6 A patient with a naso-orbito-ethmoidal fracture accompanied by defects in the orbital floor and medial wall [12, 31]. **a** Preoperative computed tomography (CT) image; **b** intraoperative view of orbital reconstruction using a bioresorbable sheet and tack fixation [SuperFIXORB-MX® (OSTEO-TRANS MS®) system]; **c** CT image obtained immediately postoperatively showing adequate orbital reconstruction; **d** CT image obtained at 6 months postoperatively showing optimal mechanical, bioactive, osteoconductive, and bioresorbable characteristics; **e** CT image obtained at 12 months postoperatively showing healed bone with good function of the bioactive and osteoconductive material in the orbit

The unsintered hydroxyapatite/PLLA composite material has significant advantages because of its bioactive, osteoconductive, and biodegradable properties, and may have broad indications in maxillofacial surgery [5, 27, 28, 30–34].

3.2.5 Unsintered Hydroxyapatite/Poly-L-Lactic Acid/Polyglycolic Acid Bioactive/Resorbable Material

The third generation bioresorbable polymer, unsintered hydroxyapatite/PLLA, has excellent bone bonding and osteoconductive properties, but the degradation time is very long. In some studies, unsintered hydroxyapatite/PLLA fragments were still detected 5 years after insertion [30]. The long resorption time of unsintered hydroxyapatite/PLLA may be associated with adverse effects; therefore, there is a need to create materials with the favorable characteristics of unsintered hydroxyapatite/PLLA, but with shorter resorption times, to improve the outcomes of maxillofacial reconstruction surgery.

The unsintered hydroxyapatite/PLLA/PGA polymers are fourth generation bioresorbable bone fixation devices (Fig. 3.5). This new material consists of unsintered hydroxyapatite (10% of the weight), and the co-polymer of 88:12 PLLA:PGA. The addition of PGA reduces the degradation time of the material.

Two recent animal studies compared unsintered hydroxyapatite/PLLA/PGA with unsintered hydroxyapatite/PLLA. Ngo et al. found equal bone regeneration ability between unsintered hydroxyapatite/PLLA/PGA and unsintered hydroxyapatite/PLLA applied to critical-sized rat mandible defects in vivo [10]. The new bone formed in the unsintered hydroxyapatite/PLLA/PGA group was larger in quantity, and matured earlier, than that in the unsintered hydroxyapatite/PLLA group, but the differences did not reach statistical significance. Ishizuka et al. [37] evaluated the interaction of bioactive materials (unsintered hydroxyapatite/PLLA and unsintered hydroxyapatite/PLLA/PGA) with mandibular periosteum and measured the molecular weights of both materials after immersion in the animal's body [37]. The interaction of periosteum with the osteoconductive materials (unsintered hydroxyapatite/PLLA and unsintered hydroxyapatite/PLLA/PGA), and bone regeneration, were demonstrated using immunohistochemistry with Runx2 and periostin antibody. In addition, the molecular weight in the unsintered hydroxyapatite/PLLA/PGA group after 16 weeks was only half of that in the unsintered hydroxyapatite/PLLA group. Therefore, both unsintered hydroxyapatite/PLLA/PGA and unsintered hydroxyapatite/PLLA are appropriate choices of reconstructive material for the maxillofacial region, where many hollow anatomical structures exist. Furthermore, unsintered hydroxyapatite/PLLA/PGA is likely to have a shorter degradation time than unsintered hydroxyapatite/PLLA. Further in vivo studies and clinical trials should be performed to explore the use of this promising material.

3.3 Magnesium-Based Bioresorbable Material

Magnesium is one of the most common elements found on earth, as well as in the human body. This metal has been used in orthopedic surgery for over a century. Magnesium has high compatibility but rapidly degrades, which results in excessive gas formation and accumulation in the body [38]. Therefore, magnesium has limited clinical utility as a bone fixation device. With advances in technology, magnesium alloys have been developed to reduce impurities, improve compatibility, and increase resistance. These magnesium alloys have been used as materials for cardiac stents and orthopedic screws for more than a decade [39].

Magnesium-based materials are processed by alloying, surface coating and treatment. Aluminum, rare earth elements, and other metals have been mixed with magnesium to create alloys. The addition of these elements to the alloy not only improves the strength and ductility, but also the corrosion resistance [40–42]. Surface modification and coating provide a protective layer for the metal, which prevents corrosion. In some studies, a significant reduction in corrosion was seen with coated compared to uncoated magnesium [39, 43].

Magnesium alloys have tensile strength ranging from 140 to 550 MPa [44] and mimic the mechanical properties of cortical bone [45]. Since magnesium alloys are weaker than titanium, maxillofacial bone fixation devices made from magnesium alloys need to be thicker to compensate for the difference in strength [44]. Multiple animal and human studies have found that magnesium alloys have high biocompatibility and rarely cause complications [46–49]. Current magnesium alloys are designed to degrade more slowly to facilitate bone healing. Naujokat et al. discovered that only 9% of the total volume of the WE43 magnesium alloy osteosynthesis material appeared to have lower density on a computed tomography (CT) scan [44]. Surprisingly, new bone was formed on the surface of the magnesium device. Although the mechanism of bone regeneration is unclear, it has also been observed in other studies [50, 51]. Future studies should compare the bone regeneration ability of magnesium alloys, unsintered hydroxyapatite/PLLA, and unsintered hydroxyapatite/PLLA/PGA.

Based on animal and human studies, magnesium alloys have potential as maxillofacial bioresorbable materials [38, 47, 51–54]. To the best of our knowledge, only two studies have evaluated the use of magnesium alloys as materials for bone screws to fix the head of mandibular condyle [55, 56]. The results of these studies were promising but more clinical research is still required.

3.4 Clinical Significance

Bioresorbable osteosynthetic implants are used to stabilize fractures, osteotomies, and bone grafts of facial bones [3, 4, 13, 16, 25, 34, 57]. The use of bioresorbable osteosynthetic implants is appropriate for the mid-face due to the fact that the fracture is easily accessible, and the implant has low biomechanical stress. Furthermore, based on recent advances in implant technology, bioresorbable osteosynthetic systems can also be used for the mandible.

3.4.1 Clinical Applications in Orthognathic Surgery

In orthognathic surgery, bioresorbable implants have the advantage over titanium of not needing a second operation for removal of the implant [6, 57–60]. Le Fort I osteotomies are stabilized with four L-shaped bioabsorbable plates and secured bilaterally in the pyriform aperture and zygomatic buttress [6, 13, 16]. Good outcomes have also been reported after the use of a biodegradable mesh [6, 13]. Segmental Le Fort I osteotomy is stabilized via standard bioresorbable fixation [34]. A systematic review of the effectiveness of bioresorbable fixation in orthognathic surgery by Fedorowicz et al. [61] reported adverse effects, mainly in the posterior maxillary region [21, 61]. Infection was associated with loose screws and wound dehiscence. A recent meta-analysis reported similar complication rates between bioresorbable and conventional titanium fixation in two trials of Le Fort I orthognathic surgery and three trials of the bimaxillary Le Fort I osteotomy plus BSSRO [25].

The use of bioresorbable plates in BSSRO has been well-described [57, 58]. Standard techniques to stimulate osteosynthesis during BSSRO include triangular placement of bicortical screws or the use of monocortical screws and two PLLA and unsintered hydroxyapatite/PLLA mini-plates [6, 25]. The two plates are applied above and below the inferior alveolar canal. Regardless of the fixation method, mandibular setback causes more instability than mandibular advancement [6, 61–65]. BSSRO with fixation for mandibular setback using a mini-resorbable plate leads to segment mobility during the early postoperative period [63, 65]. A 0.7-mm-thick unsintered hydroxyapatite/PLLA mesh can also be applied after BSSRO, especially when major segmental movements have been performed [63, 65]. Osteosynthesis using unsintered hydroxyapatite/PLLA devices during orthognathic surgery is advantageous because of the rigidity, osteoconductivity, and bone-bonding capacity of these devices [5, 32, 33, 62]. Furthermore, a recent meta-analysis showed no significant difference between bioresorbable and titanium plate fixation groups after BSSRO [25], and there was also a similar rate of complications, including infections, temporomandibular disorders, paresthesia, palpability, dehiscence, material-related complications, exposure, and relapse [25]. The safety profile of bioresorbable and titanium fixation devices is similar after bimaxillary operations, BSSRO, and Le Fort I operations, except for lower palpability and the lack of requirement for second surgery with the bioresorbable devices [6, 16, 25, 64, 66].

3.4.2 Clinical Application to Maxillofacial Trauma Surgery

The application of bioresorbable osteosynthesis plates during maxillofacial trauma surgery has been well-documented [3, 9, 15, 22, 29]. Similar stability of mid-facial fractures can be achieved using unsintered hydroxyapatite/PLLA and PLLA plates with titanium plates [3, 9, 13, 15, 25, 29, 66]. Bos et al. was the first to study the use of biodegradable plates and screws during maxillofacial surgery for fixation of zygomatic fractures [67]. The use of biodegradable plates and screws has also been extended to craniomaxillofacial fracture surgeries of the mandible or mid-face, and to three-dimensional orbital trauma reconstruction [7, 9, 15, 22, 29, 68]. The resorbable system can be used for rigid internal fixation under conditions where muscular and stress forces are not a determining factor in fragment displacement, including mid-face and whole mandibular fractures [7–9, 15, 22, 29, 68]. In children, effective use of resorbable PGA and PLLA plates combined with brief postoperative intermaxillary fixation for mandibular fractures was reported by Stanton et al. [69]. However, the stability of fixation, duration of degradation, and rate of complications (e.g., foreign-body reactions) need to be explored further. Park et al. suggested that resorbable plates and screws should be selected carefully on an individual patient basis depending on the fracture site and presence of infection [70]. The use of biodegradable plates is recommended for maxillofacial fractures with minimal load [9, 16, 31, 33]. In addition, degradation of these devices, such as PLLA into carbon dioxide and water, may take up to 3 years [17–19]. Therefore, resorbable plates and screws may be used for uncomplicated fractures of the mid-face and mandible after considering the load of mastication at surgical sites [25, 70, 71].

A recent meta-analysis of five trials of maxillofacial mid-face and mandibular fracture fixation revealed a significantly lower complication rate in the bioresorbable than titanium group [13, 25, 72–75]. Subgroup analyses according to complications revealed that palpability was more common in the titanium than bioresorbable group. The rates of infection, paresthesia, foreign-body reactions, dehiscence, and malocclusion, and of complications related to materials, exposure, and mobility, were not higher in the bioresorbable group compared to the titanium group [25, 74]. However, metal ions were detected near the implant site, suggesting leakage of the metal into body fluids [5, 13, 25, 30, 72–75]. The bioresorbable fixation system has a favorable safety profile, similar to that of titanium fixation. Next generation unsintered hydroxyapatite/PLLA bioactive/bioresorbable plates are suitable for internal bone fixation of maxillofacial fractures because of their durability, bioactivity, and osteoconductive potential [5, 27, 28, 30–33].

The bone regeneration ability of unsintered hydroxyapatite/PLLA and unsintered hydroxyapatite/PLLA/PGA is particularly useful for reconstructing the hollow structures of the face (i.e., orbits and paranasal sinuses) (Fig. 3.6). Traumatic injuries of the orbital floor and walls are usually reconstructed using biomaterials. The traditional materials used for reconstruction are autogenous bone from any other site, non-resorbable materials like titanium, polymers such as porous polyethylene, and resorbable materials. Although autogenous bone is the ideal choice for orbital reconstruction, a donor site is required, there is a risk of associated complications, and the resorption of autogenous bone is unpredictable [76]. Titanium mesh is very difficult to remove [76]. Moreover, if the orbital contents are not reduced carefully, the sharp edge of the material can cause injury, and there is a risk of orbital adherence syndrome [26, 77]. Porous polyethylene is not visible on radiologic imaging and is also expensive [78]. In addition, non-resorbable implants are permanent and may cause delayed complications. Non-osteoconductive resorbable reconstruction materials, on the other hand, should only be used for small orbital defects; otherwise, the defect will re-appear when the material is resorbed. Therefore, bioresorbable osteoconductive materials, such as unsintered hydroxyapatite/PLLA and unsintered hydroxyapatite/PLLA/PGA, are the ideal choice for reconstruction after orbital injury or a defect-type lesion. The newly formed bone fuses with the native bone and replaces the material that is resorbed [10, 11, 37]. Various studies have demonstrated the effectiveness of these materials for treating orbital defects [29, 31, 79, 80].

Therefore, bioresorbable fixation systems appear to be a reasonable option for maxillofacial fracture fixation, but large-scale randomized, prospective trials are required to confirm the safety of this approach [25, 57, 81].

3.5 Conclusions and Future Perspectives

There have been significant technological advances in the bioresorbable osteosynthesis plates used in oral and maxillofacial surgery. The development of new systems, and improvement of old ones, has led to several advantages over titanium metal plate systems for both patients and surgeons. We presented an overview of the currently available bioresorbable osteosynthesis materials and osteofixation systems.

The use of bioresorbable osteosynthetic materials is associated with many complications, but these materials have some advantages over titanium fixation, including the lack of requirement for a second surgery, the possibility of radiological assessment, and improved bone healing and remodeling. Furthermore, based on recent studies, bioabsorbable osteosynthesis systems are reliable for osteofixation in various maxillofacial surgeries, including fragment fixation in orthognathic and trauma surgeries. The use of bioresorbable devices is associated with stable fixation, uneventful bone healing, optimal remodeling, and skeletal stability similar to that of titanium devices.

The third- and fourth-generation bioresorbable materials may be used in maxillofacial surgeries because of their optimal mechanical, bioactive, osteoconductive, and bioresorbable characteristics. Although magnesium alloys have several favorable characteristics, more trials are needed to confirm their feasibility for use as a reconstruction material.

Future studies of materials for maxillofacial osteosynthetic plates should focus on reducing foreign body reactions and enhancing biocompatibility, mechanical strength, bioactivity, the bioresorption rate, and customizability.

Acknowledgements This work was supported in part by a Grant-in-Aid for Scientific Research from JSPS KAKENHI to T.K. (#21K10042).

References

1. Miloro M, Ghali GE, Larsen PE et al (eds) (2012) Peterson's principles of oral and maxillofacial surgery. People's Medical Publishing House (PMPH), Connecticut
2. Luo M, Yang X, Wang Q et al (2018) Skeletal stability following bioresorbable versus titanium fixation in orthognathic surgery: a systematic review and meta-analysis. Int J Oral Maxillofac Surg 47:141–151
3. Sukegawa S, Kanno T, Katase N et al (2016) Clinical evaluation of an unsintered hydroxyapatite/poly-l-lactide osteoconductive composite device for the internal fixation of maxillofacial fractures. J Craniofac Surg 27:1391–1397
4. Al-Moraissi EA, Ellis E 3rd (2015) Biodegradable and titanium osteosynthesis provide similar stability for orthognathic surgery. J Oral Maxillofac Surg 73:1795–1808
5. Landes C, Ballon A, Ghanaati S et al (2014) Treatment of malar and midfacial fractures with osteoconductive forged unsintered hydroxyapatite and poly-L-lactide composite internal fixation devices. J Oral Maxillofac Surg 72:1328–1338
6. Park YW (2015) Bioabsorbable osteofixation for orthognathic surgery. Maxillofac Plast Reconstr Surg 37:6
7. Yolcu Ü, Alan H, Malkoç S et al (2015) Cytotoxicity evaluation of bioresorbable fixation screws on human gingival fibroblasts and mouse osteoblasts by real-time cell analysis. J Oral Maxillofac Surg 73:1562.e1-10
8. Bali RK, Sharma P, Jindal S et al (2013) To evaluate the efficacy of biodegradable plating system for fixation of maxillofacial fractures: a prospective study. Natl J Maxillofac Surg 4:167–172
9. Sukegawa S, Kanno T, Nagano D et al (2016) The clinical feasibility of newly developed thin flat-type bioresorbable osteosynthesis devices for the internal fixation of zygomatic fractures: is there a difference in healing between bioresorbable materials and titanium osteosynthesis? J Craniofac Surg 27:2124–2129
10. Ngo HX, Dong QN, Bai Y et al (2020) Bone regeneration capacity of newly developed uncalcined/unsintered hydroxyapatite and poly-l-lactide-co-glycolide sheet in maxillofacial surgery: an *in vivo* study. Nanomaterials 11:22
11. Dong QN, Kanno T, Bai Y et al (2019) Bone regeneration potential of uncalcined and unsintered hydroxyapatite/poly l-lactide bioactive/osteoconductive sheet used for maxillofacial reconstructive surgery: an in vivo study. Materials 12:2931
12. Kanno T, Sukegawa S, Furuki Y et al (2018) Overview of innovative advances in bioresorbable plate systems for oral and maxillofacial surgery. Jpn Dent Sci Rev 54:127–138
13. Schumann P, Lindhorst D, Wagner ME et al (2013) Perspectives on resorbable osteosynthesis materials in craniomaxillofacial surgery. Pathobiology 80:211–217
14. Kumar CR, Sood S, Ham S (2005) Complications of bioresorbable fixation systems in pediatric neurosurgery. Childs Nerv Syst 21:205–210
15. Pina S, Ferreira JMF (2012) Bioresorbable plates and screws for clinical applications: a review. J Healthcare Eng. https://doi.org/10.1260/2040-2295.3.2.243
16. van Bakelen NB, Buijs GJ, Jansma J et al (2014) Decision-making considerations in application of biodegradable fixation systems in maxillofacial surgery—a retrospective cohort study. J Craniomaxillofac Surg 42:417–422
17. Pihlajamäki H, Böstman O, Hirvensalo E et al (1992) Absorbable pins of self-reinforced poly-L-lactic acid for fixation of fractures and osteotomies. J Bone Joint Surg Br 74:853–857
18. Bergsma JE, de Bruijn WC, Rozema FR et al (1995) Late degradation tissue response to poly(L-lactide) bone plates and screws. Biomaterials 16:25–31

19. Bergsma EJ, Rozema FR, Bos RR et al (1993) Foreign body reactions to resorbable poly(L-lactide) bone plates and screws used for the fixation of unstable zygomatic fractures. J Oral Maxillofac Surg 51:666–670
20. Suuronen R, Haers PE, Lindqvist C et al (1999) Update on bioresorbable plates in maxillofacial surgery. Facial Plast Surg 15:61–72
21. Sukegawa S, Kanno T, Matsumoto K et al (2018) Complications of a poly-L-lactic acid and polyglycolic acid osteosynthesis device for internal fixation in maxillofacial surgery. Odontology 106:360–368
22. Koike T, Kanno T, Sekine J (2015) A case of naso-orbital-ethmoid fracture following unexpected airbag deployment. J Oral Maxillofac Surg Med Pathol 27:522–524
23. Wiltfang J, Merten HA, Schultze-Mosgau S et al (2000) Biodegradable miniplates (LactoSorb): long-term results in infant minipigs and clinical results. J Craniofac Surg 11:239–243; discussion 244–245
24. Edwards RC, Kiely KD, Eppley BL (2001) The fate of resorbable poly-L-lactic/polyglycolic acid (LactoSorb) bone fixation devices in orthognathic surgery. J Oral Maxillofac Surg 59:19–25
25. Yang L, Xu M, Jin X et al (2013) Complications of absorbable fixation in maxillofacial surgery: a meta-analysis. PLoS One 8:e67449
26. Young SM, Sundar G, Lim TC et al (2017) Use of bioresorbable implants for orbital fracture reconstruction. Br J Ophthalmol 101:1080–1085
27. Shikinami Y, Matsusue Y, Nakamura T (2005) The complete process of bioresorption and bone replacement using devices made of forged composites of raw hydroxyapatite particles/poly l-lactide (F-u-HA/PLLA). Biomaterials 26:5542–5551
28. Shikinami Y, Hata K, Okuno M (1997) Ultra-high strength resorbable implants for oral and maxillofacial surgery made from composites of bioactive ceramic particles/polylactide. Int J Oral Maxillofac Surg 26:37
29. Kanno T, Karino M, Yoshino A et al (2017) Feasibility of single folded unsintered hydroxyapatite particles/poly-l-lactide composite sheet in combined orbital floor and medial wall fracture reconstruction. J Hard Tissue Biol 26:237–244
30. Sukegawa S, Kanno T, Kawai H et al (2015) Long-term bioresorption of bone fixation devices made from composites of unsintered hydroxyapatite particles and poly-l-lactide. J Hard Tissue Biol 24:219–224
31. Kanno T, Tatsumi H, Karino M et al (2016) Applicability of an unsintered hydroxyapatite particles/poly-l-lactide composite sheet with tack fixation for orbital fracture reconstruction. J Hard Tissue Biol 25:329–334
32. Sukegawa S, Kanno T, Manabe Y et al (2017) Biomechanical loading evaluation of unsintered hydroxyapatite/poly-l-lactide plate system in bilateral sagittal split ramus osteotomy. Materials 10:764
33. Sukegawa S, Kanno T, Koyama Y et al (2017) Precision of post-traumatic orbital reconstruction using unsintered hydroxyapatite particles/poly-l-lactide composite bioactive/resorbable mesh plate with and without navigation: a retrospective study. J Hard Tissue Biol 26:274–280
34. van Bakelen NB, Buijs GJ, Jansma J et al (2013) Comparison of biodegradable and titanium fixation systems in maxillofacial surgery: a two-year multi-center randomized controlled trial. J Dent Res 92:1100–1105
35. Shikinami Y, Okuno M (1999) Bioresorbable devices made of forged composites of hydroxyapatite (HA) particles and poly-L-lactide (PLLA): Part I. Basic characteristics. Biomaterials 20:859–877
36. Kokubo T (ed) (2008) Bioceramics and their clinical applications. Woodhead Publishing, England
37. Ishizuka S, Dong QN, Ngo HX et al (2021) Bioactive regeneration potential of the newly developed uncalcined/unsintered hydroxyapatite and poly-l-lactide-co-glycolide biomaterial in maxillofacial reconstructive surgery: an in vivo preliminary study. Materials 14:2461
38. Chaya A, Yoshizawa S, Verdelis K et al (2015) In vivo study of magnesium plate and screw degradation and bone fracture healing. Acta Biomater 18:262–269

39. Schaller B, Saulacic N, Imwinkelried T et al (2016) In vivo degradation of magnesium plate/screw osteosynthesis implant systems: soft and hard tissue response in a calvarial model in miniature pigs. J Craniomaxillofac Surg 44:309–317
40. Xin Y, Hu T, Chu PK (2011) In vitro studies of biomedical magnesium alloys in a simulated physiological environment: a review. Acta Biomater 7:1452–1459
41. Peng Q, Huang Y, Zhou L et al (2010) Preparation and properties of high purity Mg-Y biomaterials. Biomaterials 31:398–403
42. Pardo A, Merino MC, Coy AE et al (2008) Corrosion behaviour of magnesium/aluminium alloys in 3.5wt.% NaCl. Corros Sci 50:823–834
43. Zhuang J, Jing Y, Wang Y et al (2016) Degraded and osteogenic properties of coated magnesium alloy AZ31; An experimental study. J Orthop Surg Res 11:30
44. Naujokat H, Seitz JM, Açil Y et al (2017) Osteosynthesis of a cranio-osteoplasty with a biodegradable magnesium plate system in miniature pigs. Acta Biomater 62:434–445
45. Torroni A, Xiang C, Witek L et al (2017) Biocompatibility and degradation properties of WE43 Mg alloys with and without heat treatment: in vivo evaluation and comparison in a cranial bone sheep model. J Craniomaxillofac Surg 45:2075–2083
46. Atkinson HD, Khan S, Lashgari Y et al (2019) Hallux valgus correction utilising a modified short scarf osteotomy with a magnesium biodegradable or titanium compression screws—a comparative study of clinical outcomes. BMC Musculoskelet Disord 20:334
47. Gigante A, Setaro N, Rotini M et al (2018) Intercondylar eminence fracture treated by resorbable magnesium screws osteosynthesis: a case series. Injury 49(Suppl 3):S48–S53
48. Grün NG, Holweg P, Tangl S et al (2018) Comparison of a resorbable magnesium implant in small and large growing-animal models. Acta Biomater 78:378–386
49. Jähn K, Saito H, Taipaleenmäki H et al (2016) Intramedullary Mg_2Ag nails augment callus formation during fracture healing in mice. Acta Biomater 36:350–360
50. Oshibe N, Marukawa E, Yoda T et al (2019) Degradation and interaction with bone of magnesium alloy WE43 implants: a long-term follow-up in vivo rat tibia study. J Biomater Appl 33:1157–1167
51. Schaller B, Saulacic N, Beck S et al (2017) Osteosynthesis of partial rib osteotomy in a miniature pig model using human standard-sized magnesium plate/screw systems: effect of cyclic deformation on implant integrity and bone healing. J Craniomaxillofac Surg 45:862–871
52. Kong X, Wang L, Li G et al (2018) Mg-based bone implants show promising osteoinductivity and controllable degradation: a long-term study in a goat femoral condyle fracture model. Mater Sci Eng C Mater Biol Appl 86:42–47
53. Kose O, Turan A, Unal M et al (2018) Fixation of medial malleolar fractures with magnesium bioabsorbable headless compression screws: short-term clinical and radiological outcomes in eleven patients. Arch Orthop Trauma Surg 138:1069–1075
54. Plaass C, von Falck C, Ettinger S et al (2018) Bioabsorbable magnesium versus standard titanium compression screws for fixation of distal metatarsal osteotomies—3 year results of a randomized clinical trial. J Orthop Sci 23:321–327
55. Leonhardt H, Ziegler A, Lauer G et al (2021) Osteosynthesis of the mandibular condyle with magnesium-based biodegradable headless compression screws show good clinical results during a 1-year follow-up period. J Oral Maxillofac Surg 79:637–643
56. Leonhardt H, Franke A, McLeod NMH et al (2017) Fixation of fractures of the condylar head of the mandible with a new magnesium-alloy biodegradable cannulated headless bone screw. Br J Oral Maxillofac Surg 55:623–625
57. Gareb B, van Bakelen NB, Buijs GJ et al (2017) Comparison of the long-term clinical performance of a biodegradable and a titanium fixation system in maxillofacial surgery: a multicenter randomized controlled trial. PLoS One 12:e0177152
58. van Bakelen NB, Boermans BD, Buijs GJ et al (2014) Comparison of the long-term skeletal stability between a biodegradable and a titanium fixation system following BSSO advancement—a cohort study based on a multicenter randomised controlled trial. Br J Oral Maxillofac Surg 52:721–728

59. Turvey TA, Bell RB, Tejera TJ et al (2002) The use of self-reinforced biodegradable bone plates and screws in orthognathic surgery. J Oral Maxillofac Surg 60:59–65
60. Kulkarni RK, Moore EG, Hegyeli AF et al (1971) Biodegradable poly(lactic acid) polymers. J Biomed Mater Res 5:169–181
61. Fedorowicz Z, Nasser M, Newton JT et al (2007) Resorbable versus titanium plates for orthognathic surgery. Cochrane Datab Syst Rev. https://doi.org/10.1002/14651858.CD006204.pub2
62. Landes CA, Ballon A, Tran A et al (2014) Segmental stability in orthognathic surgery: hydroxyapatite/Poly-l-lactide osteoconductive composite versus titanium miniplate osteosyntheses. J Craniomaxillofac Surg 42:930–942
63. Ko EW, Huang CS, Lo LJ et al (2013) Alteration of masticatory electromyographic activity and stability of orthognathic surgery in patients with skeletal class III malocclusion. J Oral Maxillofac Surg 71:1249–1260
64. Paeng JY, Hong J, Kim CS et al (2012) Comparative study of skeletal stability between bicortical resorbable and titanium screw fixation after sagittal split ramus osteotomy for mandibular prognathism. J Craniomaxillofac Surg 40:660–664
65. Ueki K, Marukawa K, Shimada M et al (2006) Maxillary stability following Le Fort I osteotomy in combination with sagittal split ramus osteotomy and intraoral vertical ramus osteotomy: a comparative study between titanium miniplate and poly-L-lactic acid plate. J Oral Maxillofac Surg 64:74–80
66. Sukegawa S, Kanno T, Shibata A et al (2015) Use of templates and self-tapping metal screws for temporary fixation of a resorbable plate system. Ann Maxillofac Surg 5:231–233
67. Bos RR, Boering G, Rozema FR et al (1987) Resorbable poly(L-lactide) plates and screws for the fixation of zygomatic fractures. J Oral Maxillofac Surg 45:751–753
68. Singh M, Singh RK, Passi D et al (2016) Management of pediatric mandibular fractures using bioresorbable plating system—efficacy, stability, and clinical outcomes: our experiences and literature review. J Oral Biol Craniofac Res 6:101–106
69. Stanton DC, Liu F, Yu JW et al (2014) Use of bioresorbable plating systems in paediatric mandible fractures. J Craniomaxillofac Surg 42:1305–1309
70. Park CH, Kim HS, Lee JH et al (2011) Resorbable skeletal fixation systems for treating maxillofacial bone fractures. Arch Otolaryngol Head Neck Surg 137:125–129
71. Buijs GJ, van der Houwen EB, Stegenga B et al (2007) Mechanical strength and stiffness of biodegradable and titanium osteofixation systems. J Oral Maxillofac Surg 65:2148–2158
72. Bhatt K, Roychoudhury A, Bhutia O et al (2010) Equivalence randomized controlled trial of bioresorbable versus titanium miniplates in treatment of mandibular fracture: a pilot study. J Oral Maxillofac Surg 68:1842–1848
73. Lee HB, Oh JS, Kim SG et al (2010) Comparison of titanium and biodegradable miniplates for fixation of mandibular fractures. J Oral Maxillofac Surg 68:2065–2069
74. Dorri M, Nasser M, Oliver R (2009) Resorbable versus titanium plates for facial fractures. Cochrane Datab Syst Rev. https://doi.org/10.1002/14651858.CD007158.pub2
75. Meningaud JP, Poupon J, Bertrand JC et al (2001) Dynamic study about metal release from titanium miniplates in maxillofacial surgery. Int J Oral Maxillofac Surg 30:185–188
76. Kim HK, Baek WI, Bae TH et al (2015) Usefulness of subciliary approach by using lacrimal sac stripping for large isolated medial orbital fracture reconstruction. Ann Plast Surg 75:170–173
77. Chou C, Kuo YR, Chen CC et al (2017) Medial orbital wall reconstruction with porous polyethylene by using a transconjunctival approach with a caruncular extension. Ann Plast Surg 78:S89–S94
78. Park H, Kim HS, Lee BI (2015) Medial wall orbital reconstruction using unsintered hydroxyapatite particles/poly l-lactide composite implants. Arch Craniofac Surg 16:125–130
79. Dong QN, Karino M, Koike T et al (2020) Navigation-assisted isolated medial orbital wall fracture reconstruction using an u-ha/plla sheet via a transcaruncular approach. J Invest Surg 33:644–652
80. Kohyama K, Morishima Y, Arisawa K et al (2018) Immediate and long-term results of unsintered hydroxyapatite and poly L-lactide composite sheets for orbital wall fracture reconstruction. J Plast Reconstr Aesthet Surg 71:1069–1075

81. van Bakelen NB, Vermeulen KM, Buijs GJ et al (2015) Cost-effectiveness of a biodegradable compared to a titanium fixation system in maxillofacial surgery: a multicenter randomized controlled trial. PLoS One 10:e0130330.

Quang Ngoc Dong Dr. Quang graduated with an undergraduate degree in Dentistry in 2011 from Hanoi Medical University. After that, he started working as an Oral and Maxillofacial surgeon in the National Hospital of Odonto-Stomatology in Hanoi, Vietnam from September 2011. From 2014 to 2015, Dr. Quang studied in The University of Manchester, United Kingdom and graduated with the master's degree in Oral and Maxillofacial Surgery. Since September 2016 to March 2021, Dr. Quang attended the Ph.D. program and Clinical Fellowship in the Department of Oral and Maxillofacial Surgery, Shimane University Faculty of Medicine in Japan. His research interest includes the clinical application of bioactive and osteoconductive biomaterial in Oral and Maxillofacial surgery and the use of computer-assisted surgical planning and computer-assisted fabrication of surgical guide in maxillofacial reconstruction and orthognathic surgery.

Takahiro Kanno Professor Kanno graduated from School of Dentistry and Graduate School of Dentistry, Kyushu Dental College, Kitakyushu, Fukuoka, Japan in 2001 (DDS) and be awarded PhD degree (Maxillofacial Surgery; Director, Chief & Mentor: Prof. Tetsu Takahashi) from Graduate School of Kyushu Dental College in 2005. He worked at The Department of Oral and Maxillofacial Surgery of Kyushu Dental College (2001–2005) and at The Division of Oral-Maxillofacial Surgery of Kagawa Prefectural Central Hospital (2005–2012). He studied abroad for clinical fellowships (AO fellow and IBRA fellow) at The Department of Oral and Maxillofacial Surgery, Ludwig-Maximilians-Universität Munich, Germany (2006) and at The Department of CranioMaxillofacial Surgery, University of Bern, Switzerland (2007). He is now employed by Shimane University Faculty of Medicine as a fulltime medical faculty (2012-present). He is the Chair Professor and the Director of The Department of Oral and Maxillofacial Surgery, the Chief Surgeon of Maxillofacial Trauma Center. He has more than 200 scientific papers published, among them 130 published in international journals. Specialty: Surgical treatment of oral, head and neck cancer and reconstruction, Maxillofacial trauma, Orthognathic surgery and Oral and maxillofacial implant surgery.

Chapter 4
Tissue Engineering Strategies for Craniomaxillofacial Surgery: Current Trends in 3D-Printed Bioactive Ceramic Scaffolds

Lukasz Witek, Vasudev Vivekanand Nayak, Christopher M. Runyan, Nick Tovar, Sharbel Elhage, James C. Melville, Simon Young, David H. Kim, Bruce N. Cronstein, Roberto L. Flores, and Paulo G. Coelho

Abstract The advancements in craniomaxillofacial surgical reconstruction have led to improvements in patient-specific treatment care. One of the primary goals is to restore form and function by balancing aesthetic, functional and physiological demands while minimizing morbidity to the patient. Although the promise of bone tissue engineering has yet to be fully achieved, clinically, translational developments have been evaluated with favorable results. This chapter reviews the recent

---

L. Witek (✉) · V. V. Nayak · N. Tovar · P. G. Coelho
Department of Biomaterials, New York University College of Dentistry, New York, NY, USA
e-mail: lukasz.witek@nyu.edu

L. Witek
Department of Biomedical Engineering, New York University Tandon School of Engineering, Brooklyn, NY, USA

V. V. Nayak · P. G. Coelho
Department of Mechanical and Aerospace Engineering, New York University Tandon School of Engineering, Brooklyn, NY, USA

C. M. Runyan · S. Elhage
Department of Plastic Surgery, Wake Forest University School of Medicine, Winston-Salem, NC, USA

N. Tovar
Department of Oral and Maxillofacial Surgery, NYU Langone Medical Center and Bellevue Hospital Center, New York, NY, USA

J. C. Melville · S. Young
Department of Oral & Maxillofacial Surgery, The University of Texas Health Science Center at Houston School of Dentistry, Houston, TX, USA

D. H. Kim
Department of Surgery, Division of Plastic and Reconstructive Surgery, Lewis Katz School of Medicine, Temple University, Philadelphia, PA, USA

B. N. Cronstein
Department of Medicine, New York University Grossman School of Medicine, New York, NY, USA

© The Author(s), under exclusive license to Springer Nature Singapore Pte Ltd. 2022

A. H. Choi and B. Ben-Nissan (eds.), *Innovative Bioceramics in Translational Medicine II*, Springer Series in Biomaterials Science and Engineering 18,
https://doi.org/10.1007/978-981-16-7439-6_4

and current advances in biomaterials science combined with 3D printing combined with bioactive molecule osteogenic stimulation, and their integration toward the development of devices capable of facilitating craniomaxillofacial bony restoration.

Keywords Craniomaxillofacial Surgery · 3D Printing · Additive Manufacturing · Adenosine · Dipyridamole (DIPY) · Bone Tissue Engineering · Recombinant Human Bone Morphogenic Protein (rhBMP-2)

4.1 Introduction

The principles of craniomaxillofacial surgery were established with the goals of restoring form and function by balancing aesthetic, functional and physiological demands while minimizing morbidity to the patient. Surgeons, specifically plastic and reconstructive surgeons, routinely operate on a breadth of conditions, without being defined by an anatomical area, to address and treat the specific needs of the patient from head to toe through a variety of reconstructive techniques. Plastic surgeons are oftentimes a part of a team of specialists, including dentists, oral and maxillofacial surgeons, neurosurgeons, orthopedic surgeons, and otolaryngologists. While each anatomical realm has its own set of considerations, the common surgical principle is the replacement of "*like with like*". Therefore, the use of autogenous tissue with analogous histologic and mechanical properties is often preferable over prosthetic replacements. In craniomaxillofacial reconstruction, for example, autogenous tissue is preferred for the restoration of congenital, oncologic, traumatic, and iatrogenic deformities. To date, limitations to the gold standard of autogenous tissue reconstruction remain such as size, shape, and stock of donor tissue, coupled with extended operative time and donor site morbidity. However, these challenges present an opportunity for modern and innovative approaches to craniomaxillofacial reconstruction, such as the application for additive manufacturing, specifically three-dimensional (3D) printing.

R. L. Flores · P. G. Coelho
Hansjörg Wyss Department of Plastic Surgery, New York University Grossman School of Medicine, New York, NY, USA

4.2 Direct Ink Writing for Craniomaxillofacial Applications

Three-dimensional printing (3DP) technology can be leveraged to reconstruct craniomaxillofacial bone defects by designing and manufacturing customized patient models, surgical templates and cutting guides, as well as patient-specific implants and tissue engineering devices. Bony defects or deformities in the craniomaxillofacial skeleton can be congenital, infectious, neoplastic, traumatic, or iatrogenic in etiology. Approximately one-third of all birth defects affect the craniomaxillofacial region. Examples include bony defects of the mid-face such as alveolar clefts characterized by an anterior alveolar defect of the maxilla, as well as cranial deformities such as craniosynostosis or premature fusion of one or more of the cranial sutures. Additionally, edentulism, a condition that more commonly affects the aging and elderly population, is a leading contributor of alveolar bone deficiency. Etiologies for edentulism may be linked to genetic factors, periodontal disease, or trauma. Ensuing tooth loss and the lack of mechanical stress on the alveolar ridge result in physiological alveolar bone resorption and ridge atrophy. Bone loss in the alveolar region typically leads to one of the three primary types of ridge deficiencies: horizontal, vertical, or a complex configuration, where both vertical and horizontal component dimensions should be reestablished for adequate restoration of masticatory function through dental implants.

There are a variety of different biomaterials which can be employed for bone tissue engineering. Bioactive calcium phosphates are most commonly selected due to their well-documented safety profile and biocompatibility [1, 2]. Specifically, calcium-phosphate ($CaPO_4$) bioactive ceramics, i.e., hydroxyapatite and β-tri-calcium phosphate (β-TCP), have been identified as suitable alternatives to autogenous bone grafts [3] due to their similarities to bone. Hydroxyapatite, which is the predominant inorganic component of bone, has been most commonly utilized ceramic. Although hydroxyapatite is both biocompatible and osseoconductive, its degradation kinetics (~2% per year in vivo [4]) render it an unfavorable material for replacement of scaffold material by regenerated bone. As a result, β-TCP ceramic was developed [5] and similar to hydroxyapatite, β-TCP is both biocompatible and osseoconductive. Furthermore, β-TCP has the capacity to regenerate bone while being resorbed over a 6-to-18-month period depending on overall macro, micro, and nanoporosity [4, 5].

Due to inferior bulk form configurations and degradation kinetics of bioactive ceramics, an impetus was necessary for altering and improving construct designs in an effort to improve performance and facilitate bone regeneration. While most bioactive ceramics exist in powder form or in prefabricated bulk shapes, fabrication of patient-specific bone grafting scaffolds with fit and fill designs has remained a challenge. However, with the advent of 3D printing technology and its subsequent evolution, equipment capable of printing viscous colloidal gels were developed [6, 7], ultimately allowing the fabrication of personalized devices that are capable of fitting and filling bony defects of different size and complex tri-dimensional shape. Through the utilization of a viscous colloidal ink system in conjunction with 3D printing, β-TCP can be printed into patient specific devices matching volumetric configuration of the bone defect. Such capabilities allow for maximal interaction with the interface of the bone defect for a prompt and complete osteointegration. Most commonly, latticed-based structures with tailored surface texture facilitate the osseo-conduction, vascular in-growth, and scaffold degradation/absorption. With additive manufacturing or 3D printing technology, such customized modifications to scaffold design can be systematically incorporated to tailor the device at various scales (i.e., macro-, meso-, micro- and nanometer) levels [8–11].

4.3 Personalized Fabrication of Scaffolds

The utilization of 3D printing technologies coupled with medical imaging (i.e., computed tomography (CT), magnetic resonance imaging (MRI), ultrasound or, in some cases, high-resolution 3D photographs) to acquire volumetric data of the patient and area of interest/anatomy may be utilized for the fabrication of not only surgical guides/templates but also patient specific, fit and fill scaffolds for bone regeneration [12–15].

After the volumetric data is acquired, it is subsequently exported to a data-editing platform for processing, refinement, and isolation of the defect site. Digital data-editing and surface refinement are finalized, the digitally reconstructed fit and fill solid is then exported into a stereolithography (.STL) format that is digitally sliced and meshed into a multilayered three-dimensional (3D) object, yielding a final scaffold composed of ~5–7 layers per millimeter (mm) (Fig. 4.1). During the digital processing steps, the engineer or end user tailor the scaffold's attributes, such as strut/rod and pore size. Once the final digital model is completed, a tool path is saved in a code specified to the printer that is exported to the Direct Ink Writing printer (Fig. 4.2).

The Direct Ink Writing (DIW) printing process, commonly referred to as robo-casting, is the process of assembling a bioactive ceramic colloidal gel in a layer-by-layer fashion [16]. The DIW technology utilizes a 3D printer assembled with a stationary platform and a moveable gantry. The DIW equipment follows a tool path

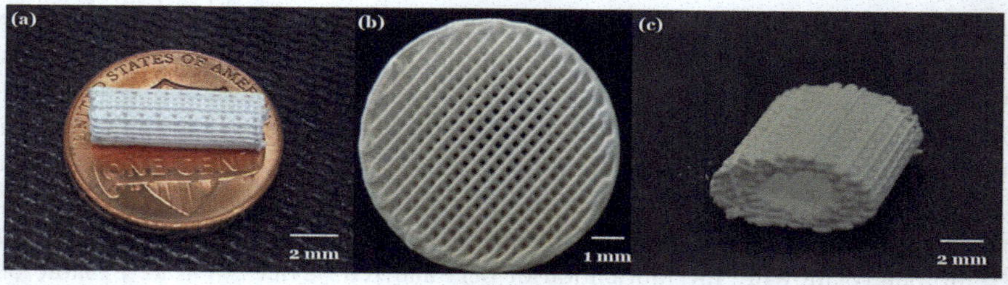

Fig. 4.1 Digital images of different scaffold configurations used for regeneration of boney defects of **a** radial; **b** calvaria; and **c** full-thickness mandibular defect

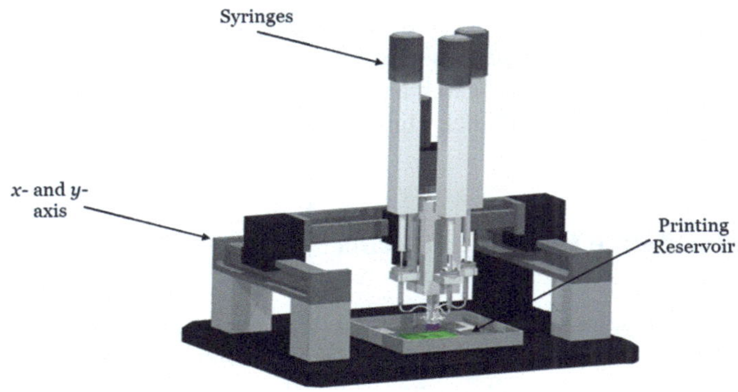

Fig. 4.2 Schematic of Direct-Ink Write 3D printer (3D Inks LLC, Tulsa, OK) used to assemble scaffolds. The machine is composed of 3 syringe pumps, along with a x-, y-, z-axis gantry. The printer deposits the primary material and secondary support structure in a reservoir located directly below the printing stage

from the saved programming code file and 'extrudes' successive layers (z-axis), in the x- and y-axis, of colloidal gel to fabricate the tailored lattice scaffold. Notably, the DIW process coupled with the bioactive ceramic colloidal inks does require a post-processing step of sintering, which involves heating the construct to 1100 °C to densify the structure in order to achieve sufficient mechanical properties for surgical handling (Fig. 4.3).

The DIW method presents itself as one of the most favorable to the biomedical field, due to its ability to not only tailor the colloidal ink suspensions to closely mimic properties of native bone [7, 17], but also simultaneously design patient-specific fit and fill scaffolds. A recent literature search revealed many publications

Fig. 4.3 Lateral view of 3D-printed β-TCP scaffold being held with surgical forceps for alveolar ridge reconstruction in a pre-clinal rabbit model

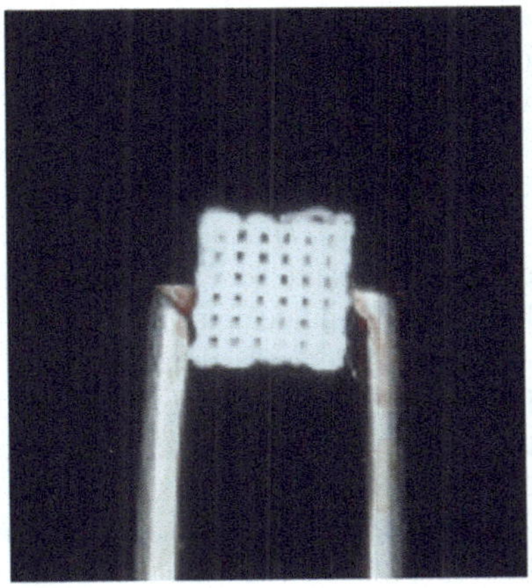

demonstrating successful printing of volumetrically complex lattice structures for various defects, but there is a paucity of relevant in vivo testing of such constructs. Of note, there has been more research and publications made available about these bioactive ceramic scaffolds and their clinically relevant applications [14, 15, 18].

4.4 Regenerative Pharmaceuticals

On a molecular level, bone regeneration is a multifaceted process, which is continuously remodeling, and in addition, there is osseoinduction with the commitment of osteogenic progenitor cells to differentiate [3, 19]. Combining patient-specific 3D-printed scaffolds, with organized pores augmented with osseoinductive pharmaceuticals has the potential to increase osteoblast activity and decrease osteoclast activity, resulting in increased formation of new bone [19, 20]. Therefore, the implementation of in vivo studies provides important insight into the role of agonists, (e.g., adenosine) with respect to bone regeneration. Adenosine's effects are systemic because its receptors can be found in nearly all tissues. Unfortunately, due to its limited extracellular presence and a restricted regulatory pathway ensuring a reduced half-life, non-stressed cellular environments do not accumulate adequate extracellular adenosine concentrations to activate adenosine receptors. Therefore, the ideal approach to adenosine-induced bone regeneration would target the local delivery pathway that can facilitate extracellular concentrations to activate adenosine receptors, ultimately inducing bone formation.

The application of targeting pathways at specific sites has become feasible with application of 3D-printed bioactive ceramic scaffold carriers enabling the delivery of osseoinductive pharmaceuticals. The commercially available and FDA-approved osteogenic agent recombinant human bone morphogenic protein (rhBMP-2), which is commonly delivered locally, results in increased osteogenesis when compared to scaffold alone [20–22]. Unfortunately, rhBMP-2 has been associated with significant disadvantages such as ectopic and exuberant bone formation and in pediatric patients, with premature suture closure in the craniomaxillofacial skeleton.

Alternatively, activators of purinergic receptors have exhibited favorable bone regenerative potential without the adverse side effects associated with rhBMP-2 (Fig. 4.4) [23–27]. Adenosine is now garnering attention due to its osteogenic properties. Adenosine serves as a marker of metabolic status at the cellular level and is known to attenuate activity for a wide array of cell types as a protective mechanism. Adenosine release is stimulated in sepsis [28], is implicated in neutrophil suppression [29] and induces a protective vasodilatory and negative inotropic effect on stressed cardiac vessels and tissue. This mechanism forms the basis of anti-platelet agents and those used in cardiac stress testing. Alternative approaches to adenosine receptor activation, such as pharmacological manipulation, have been previously explored and targeting this pathway has the capacity to achieve the necessary adenosine concentrations to affect its receptors without the need to induce a stressful cellular environment. The most notable pharmacologic agent is Dipyridamole (DIPY), an adenosine A_{2A}

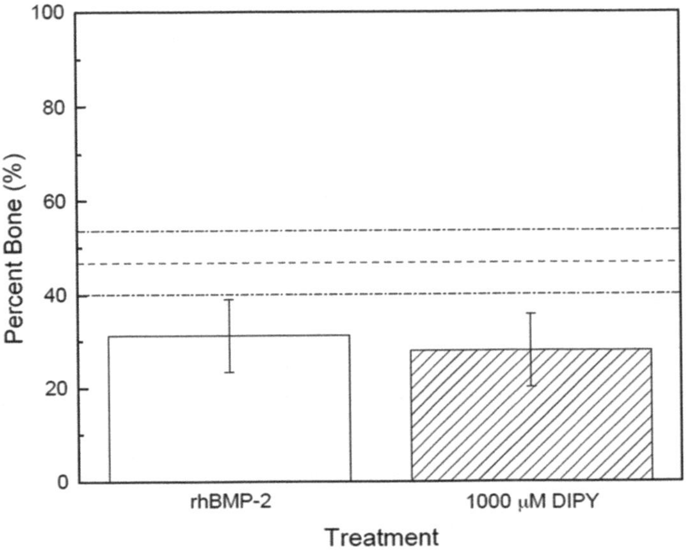

Fig. 4.4 Volumetric analysis of calvarial bone volume fraction regenerated by scaffolds treated with rhBMP-2 and 1000 μM DIPY at 8 weeks. Native bone porosity is denoted by dashed line; black dashed lines are 95% confidence intervals

receptor indirect agonist. DIPY functions via the Type 1 equilibrative nucleoside transporter (ENT1) to block adenosine reuptake into the cell and thereby results in extracellular accumulation (Fig. 4.5) [30–32]. Dipyridamole has recently been shown to further stimulate osteoblast function while limiting osteoclast formation; these principles have recently been reproduced in highly translational preclinical models.

For Dipyridamole (DIPY), the dosages needed to induce bone regeneration are much lower when delivered via scaffold compared to those via oral administration. Furthermore, the dose needed to achieve bone regeneration is lower than established systemic doses for pediatric patients [18]. Our hypothesis was validated that adenosine receptor activation augments the bone regenerative capacity of bioactive ceramic 3D-printed scaffolds; our group utilized the same scaffold material and design on mice critical sized calvarial defects (Fig. 4.6). The results of the study showed significantly increased amounts of bone formation for scaffolds immersed in dipyridamole prior to implantation, relative to the plain scaffold group. These strategies were further explored and confirmed in large translational sheep calvarial model (Fig. 4.7) [18]. This result among others has led to studies assessing the regenerative potential of these principles in various shaped defects in skeletally immature and mature translational models.

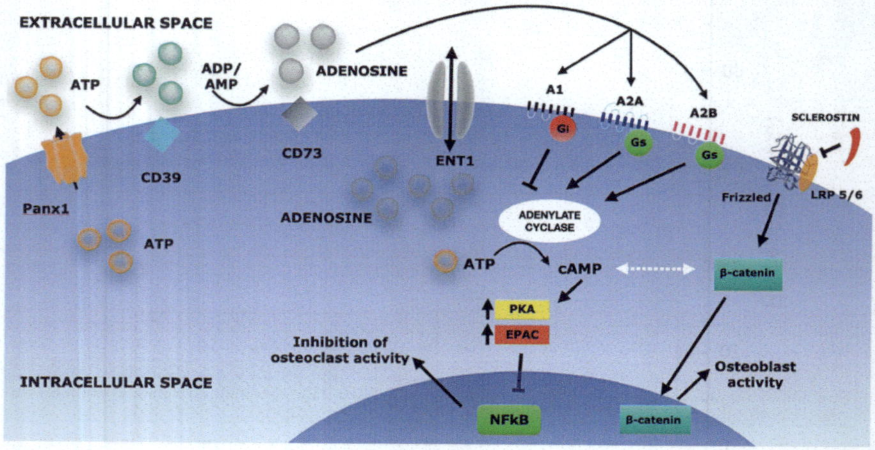

Fig. 4.5 Mechanism by which facilitates osteogenesis. Dipyridamole inhibits equilibrative nucleoside transporter 1 (ENT-1) leading to increased osteoblast activity and decreased osteoclast activity via increased $A_{2A}R$ signaling in an adenosine-dependent fashion. (Reprint with permission from [33])

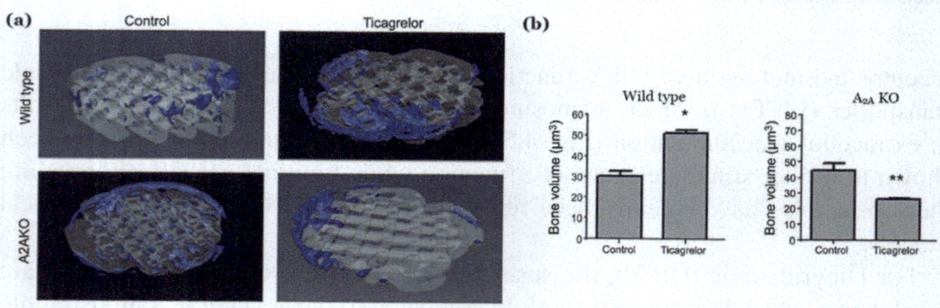

Fig. 4.6 Volumetric 3-D analysis of bone formation in a collagen-coated scaffold. **a** MicroCT images indicating new bone formation (blue) and the scaffold (gray). **b** Volume of new bone formation in the scaffolds at the site of trephination are presented as the means ± SEM. (Reprint with permission from [23])

4.5 Translational Evidence: Preclinical Models

Our group successfully utilized 3D-printed bioactive ceramic scaffolds in translational models for regeneration of critical-sized, full-thickness defects in both skeletally immature and mature rabbit and pig models [14, 34]. The results unequivocally showed that scaffolds seamlessly fit and filled critical-sized defects and bridged the defects with newly formed bone as early as 8 weeks in vivo [14, 35].

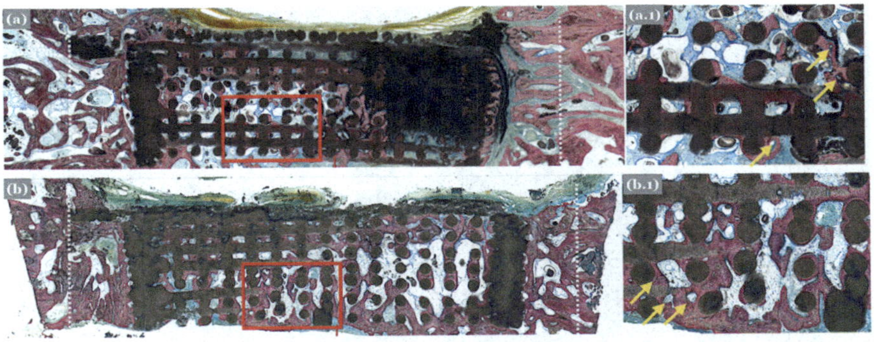

Fig. 4.7 Representative histologic micrographs from animals in the **a** CTRL and **b** DIPY group after 6-weeks of healing. (a.1 and b.1) Close-up depicting significant bony infill within the space between the defect margin and wall of the scaffold, with several osteons highlighted (Yellow Arrows) that provide evidence of lamellar reorganization. Osteon development is evident and more prevalent in the (b.1) DIPY group compared to (a.1) control. (Adapted and reprint with permission from [18])

The challenges for regenerating the skeletally immature bone are different from that of a skeletally mature bone. For example, alveolar cleft defects of the primary palate are associated with structural instability of the maxillary arch, an inability to support tooth eruption, and facial asymmetry. The current bone tissue engineering-based treatment modalities must take into consideration the developing craniomaxillofacial skeleton and thus cannot utilize treatment approach that may potentially induce asymmetrical growth.

Based on the promising preclinical results seen in experiments reconstructing calvaria and mandibular defects, the application of dipyridamole-loaded bioceramic scaffolds has also been investigated in critical-sized alveolar defects in translational skeletally immature rabbit and pig models, where scaffolds were coated with cross-linked bovine collagen to function as the carrier for localized DIPY delivery (Fig. 4.8). The DIPY augmented scaffolds were implanted and evaluated over 2-to-18-month periods. These preclinical experiments demonstrated the regenerative potential of 3D-printed scaffolds with dipyridamole to induce bone growth in the alveolar and calvarial bone of rabbits, with bone formation spanning the length of the established critical-sized defect. The long-term experiments, at 6- and 18-months, revealed regenerated alveolar bone is denser than unoperated bone [36], but a decrease in regenerated bone volume percentage from 6 to 18 months likely due to physiologic remodeling, culminating with similar bone density between regenerated and native contralateral alveolar bone. In addition, regenerated alveolar bone was functionally analogous to native bone in terms of mechanical properties (Fig. 4.9). This morphologic difference yet mechanical similarity has been noted in mandibular defect models [37].

In a similar study published by our group, no significant difference was noted in bone volume between the calvaria and alveolar defects repaired with 3DP-DIPY bioceramic scaffolds and autologous bone graft at 6 months (Fig. 4.10 a anb b)

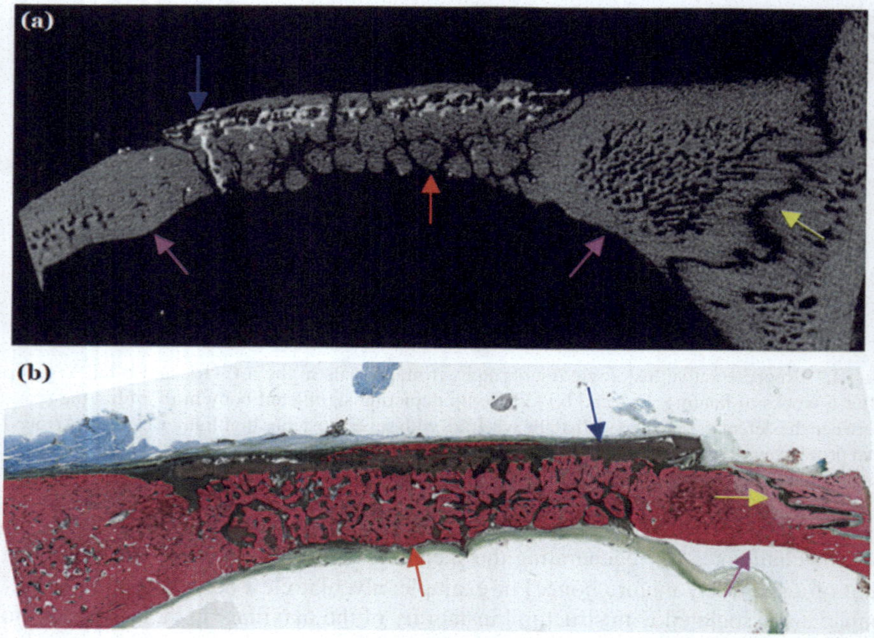

Fig. 4.8 **a** 2D microCT cross-section and **b** histological micrograph showing the remaining scaffold (blue arrow) and regenerated bone (red bone). Native bone (pink arrows) surrounding the scaffold from the periphery in a pre-clinical rabbit model

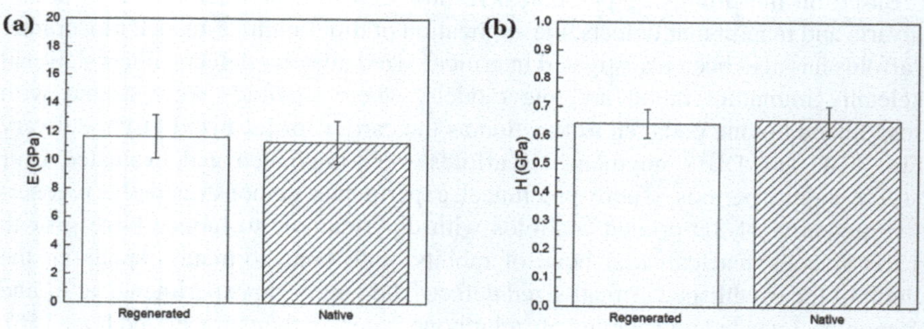

Fig. 4.9 **a** Elastic modulus (E) and **b** Hardness (H) of alveolar scaffold-regenerated bone yielded no statistical differences when compared to that of native control. (Mean ± 95% CI) [36]

[36]. Furthermore, regenerated bone density was significantly increased compared to native contralateral unoperated bone density yet with similar mechanical properties. Therefore, these results imply that regenerated alveolar bone tissue eventually remodels to native bone density without detriment to function when given enough time. Degradation rate and bone tissue properties were different between calvarial

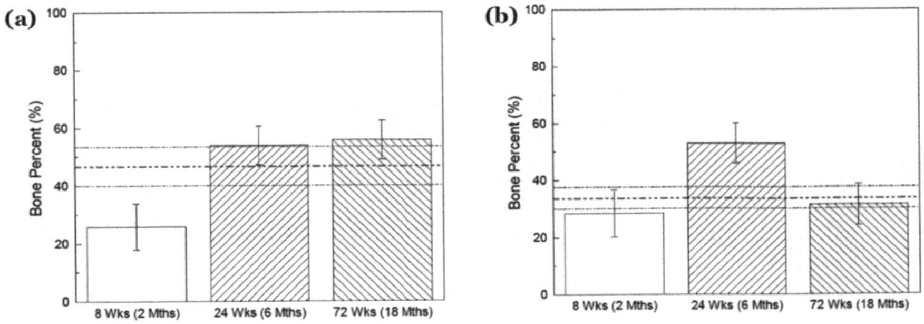

Fig. 4.10 Volumetric analysis of **a** calvaria and **b** alveolus bone yielding comparable bone volume fraction regenerated by scaffold compared to native bone (red dashed line; black dashed lines are 95% confidence intervals) at 18 months in both ($p = 0.064$ and $p = 0.337$, respectively). Error bars are 95% confidence intervals

bone and alveolar bone, highlighting the metabolic and mechanical heterogeneity in the craniomaxillofacial skeleton that is important to consider when applying tissue engineering principles to craniomaxillofacial reconstruction.

This most recent long-term animal study of immature rabbits through the time of facial maturity reported a degradation kinetic profile [38]. The study demonstrated an acceleration of β-TCP degradation to 55–90%/year through 3D printing augmented with DIPY (Fig. 4.11). The qualitative assessment indicated that the absorbed β-TCP was replaced by vascularized, organized bone, with histologic and mechanical

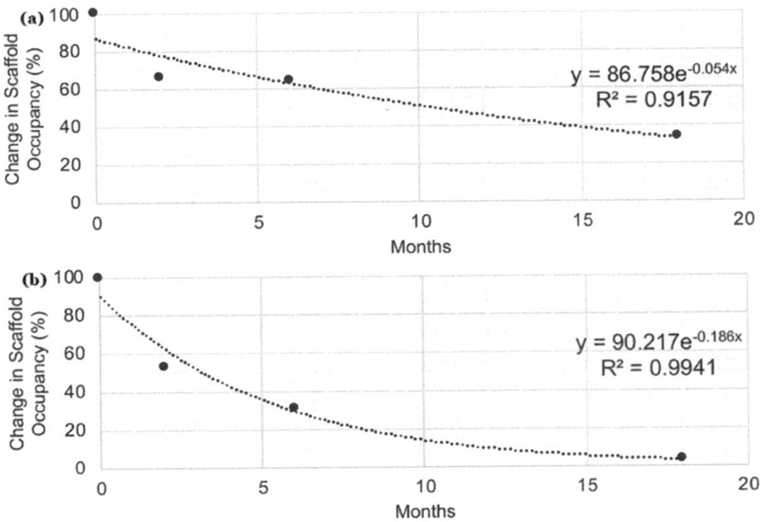

Fig. 4.11 Degradation kinetics analysis of scaffold over 18 months used to calculate annual degradation rate of β-TCP in calvarium and alveolus (54.6% and 90.5%, respectively). Change in Scaffold occupancy can be calculated using the following equation [38]

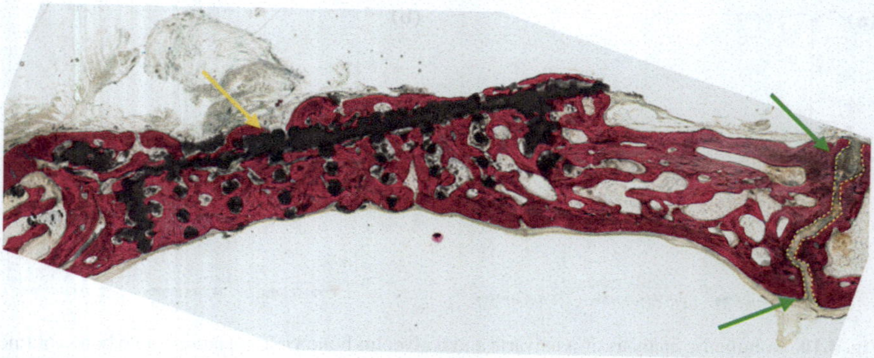

Fig. 4.12 Histological micrograph of 1000 μMDIPY scaffold after 24 weeks of healing in a rabbit calvaria. The yellow arrow points to the scaffold material, black, remaining. The green arrows, far right, point to the suture, with the dashed yellow line illustrating that it remained patent

properties similar to native bone and without damage noted to the growing suture (Fig. 4.12). Most importantly, craniomaxillofacial growth appeared to be comparable between 3D-printed β-TCP scaffold and autologous bone graft control - with no readily evident increase in asymmetry, ectopic bone formation, or morbidity or mortality associated with DIPY-coated 3D-printed β-TCP scaffold-treated defects.

$$Scaffold\ Occupancy = \frac{\%Scaffold\ at\ time,\ t}{\%Scaffold\ at\ time\ minus\ 1,\ (t-1)}$$

4.6 Conclusions and Future Directions

3D printing has the ability to create patient specific devices/scaffolds based on clinical imaging offering a valuable approach to personalized fit-and-fill bony defects (Fig. 4.13) [39]. Traditionally scaffolds were composed of bulk material such as hydroxyapatite [40], but 3D printing allows for the design of porous scaffolds that act as temporary void fillers similar to bulk scaffolds with the added benefit of pores that behave as healing chambers, guiding osteoconduction from the walls of the defect into pores via an intramembranous-like healing pathway and filling the scaffold lattice structure with woven bone. With the rapidly evolving clinical setting, new biomaterials and pharmacologic advances are likely to continue to be introduced to clinician and surgeons. Therefore, a concrete understanding of the array of clinically available regenerative treatment modalities, with translational potential, will facilitate further innovation.

Clinicians, engineers, and surgeons have all witnessed striking clinical and scientific innovations for craniofacial reconstruction. This has been in large part

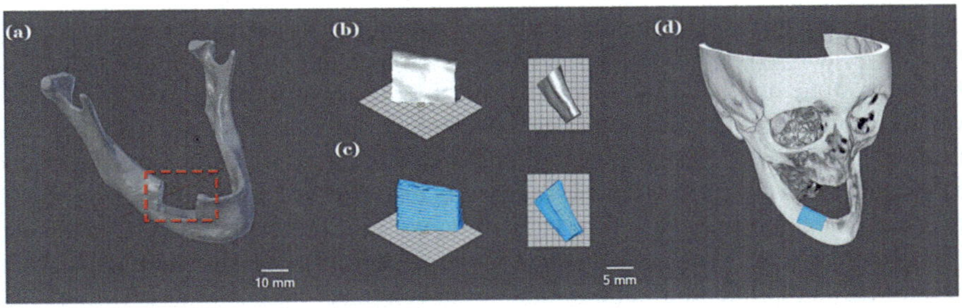

Fig. 4.13 A CT rendering of a patient's isolated mandible with the defect (region of interest) denoted in red dashed box, **b** the isolated volumetric section from the CT data, highlighted in red box in part **a** of figure, and **c** final converted multilayered scaffold rendition; **d** Volumetric rendering of a patient's scan with the custom 3D printed scaffold virtually placed at the site of defect. (Adapted and reprint with permission from [39])

due to the introduction of reconstructive microvascular surgery, autogenous non-vascularized bone grafting techniques, distraction osteogenesis, osteoconductive biomaterials, osteoinductive growth factors, computer-aided design and manufacturing (CAD/CAM), virtual surgical planning (VSP) coupled with three-dimensional (3D) printing technology [41–44]. This expansion of the reconstructive surgical armamentarium has allowed surgeons to have a myriad of techniques and technologies that address the functional and aesthetic needs of patients regardless of anatomic and/or medical challenges at their disposal. Such advances in the field have enabled craniofacial reconstruction to be categorized into four categories: (1) soft tissue pedicled flaps; (2) non-vascularized soft and hard tissue grafts; (3) soft and hard tissue vascularized grafts; and (4) alloplastic reconstructions with prosthetic appliances. On occasion, composite techniques can be used, such as the staged reconstruction of a defect where soft tissue is first added to a defect site followed by bone at a later time. A newer, more modern, technique using a hybrid of vascularized soft and hard tissue with nonvascular tissue engineered allogenic grafts, while reported in a limited number of cases has shown favorable outcomes [45].

With the advances in bone tissue engineering, there has been a paradigm shift, which has occurred over the past 10 years regarding the role and contribution of mesenchymal stem or progenitor cells (MSCs) toward bone tissue engineering. Previous preclinical large animal models using bone tissue engineering devices have demonstrated increased bone formation with the addition of both adipose stem cells (ASCs) and bone marrow-derived stem cells (BMSCs) [46, 47].

In 2010, Muschler et al. identified four strategies for the use of osteogenic cells with bone tissue engineering devices, including: (1) targeting local MSCs to a local site using osteoconductive scaffolds, (2) transplantation of MSCs with or without processing, (3) homing of MSCs into a wound site from regional tissues or systemic circulation using bioactive factors, targeting with selective binding or magnetic fields, and (4) ex vivo modification of MSCs to enhance their performance followed by introduction using the other listed strategies [48]. However these strategies presume that the contribution of MSCs towards bone formation is largely direct, meaning that the MSCs themselves differentiate into the heterogenous cell types which comprise trabecular bone and bone marrow. After all, the initial promising aspect of MSCs was their multipotency and ability to differentiate into various mesenchymal cell lineages [49]. However, studies examining the cell fate of MSCs in animal models demonstrate very little engraftment, suggesting instead that the contribution of MSCs to tissue formation may be indirect [50–53]. We propose that it may be adequate to provide constructs with osteogenic factors to recruit local MSCs rather than to directly include progenitor cells in bone tissue engineering constructs. Ultimately, for the reconstructive craniomaxillofacial surgeon omitting MSCs greatly reduces regulatory hurdles encountered when attempting clinical translation.

Tissue engineering for the craniomaxillofacial bone region is and has been the vision of surgeons for a long time. While the clinical potential of tissue engineering in craniomaxillofacial regeneration is yet to be clinically deployed in a large scale, promising patient specific constructs are under actively investigation in large preclinical translational models. As described, 3D-printed bioactive ceramic scaffolds with deliberate geometries for osseoconduction and osseoinduction provide regenerative cells with physical pathways to communicate with one another and fill defects. Additionally, for the first time, significant bone healing responses to bones such as the calvaria and mandible are being reported, where $A_{2A}R$ activation appears to stimulate local durable tissue which contributes to bone regeneration [15].

Acknowledgements The work presented in this chapter was supported by NIH/NIAMS (MPI- Cronstein, Coelho) 5R01AR068593-02, 3R01AR068593- 02S1, 5R01AR068593-03, & 3R01AR068593- 03S1, NIH/NICHD (MPI- Coelho, Flroes, Crosntein) R21HD090664-01, and DoD (MPI- Rodriguez, Coelho) W81XWH-16-1-0772. Dr. Coelho is a co-inventor of the 3D printing technology presented in this chapter (Patent #: US20150150681A1).

References

1. Inzana JA, Olvera D, Fuller SM et al (2014) 3D printing of composite calcium phosphate and collagen scaffolds for bone regeneration. Biomaterials 35:4026–4034
2. Simon JL, Michna S, Lewis JA et al (2007) *In vivo* bone response to 3D periodic hydroxyapatite scaffolds assembled by direct ink writing. J Biomed Mater Res A 83:747–758
3. Roberts TT, Rosenbaum AJ (2012) Bone grafts, bone substitutes and orthobiologics: the bridge between basic science and clinical advancements in fracture healing. Organogenesis 8:114–124
4. Moore WR, Graves SE, Bain GI (2001) Synthetic bone graft substitutes. ANZ J Surg 71:354–361
5. Kivrak N, Taş AC (1998) Synthesis of Calcium hydroxyapatite-tricalcium phosphate (ha-tcp) composite bioceramic powders and their sintering behavior. J Am Ceram Soc 81:2245–2252
6. Witek L, Shi Y, Smay J (2017) Controlling calcium and phosphate ion release of 3D printed bioactive ceramic scaffolds: An *in vitro* study. J Adv Ceram 6:157–164
7. Witek L, Smay J, Silva NRFA et al (2013) Sintering effects on chemical and physical properties of bioactive ceramics. J Adv Ceram 2:274–284
8. Coelho PG, Jimbo R (2014) Osseointegration of metallic devices: current trends based on implant hardware design. Arch Biochem Biophys 561:99–108

9. Jimbo R, Anchieta R, Baldassarri M et al (2013) Histomorphometry and bone mechanical property evolution around different implant systems at early healing stages: an experimental study in dogs. Implant Dent 22:596–603
10. Wilson CE, de Bruijn JD, van Blitterswijk CA et al (2004) Design and fabrication of standardized hydroxyapatite scaffolds with a defined macro-architecture by rapid prototyping for bone-tissue-engineering research. J Biomed Mater Res A 68:123–132
11. Steigenga JT, Al-Shammari KF, Nociti FH et al (2003) Dental implant design and its relationship to long-term implant success. Implant Dent 12:306–317
12. Bekisz JM, Liss HA, Maliha SG et al (2019) In-house manufacture of sterilizable, scaled, patient-specific 3d-printed models for rhinoplasty. Aesthet Surg J 39:254–263
13. Flores RL, Liss H, Raffaelli S et al (2017) The technique for 3D printing patient-specific models for auricular reconstruction. J Craniomaxillofac Surg 45:937–943
14. Lopez CD, Diaz-Siso JR, Witek L et al (2018) Three dimensionally printed bioactive ceramic scaffold osseoconduction across critical-sized mandibular defects. J Surg Res 223:115–122
15. Lopez CD, Witek L, Torroni A et al (2018) The role of 3D printing in treating craniomaxillofacial congenital anomalies. Birth Defects Res 110:1055–1064
16. Cesarano J III, Baer TA, Calvert P (1997) Recent developments in freeform fabrication of dense ceramics from slurry deposition. https://doi.org/10.2172/554831. Accessed 29 Jun 2021
17. Zhang B, Song J (2018) 3D-printed biomaterials for guided tissue regeneration. Small Methods 2:1700306
18. Bekisz JM, Flores RL, Witek L et al (2018) Dipyridamole enhances osteogenesis of three-dimensionally printed bioactive ceramic scaffolds in calvarial defects. J Craniomaxillofac Surg 46:237–244
19. Herford AS, Nguyen K (2015) Complex bone augmentation in alveolar ridge defects. Oral Maxillofac Surg Clin North Am 27:227–244
20. Teven CM, Fisher S, Ameer GA et al (2015) Biomimetic approaches to complex craniofacial defects. Ann Maxillofac Surg 5:4–13
21. Hidalgo DA, Rekow A (1995) A review of 60 consecutive fibula free flap mandible reconstructions. Plast Reconstr Surg 96:585–596
22. Hidalgo DA (1994) Condyle transplantation in free flap mandible reconstruction. Plast Reconstr Surg 93:770–781. (discussion 782–3)
23. Mediero A, Wilder T, Reddy VS et al (2016) Ticagrelor regulates osteoblast and osteoclast function and promotes bone formation *in vivo* via an adenosine-dependent mechanism. FASEB J 30:3887–3900
24. Mediero A, Wilder T, Perez-Aso M et al (2015) Direct or indirect stimulation of adenosine A2A receptors enhances bone regeneration as well as bone morphogenetic protein-2. FASEB J 29:1577–1590
25. Mediero A, Cronstein BN (2013) Adenosine and bone metabolism. Trends Endocrinol Metab 24:290–300
26. Mediero A, Frenkel SR, Wilder T et al (2012) Adenosine A2A receptor activation prevents wear particle-induced osteolysis. Sci Transl Med 4:135ra65.
27. Costa MA, Barbosa A, Neto E et al (2011) On the role of subtype selective adenosine receptor agonists during proliferation and osteogenic differentiation of human primary bone marrow stromal cells. J Cell Physiol 226:1353–1366
28. Martin C, Leone M, Viviand X et al (2000) High adenosine plasma concentration as a prognostic index for outcome in patients with septic shock. Crit Care Med 28:3198–3202
29. Cronstein BN, Kramer SB, Weissmann G et al (1983) A new physiological function for adenosine: regulation of superoxide anion production. Trans Assoc Am Phys 96:384–391
30. Monagle P, Chan AKC, Goldenberg NA et al (2012) Antithrombotic therapy in neonates and children: antithrombotic therapy and prevention of thrombosis, 9th ed: American College of Chest Physicians Evidence-Based Clinical Practice Guidelines. Chest 141(2 Suppl):e737S-e801S
31. Patrono C, Coller B, Dalen JE et al (1998) Platelet-active drugs: the relationships among dose, effectiveness, and side effects. Chest 114(5 Suppl):470S-488S

32. FitzGerald GA (1987) Dipyridamole. N Engl J Med 316:1247–1257
33. Lopez CD, Coelho PG, Witek L et al (2019) Regeneration of a pediatric alveolar cleft model using three-dimensionally printed bioceramic scaffolds and osteogenic agents: comparison of dipyridamole and rhBMP-2. Plast Reconstr Surg 144:358–370
34. Tovar N, Witek L, Atria P et al (2018) Form and functional repair of long bone using 3D-printed bioactive scaffolds. J Tissue Eng Regen Med 12:1986–1999
35. Maliha SG, Lopez CD, Coelho PG et al (2020) Bone tissue engineering in the growing calvaria using dipyridamole-coated, three-dimensionally-printed bioceramic scaffolds: construct optimization and effects on cranial suture patency. Plast Reconstr Surg 145:337e–347e
36. Wang MM, Flores RL, Witek L et al (2019) Dipyridamole-loaded 3D-printed bioceramic scaffolds stimulate pediatric bone regeneration *in vivo* without disruption of craniofacial growth through facial maturity. Sci Rep 9:18439
37. Yuan J, Cui L, Zhang WJ et al (2007) Repair of canine mandibular bone defects with bone marrow stromal cells and porous beta-tricalcium phosphate. Biomaterials 28:1005–1013
38. Shen C et al (2021) Transforming the degradation rate of β-tricalcium phosphate bone replacement using 3-dimensional printing. Ann Plastic Surg. https://doi.org/10.1097/sap.0000000000002965. PMID: 34611100
39. Witek L, Colon RR, Wang MM et al (2019) Tissue-engineered alloplastic scaffolds for reconstruction of alveolar defects. In: Mozafari M, Sefat F, Atala A (eds) Handbook of tissue engineering scaffolds, vol 1. Woodhead Publishing, United Kingdom, pp 505–520
40. Sheikh Z, Najeeb S, Khurshid Z et al (2015) Biodegradable materials for bone repair and tissue engineering applications. Materials 8:5744–5794
41. Marx RE, Tursun R (2013) A qualitative and quantitative analysis of autologous human multipotent adult stem cells derived from three anatomic areas by marrow aspiration: tibia, anterior ilium, and posterior ilium. Int J Oral Maxillofac Implants 28:e290–e294
42. Herford AS (2009) rhBMP-2 as an option for reconstructing mandibular continuity defects. J Oral Maxillofac Surg 67:2679–2684
43. Hirsch DL, Garfein ES, Christensen AM et al (2009) Use of computer-aided design and computer-aided manufacturing to produce orthognathically ideal surgical outcomes: a paradigm shift in head and neck reconstruction. J Oral Maxillofac Surg 67:2115–2122
44. Sacco AG, Chepeha DB (2007) Current status of transport-disc-distraction osteogenesis for mandibular reconstruction. Lancet Oncol 8:323–330
45. Kim BB, Zaid WY, Park EP et al (2014) Hybrid mandibular reconstruction technique: Preliminary case series of prosthetically-driven vascularized fibula free flap combined with tissue engineering and virtual surgical planning. J Oral Maxillofac Surg 72:e6–e7
46. Bajuri MY, Selvanathan N, Dzeidee Schaff FN et al (2021) Tissue-engineered hydroxyapatite bone scaffold impregnated with osteoprogenitor cells promotes bone regeneration in sheep model. Tissue Eng Regen Med 18:377–385
47. Liu G, Zhang Y, Liu B et al (2013) Bone regeneration in a canine cranial model using allogeneic adipose derived stem cells and coral scaffold. Biomaterials 34:2655–2664
48. Muschler GF, Raut VP, Patterson TE et al (2010) The design and use of animal models for translational research in bone tissue engineering and regenerative medicine. Tissue Eng Part B Rev 16:123–145
49. Pittenger MF, Mackay AM, Beck SC et al (1999) Multilineage potential of adult human mesenchymal stem cells. Science 284:143–147
50. Toma C, Wagner WR, Bowry S et al (2009) Fate of culture-expanded mesenchymal stem cells in the microvasculature: *in vivo* observations of cell kinetics. Circ Res 104:398–402
51. Caplan AI, Dennis JE (2006) Mesenchymal stem cells as trophic mediators. J Cell Biochem 98:1076–1084
52. François S, Bensidhoum M, Mouiseddine M et al (2006) Local irradiation not only induces homing of human mesenchymal stem cells at exposed sites but promotes their widespread engraftment to multiple organs: a study of their quantitative distribution after irradiation damage. Stem Cells 24:1020–1029

53. McBride C, Gaupp D, Phinney DG (2003) Quantifying levels of transplanted murine and human mesenchymal stem cells *in vivo* by real-time PCR. Cytotherapy 5:7–18

Lukasz Witek Dr. Witek is currently an Assistant Professor at the Department of Biomaterials and holds an affiliated position at the Department of Biomedical Engineering at NYU Tandon School of Engineering. Dr. Witek is an NYU alumnus; he earned his M.S. in Biomaterials Science under the mentorship of Dr. Paulo G. Coelho DDS, Ph.D. in 2011. Dr. Witek earned his Ph.D. in Chemical Engineering from Oklahoma State University in 2015, completing his thesis under the guidance of James E. Smay, Ph.D. Dr. Witek's research interests include utilizing 3D printing technologies for biomedical/biomaterial applications. His work focuses on extrusion-based, three-dimensional printing of bioactive ceramic scaffolds for bone regeneration.

Vasudev Vivekanand Nayak Mr. Nayak is a mechanical engineer by training with a specialization in additive manufacturing and biomaterials for soft and hard tissue regeneration. His recent work focused on determining the bactericidal efficacy of silver ions augmented into β-Tricalcium Phosphate. Mr. Nayak's doctoral research is dedicated towards engineering novel printing techniques and biocompatible shape memory polymers to cater towards the emerging need of minimally invasive surgery in orthopedics. Mr. Nayak, who is a Ph.D. candidate and currently serves as a research assistant at the Craniomaxillofacial Orthopedic Biomaterials Regenerative Applications Lab under the guidance and leadership of Drs. Lukasz Witek and Paulo G. Coelho.

Christopher M. Runyan Dr. Runyan is an Assistant Professor at the Department of Plastic and Reconstructive Surgery at Wake Forest Baptist Medical Center. Dr. Runyan is a member of the American Cleft Palate-Craniofacial Association and the American Society of Craniofacial Surgeons. His specialty areas include pediatric plastic surgery and craniofacial reconstructive surgery. His research areas include bone regeneration, cleft lip and palate, craniosynostosis, Pierre Robin Sequence, and distraction osteogenesis.

Nick Tovar Dr. Tovar is a PGY-2 Oral Maxillofacial Surgery Resident at NYU's Langone-Bellevue Hospital. Dr. Tovar previously was an Adjunct Scientist in the Department of Biomaterials at New York University College of Dentistry in New York. His primary research focuses are on the rapid repair and regeneration of damaged tissue due to trauma or disease, in an effort to develop bone graft materials and metal-based implants for the dental and orthopedic applications.

Sharbel Elhage Dr. Elhage is currently a surgery resident at the Carolinas Medical Center in Charlotte, North Carolina. He attended Clemson University and the Medical University of South Carolina.

James C. Melville Dr. Melville is an associate professor in the Bernard and Gloria Pepper Katz Department of Oral and Maxillofacial Surgery at UTHealth School of Dentistry. He is a diplomate of the American Board of Oral and Maxillofacial Surgery and a fellow of the American Association of Oral and Maxillofacial Surgeons, the American Academy of Craniomaxillofacial Surgeons, and American College of Surgeons. He has written a Springer textbook on maxillofacial reconstruction and contributed to 18 chapters in textbooks concerning oral and maxillofacial surgery, oncology, maxillofacial reconstruction, microvascular reconstruction, and tissue engineering. In addition, he is the author of more than 30 articles in peer-reviewed journals. He is a reviewer for the Journal of Oral and Maxillofacial Surgery (JOMS), Triple OOOO Journal, Journal of Cranio-Maxillofacial Surgery, Journal of Tissue Engineering, Journal of Biomedical Materials Research, Experimental Dermatology and of Journal of American Dental Association (JADA).

Simon Young Dr. Young is an Associate Professor in the Department of Oral & Maxillofacial Surgery at The University of Texas Health Science Center at Houston School of Dentistry and a board-certified oral and maxillofacial surgeon with significant experience in the treatment and reconstruction of craniomaxillofacial trauma and pathology.

Dr. Young's research efforts include the synthesis and characterization of implantable biomaterials designed to elicit in situ cell recruitment and programming. His work includes the use of materials able to simultaneously deliver multiple bioactive factors with distinct release profiles. These constructs have been used in a diverse set of applications such as the promotion of craniofacial bone regeneration and cancer immunotherapy. He has broad experience in the fields of polymer synthesis and characterization, growth factor delivery, in vivo models, characterization of bone and neovascularization, cancer immunotherapy, and implantable therapeutic cancer vaccines.

The focus of his current research is to develop novel material-based immunotherapies for the treatment of head and neck cancer and explore mechanisms of how this approach may synergize with other modalities of treatment such as chemo/radiation therapy. The paradigm of in situ cell programming using biomaterials is also being explored using layer-by-layer technology to deliver bioactive factors for the enhancement of craniofacial bone regeneration and multidomain peptide hydrogels for nerve regeneration.

David Kim Dr. Kim is currently a Plastic Surgery Fellow in the Division of Plastic and Reconstructive Surgery at the Lewis Katz School of Medicine, Temple University. He graduated from Cornell University with a Bachelor of Arts degree and earned his Doctor of Medicine at the University of Maryland School of Medicine. He completed his surgical internship and residency at Brown University, where he served as chief resident in surgery. His clinical and academic interests include craniofacial surgery and general and microsurgical reconstruction.

Bruce N. Cronstein Dr. Cronstein is a rheumatologist and the Dr. Paul R. Esserman Professor of Medicine in the Department of Medicine at NYU Grossman School of Medicine. Dr. Cronstein also serves as director of the Clinical and Translational Science Institute and chief of the Division of Translational Medicine at NYU Langone. His research focuses on the anti-inflammatory effects of adenosine and its role in wound healing and bone metabolism.

Roberto L. Flores Dr. Flores currently serves as the Joseph McCarthy Associate Professor of Reconstructive Plastic Surgery in the Hansjörg Wyss Department of Plastic Surgery at NYU Langone. He is board certified in plastic and reconstructive surgery. Dr. Flores is a pioneer in the construction of digital animation and simulation tools to teach surgeons complex reconstructive procedures of the face. His top areas of expertise include micrognathia, isolated pierre robin sequence, pierre robin sequence, and crouzon syndrome.

Paulo G. Coelho Prof. Coelho has a Doctor of Dental Surgery degree from the Pontificia Universidade Catolica do Parana in Brazil as well as a Ph.D. in Materials Engineering at the University of Alabama at Birmingham. Currently, he is a Leonard I. Linkow Professor in in the Department of Biomaterials at the NYU College of Dentistry, Mechanical and Aerospace Engineering at NYU Tandon School of Engineering, and at the Hansjörg Wyss Department of Plastic Surgery at NYU School of Medicine. Dr. Coelho's research interests include hard and soft tissue engineering for reconstructive surgery, systemic effects on tissue healing, preclinical evaluation of implantable devices, cell and tissue responses to biomaterials, and designing and testing of biomedical devices.

Chapter 5
Clinical Application of Monoclonal Antibodies: Key Technological Advances and Treatment of Osteoporosis

Sian Yik Lim

Abstract Osteoporosis is a skeletal disorder in which bone strength is decreased, leading to increased risk of fracture. In recent years, monoclonal antibodies have increasingly been used to treat human diseases, including osteoporosis. In this article, we discuss two monoclonal antibodies currently approved to treat osteoporosis-denosumab, and romosozumab. We aim to give the reader a basic understanding of the basic properties of monoclonal antibodies, the technology used to develop them, and their clinical application in the treatment of osteoporosis.

Keywords Osteoporosis · Monoclonal antibodies · Denosumab · Romosozumab · Skeletal disorder · RANKL · RANK · OPG · Sclerostin

5.1 Introduction

In recent years, monoclonal antibodies have increasingly been used to treat human diseases, including cancer, autoimmune disease, asthma, and osteoporosis. The total number of monoclonal antibodies approved by the United States Federal Drug Administration was 64 in 2018 with 11 new therapeutic monoclonal antibodies being approved in 2017 alone. The sales of antibody-based therapeutic agents is expected to reach approximately 170 billion in 2022, consisting of 20% of global pharmaceutical market [1]. Monoclonal antibodies now represent one of the most critical advances in pharmacological development in the past several decades.

Osteoporosis is a skeletal disorder in which bone strength is decreased, leading to increased risk of fracture [2]. Osteoporosis, and related fragility fractures are associated with significant cost, morbidity and mortality. The treatment of osteoporosis focuses on reducing fracture risk. In this article, we discuss two monoclonal antibodies currently approved to treat osteoporosis: denosumab, and romosozumab. We aim to give the reader a basic understanding of the basic properties of monoclonal

S. Y. Lim (✉)
Straub Clinic, Hawaii Pacific Health, Honolulu, HI, USA

Pali Momi Medical Center, Hawaii Pacific Health, Honolulu, HI, USA

© The Author(s), under exclusive license to Springer Nature Singapore Pte Ltd. 2022

A. H. Choi and B. Ben-Nissan (eds.), *Innovative Bioceramics in Translational Medicine II*, Springer Series in Biomaterials Science and Engineering 18,
https://doi.org/10.1007/978-981-16-7439-6_5

antibodies, the technology used to develop them, and their clinical application in the treatment of osteoporosis.

5.1.1 *Antibody—What is It?*

An antibody is a protein molecule synthesized by the immune system in response to an antigen. Figure 5.1 shows the structure of an antibody. Antibodies consist of four polypeptides. Two heavy chains and light chains join to form a Y-shaped molecule. Antibodies circulate in the blood, recognize foreign antigens such as bacteria and viruses, triggering an immune response, leading to the neutralization of bacteria and viruses [3]. An antigen is any substance that elicits the production of an antibody by the immune system. Antigens are large molecules, usually consisting of proteins or polysaccharides. An epitope is a part of an antigen molecule to which an antibody attaches itself. A single antigen is a large molecule and may contain several distinct epitopes.

Antibodies consist of a variable region, and a constant region (Fig. 5.1). The variable region contributes to the antigen binding site and determines an antibody's affinity to a specific antigen. When antibodies bind to antibodies through the variable region, they exert direct effects (inhibition of a toxin/enzyme of a pathogen). The constant region determines the antibody subtype. The constant region of the antibody interacts with accessory molecules to mediate indirect effector functions including antibody-dependent activation cellular cytotoxicity or complement cytotoxicity.

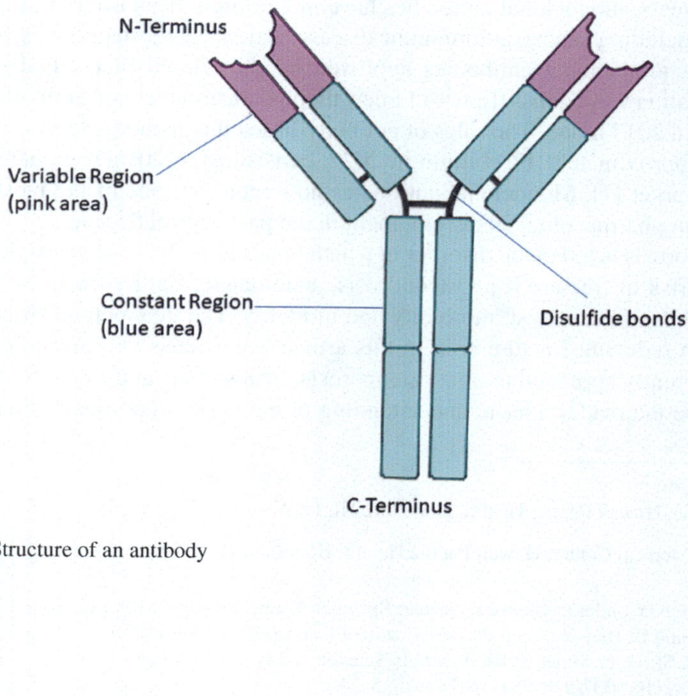

Fig. 5.1 Structure of an antibody

Table 5.1 Difference between monoclonal antibodies and polyclonal antibodies

	Monoclonal antibodies	Polyclonal antibodies
B cell source	Single B cell clone	Different B cell lineages
Antibodies	Homogeneous antibodies	Heterogeneous antibodies
Target	Single epitope of antigen	Different epitopes of a single antigen

5.1.2 B Cells—The Immune Response and Formation of Immunoglobulins

Each individual human possesses numerous clonally derived lymphocytes. The clonal selection theory states that each B cell clone arises from a single precursor and can recognize and respond to a distinct antigenic determinant. An antigen selects a specific preexisting clone and activates it, leading to the proliferation of a single cell line determined by the antigen. Memory cells remain in the circulation and proliferate when the body is exposed again to the antigen [4, 5].

The polyclonal B cell response is the typical way the human immune system responds to antigens. This response ensures that a single antigen is attacked at different overlapping parts of the antigen (epitopes). In this response, various B cell clones develop antibodies against distinct epitopes on a single antigen, leading to polyclonal antibodies. Polyclonal antibodies, to be differentiated from monoclonal antibodies, are a collection of antibodies that react towards a particular antigen, each identifying a different epitope (Table 5.1).

5.1.3 Monoclonal Antibodies: What Are They?

Monoclonal antibodies are identical antibodies produced from a single clone of genetically identical B cells. Monoclonal antibodies have high specificity to an epitope. In treating human disease, monoclonal antibodies are desirable because of their unique ability to target specific disease-related molecules.

Several advantages of monoclonal antibodies for the treatment of human disease as compared to sera-derived polyclonal antibodies include: (1) lower lot-to-lot variability, (2) low risk of pathogen transmission, (3) more significant activity per mass of protein due to specificity against the desired target, (4) lower rates of immunological complications such as serum sickness and immediate hypersensitivity [6].

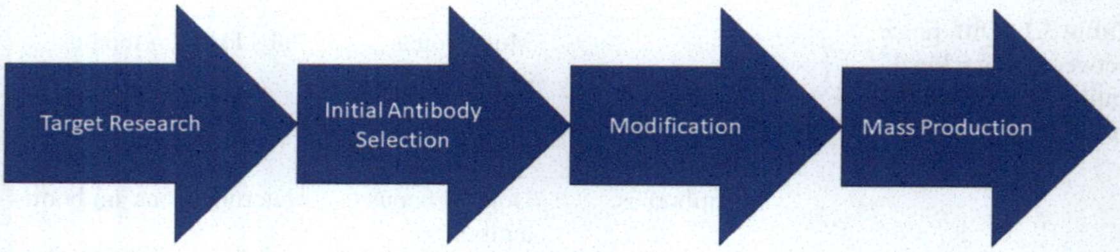

Fig. 5.2 Steps in developing monoclonal antibodies for treatment of human disease

5.2 Production of Monoclonal Antibody and Key Technological Advances

5.2.1 *Overview*

The production of monoclonal antibodies is a complex process. This section describes key concepts involved in producing monoclonal antibodies that are essential in understanding antibody treatment in osteoporosis, summarized in Fig. 5.2. Developing a monoclonal antibody requires prior research into the disease process's molecular pathways to identify targets for monoclonal antibodies. After identifying the target, the first step is antibody selection. This step involves creating/finding an antibody that has high affinity to the target antigen. One method to achieve this includes immunizing an animal with the target antigen. Other more advanced techniques include the screening of a library of antibodies developed through phage display technology. Once an antibody has been chosen, the second step involves modifications made to the antibody to generate a monoclonal antibody with the desired specificity, pharmacokinetics, and properties. Modification includes processes such as humanization and the development of antibody fragments to enhance pharmacokinetic properties and efficiency. In the third step, monoclonal antibodies are mass-produced in large quantities to treat disease. The initial technology used to achieve this was to create a hybridoma (see subsequent section). Other methods of producing immortal cell culture lines include the usage of Chinese hamster ovary cells.

5.2.2 *Hybridoma*

Plasma cell lines could not be directly used to produce monoclonal antibodies in vitro because plasma cell lines could only replicate a limited number of times before dying. The advancement of hybridoma technology was a significant step in developing monoclonal antibodies to treat human disease. In 1975, Georges Kohler and Cesar Milstein published their work titled "Continuous cultures of fused cells secreting antibody of predefined specificity" describing the hybridoma technique of production of monoclonal antibodies [7].

In the hybridoma technique, monoclonal antibodies are developed in a laboratory by first vaccinating mice with a target antigen. The target antigen stimulates B cells in the spleen of mice to produce specific antibodies. Subsequently, the B cells in the spleen are isolated. B cells that produce specific antibodies are fused with tumor cells with a strong ability to replicate, forming a hybridoma cell. Hybridoma cells, produce antibodies and are "immortal". Cultures to be cloned are subsequently selected for production of monoclonal antibodies. Development of hybridoma cells made unlimited production of monoclonal antibodies with predetermined specificity possible. Georges Kohler, and Cesar Milstein were awarded the 1984 Nobel Prize in Physiology for their work developing hybridoma cells.

5.2.3 Humanization of the Murine Antibody

However, murine antibodies are limited by (a) increased immunogenic potential; and (b) decreased efficacy. Therefore, subsequent efforts were made to transform murine antibodies to antibodies that were more similar to human antibodies. In the 1990s, chimeric antibodies were developed by combining sequences of the murine variable domain with the human constant region. In chimeric antibodies, 66% of the antibody structure is of human immunoglobulin, while 33% is of mouse immunoglobulin. The first chimeric monoclonal antibody, Abciximab, was approved in 1994 was approved to treat heart disease. Subsequently, humanized antibodies were developed by the complementary-determining region grafting technique. Mice with an unmodified genome are used to create humanized monoclonal antibodies. After immunization, target DNA is extracted, complementary-determining region loops responsible for antigen binding are transplanted into the human variable domain framework sequence. In humanized monoclonal antibodies, only the complementary determining region (CDR) is of mouse sequence origin. Humanized antibodies only contain 6–10% mouse proteins, while the 90–94% is human proteins [8]. Romosozumab, one of the monoclonal antibodies described extensively in this chapter, is a humanized monoclonal antibody.

The development of fully human antibodies was made possible by the discovery of phage display by Sir Gregory P Winter and transgenic mice engineered to produce human antibodies. Phage display allows the development of large antibody libraries derived from the human B-cell repertoire. Phage display facilitates screening for target binding antibody fragments within an extensive library without having to screen each antibody individually. Transgenic mice are engineered to produce human antibodies by replacing mouse immunoglobulin G loci with human immunoglobulin loci in the germlines of mice deficient in mouse antibody production. When transgenic mice are injected with a human therapeutic target, they produce fully human monoclonal antibodies. Target DNA is then extracted, with antibody genes cloned. Monoclonal antibodies are then expressed from the Chinese hamster ovary cells. Adalimumab was the first fully human antibody approved to treat rheumatoid arthritis in 2002. Denosumab, the other monoclonal antibody described in this chapter, is a

Table 5.2 Comparison of development of humanized monoclonal antibody versus fully human monoclonal antibody

	Humanized monoclonal antibody (i.e., romosozumab)	Fully human monoclonal antibody (i.e., denosumab)
Animal used	Mouse with unmodified genome	Mouse IgG locus is replaced with human DNA locus
Antibody selection	Mouse injected with human therapeutic target	Mouse injected with human therapeutic target
Modification	Target specific DNA extracted (CDR regions from mouse antibody grafted onto human antibody framework)	Target specific DNA extracted
Mass production	Expression of monoclonal antibody using Chinese Hamster Ovary Cells	Expression of monoclonal antibody using Chinese Hamster Ovary Cells

Table 5.3 Classification of monoclonal antibodies based on the composition of mouse/human proteins

	Composition	Substem B (International Nonproprietary Names Working Group nomenclature system before 2017)	Suffix
Murine antibodies	Mouse proteins	O	Mab
Chimeric antibodies	Approximately 2/3 human proteins /approximately 1/3 mouse proteins	Xi	Mab
Humanized antibodies	Approximately 90–94% human proteins /approximately 6–10% mouse proteins	Zu	Mab
Fully human antibodies	Fully human proteins	U	Mab

fully-humanized monoclonal antibody that was developed using Xenomouse transgenic mouse technology [9]. Table 5.2 contrasts the development of humanized monoclonal antibodies as compared to the development of fully human monoclonal antibodies. Table 5.3 summarizes the classification of monoclonal antibodies based on the composition of mouse/human proteins, and the nomenclature used in their naming.

5.2.4 *Modulation of Antibody Effector Functions*

As previously described, the Fc portion of an antibody interacts with accessory molecules, leading to indirect effector functions. Manipulating the sequences and glycosylation of the hinge and constant domains can modify antibody effector functions and circulating half-life [10, 11]. Currently, every clinically used monoclonal antibodies are immunoglobulin G monoclonal antibodies. The IgG class consists of IgG1, IgG2, IgG3, and IgG4 subtypes. IgG1 and IgG3 are potent triggers of effector mechanisms, while IgG2 and IgG4 have little indirect effects. In the development of monoclonal antibodies, one of the factors affecting antibody isotype choice is the effector functions required to mediate a biological process that confers a desired disease outcome [11]. The monoclonal antibodies used for osteoporosis treatment are of IgG2 isotype because antibodies with IgG2 isotype have little indirect effects [12].

5.3 The Naming of Monoclonal Antibodies

The International Nonproprietary Names Working Group developed a specific structure to name monoclonal antibodies. Denosumab and Romosozumab were named based on the older nomenclature system before 2017, where monoclonal antibody names consist of a prefix, Substem A, Substem B, and suffix. The prefix has no specific criteria, is the unique name of each of the monoclonal antibodies. The antibody's target determines Substem A (i.e., *tu* for tumor; *os*/*so* for bone). In contrast, the sequence from which the monoclonal antibody was derived determines Substem B (i.e., *o* for mouse; *xi* for chimeric; *zu* for humanized; *u* for fully human antibodies) (Table 5.3). The suffix *mab* is a common stem for monoclonal antibodies. Denosumab (*den*) is a fully humanized monoclonal antibody that targets bone used to treat osteoporosis (prefix: *den*, Substem A: *os* (target bone); Substem B: *u* (fully humanized); *mab*: monoclonal antibody). Romosozumab (*romo*) is a humanized monoclonal antibody that targets bone used to treat osteoporosis (prefix: *romo*; Substem A: *os* (target bone), Substem B: *zu* (humanized), *mab*: monoclonal antibody). In 2017, the International Nonproprietary Names Working Group issued new guidance to discontinue the use of Substem B.

5.4 Monoclonal Antibodies in Therapeutics: General Characteristics and Pharmacokinetics

The specificity of monoclonal antibodies towards the target antigen allows precise targeting of pathways involved in disease processes. Monoclonal antibodies modulate cells directly through binding to cell surface receptors, membrane-bound proteins, growth factors, and circulating proteins [13]. Monoclonal antibodies can also have indirect effects, mediated by the Fc region of the antibody. Indirect effects of monoclonal antibodies include: (1) recruitment of natural killer cells, monocytes, and macrophages, leading to antibody-dependent cellular cytotoxicity, and (2) activation of the complement system leading to complement cytotoxicity.

Monoclonal antibodies are large complex proteins purified from living cells. Due to their large size, monoclonal antibodies cannot cross cellular membranes. Therefore, monoclonal antibodies are generally directed against extracellular targets. Monoclonal antibodies, in general, have long half-lives; hence, dosing can range from monthly to every several months. Monoclonal antibodies are usually administered intravenously and subcutaneously because the gastrointestinal tract denatures complex proteins.

After subcutaneous administration, lymphatic channels take up monoclonal antibodies, which then enter the circulation [14]. Monoclonal antibodies leave the vasculature primarily by convective transport. Convective transport depends on the blood-tissue hydrostatic gradient and the sieving effect of the paracellular pores in the vascular epithelium. Another mechanism of monoclonal antibodies leaving the vasculature is by endocytosis/pinocytosis via endothelial cells [10, 14]. Monoclonal antibodies are large-sized molecules and are too large to be cleared by the liver and kidney. The elimination of monoclonal antibodies occur through: (1) nonspecific elimination through proteolysis by reticuloendothelial system, and (2) more specific antigen mediated clearance by internalizing the monoclonal antibody-target complex into the target cell, followed by intracellular degradation [10].

Many monoclonal antibodies exhibit non-linear pharmacokinetics: at low concentrations, the drug is cleared by first-order kinetics (a constant proportion of the drug is eliminated per unit time); at high concentrations, a drug is cleared by zero-order kinetics (a constant amount of drug is eliminated per unit of time). Non-linear pharmacokinetics observed because target mediated elimination has a small capacity and hence is susceptible to saturation [14]. At low serum concentrations, rapid saturable target mediated elimination regulates the elimination rate of the antibody. However, when higher serum concentrations saturate target-mediated elimination, eliminating antibody-protein occurs more slowly via nonspecific endocytosis and other processes [14, 15]. Therefore, this leads to an increase in exposure to drug that is more than dose-proportional at higher serum concentrations. Table 5.4 summarizes the common characteristics of monoclonal antibodies.

5.5 Osteoporosis: Regulation of Bone Metabolism

Our skeleton consists of bones, which are living tissues. During human development, bones are formed through intramembranous or endochondral ossification. After formation, bone grows and changes shape by process of modeling. Bone modeling occurs where osteoclasts and osteoblasts work independently to reshape or shape bone. Bone modeling occurs primarily in childhood but continues throughout life. Modeling leads to a net increase in bone mass. Genetic factors, combined with environmental factors such as physical strain and hormonal factors, regulate

Table 5.4 Characteristics of monoclonal antibodies

Characteristics	Monoclonal antibody
Size/Structure	High molecular weight protein
Target	High specificity, targeting extracellular targets
Half-life	Long half-life–from days to weeks
Production	Culture derived
Administration	Intravenously or subcutaneously
Absorption	Lymphatic system (after subcutaneous administration)
Distribution	Limited due to large size-convective transport, endocytosis/pinocytosis via endothelial cells
Excretion	Reticuloendothelial system, antigen mediated clearance. Not through the liver or kidneys

bone modeling. Parathyroid hormone and sclerostin inhibition also stimulate bone modeling [16].

Bone remodeling renews the skeleton. The critical cells involved in bone remodeling are osteoblasts and osteoclasts within a bone remodeling unit. In the remodeling process, osteoclasts and osteoblasts work sequentially within the same bone remodeling unit. The first step of remodeling is bone resorption. During bone resorption, osteoclasts remove mature bone from the skeleton. Bone resorption occurs when osteoclasts break down an old bone area, forming a pit or a small hole (resorption pit). Ossification (laying down of new bone) then follows resorption. During ossification, osteoblasts move into the resorption pit and start filling it with bone matrix. The bone matrix is then mineralized, forming mature bone. The bone remodeling process takes place over several months: resorption is a more rapid process (2–4 weeks), while bone formation takes up to 4–6 months to complete [17, 18].

During the process of remodeling, our skeleton breaks down small areas and rebuilds them. About 10% of the skeleton is remodeled annually [18, 19]. At that rate, the entire skeleton turns over about every ten years. Remodeling adjusts bone architecture to meet changing mechanical needs, also repairing microdamage in the bone matrix [18, 20]. This process occurs throughout life, maintaining or leading to a slight decrease in bone mass. Both systemic hormones and local factors regulate the process of bone remodeling. Systemic hormones, such as parathyroid hormone, calcitonin, vitamin D, and estrogen, affect bone remodeling. Many bone elements, known as cytokines (including RANKL), growth factors, local regulator proteins (including sclerostin), are also involved in regulating bone remodeling [21].

Understanding the bone remodeling and modeling processes as well as how it is regulated has allowed targeting specific disease-related molecules, leading to osteoporosis treatment advances. The following sections describe two monoclonal antibodies used to treat osteoporosis developed based on this understanding: (1) Denosumab, a monoclonal antibody that binds to RANKL, leading to inhibition of the RANKL/RANK/OPG system; (2) Romosozumab, a monoclonal antibody that binds to sclerostin, leading to activation of the Wnt signaling pathway.

Table 5.5 Bone turnover markers

Bone formation markers	Bone resorption markers
Bone-specific alkaline phosphatase (BSAP)	N-terminal cross-linking telopeptide (NTX) of type I collagen
Procollagen type I N propeptide (P1NP)	C-terminal cross-linking telopeptide (CTX) of type I collagen

5.6 Assessment of Bone Modeling and Remodeling: Osteoporosis Medication Efficacy

Histologic examination of bone biopsies is the gold standard to assess bone formation and resorption in bone modeling and remodeling. However, the collections of bone biopsies are invasive and are not widely available. Assessment of serum/urine biomarkers is an inexpensive and easier way to understand the whole-body bone formation and resorption at a given moment in time. When osteoclasts resorb bone, they release mature collagen fragments such as C-telopeptide (CTX) and N-telopeptide (NTX). CTX and NTX can be measured in the serum and urine. Measuring the bone mature collagen fragments allows assessment of bone resorption. Osteoblasts produce bone-specific alkaline phosphatase (BSAP) when they secret active bone matrix. When osteoblasts cleave components of collagen during collagen synthesis, they release procollagen N-terminal extension peptide (P1NP). Measuring of bone alkaline phosphatase and P1NP provides information about bone formation. Table 5.5 summarizes various bone turnover markers discussed in this chapter.

The efficacy of osteoporosis medications is assessed by their ability to reduce fragility fractures. However, using fracture as an end point is associated with significant cost and requirements of large population size. Therefore, bone mineral density (BMD) is measured as a surrogate for bone strength to gain additional information about osteoporosis medication efficacy [12].

5.7 Denosumab

5.7.1 RANKL/RANK/OPG System in Skeletal Health

The RANKL/RANK/OPG system was discovered in the mid-1990s. NF-kappa-β ligand (RANKL), expressed on osteoblasts and stromal cells, regulates osteoclast formation, function, and survival. RANK Ligand is a glycoprotein that binds to RANK on the osteoclasts, leading to maturation of CFU-M to mature osteoclasts, leading to increased bone resorption. Osteoblasts and stromal stem cells secrete osteoprotegerin. Osteoprotegerin prevents the binding of RANKL from binding to

RANK, reducing osteoclast formation. Osteoprotegerin, therefore, exerts control over the effects of RANKL on osteoclast function [22]. This understanding led to efforts targeting the RANKL/RANK/OPG system to treat osteoporosis. It was postulated that by inhibiting RANKL, osteoclast formation could be inhibited, therefore allowing the reduction of bone loss in osteoporosis.

5.7.2 Development of a Monoclonal Antibody Targeting the RANKL Pathway to Treat Osteoporosis

Although osteoprotegerin was effective in vitro in inhibiting osteoclast differentiation, very high subcutaneous doses of recombinant osteoprotegerin had to be administered to suppress bone resorption in mice. The full-length molecule with the C-terminal heparin-binding domain also exhibited poor pharmacokinetic and pharmacodynamic properties [12]. This led to the development of Fc-OPG fusion molecules, which comprised human OPG fused to the human immunoglobulin Fc region, which demonstrated higher potency, were more stable in vivo, and exhibited longer half-life [23]. A Phase I clinical study evaluated RANKL inhibition's efficacy on bone resorption using recombinant Fc-OPG. The primary outcome measures being bone turnover markers of urinary N-telopeptide (NTX) [24]. At the highest dose, an 80% reduction in NTX at four days after dosing, with bone turnover suppression lasting 45 days, was noted, suggesting targeting the RANKL was a viable candidate for the treatment for osteoporosis.

However, with Fc-OPG fusion molecules, there were concerns about developing anti-OPG antibodies in humans, which might cross-react with endogenous OPG, neutralizing its activity. One patient in the previously mentioned study did develop anti-OPG antibodies after one dose of Fc-OPG fusion molecule [24]. Although clinically non-significant, this raised concern because osteoporosis is a chronic condition requiring repeated administration of the Fc-OPG fusion molecules for continuous treatment [23]. Another potential problem was the binding of OPG to TNF-related apoptosis-inducing ligand (TRAIL), which was a natural defense mechanism of the body against cancer. Although osteoprotegerin had only a low affinity for TRAIL, this was a concern because chronic treatment was needed for osteoporosis [23].

Therefore, focus turned to the development of a specific anti-RANKL antibody. AMG 162, subsequently known as denosumab, was later developed. Denosumab is a fully human IgG2 monoclonal antibody that binds to both soluble and membrane-bound RANKL with high affinity [12]. As compared to the Fc-OPG fusion molecule, denosumab does not resemble OPG in molecule structure. Consequently, this avoids the issue of the development of anti-OPG antibodies [23]. Furthermore, denosumab does not bind to other TNF family members such as TRAIL. Thus, this alleviated concerns about interference with the defense mechanism against tumorigenesis [23]. A single dose of denosumab was more potent than the Fc-OPG fusion molecule. It showed more significant decreases in bone turnover markers at lower doses and more prolonged anti-resorptive effect at equivalent doses [24].

5.7.3 *Denosumab: Efficacy in Treatment of Osteoporosis*

Denosumab is a fully human IgG2 monoclonal antibody that binds to both soluble and membrane-bound RANKL with high affinity [12]. By binding to RANKL, denosumab prevents binding of RANKL to RANK. Preventing oligomerizing of RANK leads to decreasing osteoclast maturation, function, and survival. Denosumab primarily exerts its effect by reducing bone remodeling, leading to reduced bone breakdown and increased bone mass. Denosumab is administered as a subcutaneous injection, 60 mg once every six months. Denosumab was initially approved for the treatment of osteoporosis in postmenopausal women in 2010 [25]. In subsequent paragraphs we described the clinical development of denosumab. The key findings from Phases I, II, and III trials are summarized in Table 5.6 [23, 25–31].

In a Phase I trial of healthy postmenopausal women, a single dose of denosumab resulted in dose-dependent, rapid (within 12 h), profound (up to 84%), and sustained (up to 9 months) decrease in urinary NTX [23]. The efficacy and safety of subcutaneously administered denosumab were evaluated in 412 postmenopausal women with low bone density in a Phase II trial [26]. The primary endpoint was percentage change from baseline in bone mineral density at the lumbar spine at 12 months. A total of seven dosages and two dosing intervals (every three months, every six months) were evaluated. Twelve months of denosumab treatment led to increased bone density in the lumbar spine of 3.0–6.7%. In the hip and distal third of the radius, there was a gain in bone density of 1.9–3.6% and 0.4–1.3%, respectively.

After denosumab treatment, a decrease in serum C-telopeptide from baseline was noted three days after denosumab administration. The duration of bone turnover suppression was dose-dependent [26]. Bone specific alkaline phosphatase levels were decreased with a one-month delay. These findings are consistent with denosumab's anti-remodeling activity, where both bone resorption of osteoclasts and bone formation of osteoblasts are inhibited [26]. Denosumab demonstrated non-linear pharmacokinetics [26].

Based on the data from Phase II trial, dosing of 60 mg every six months was chosen because this was the lowest dose with the least frequent administration interval, which led to the most gain in bone mineral density in all skeletal sites. Furthermore, with 60 mg every 6-month dosing, bone turnover markers started to increase toward levels observed with open comparator alendronate at the end of the dosing interval. 60 mg every 6-month dosing was considered potentially more ideal than continuous suppression at higher doses of denosumab.

The primary trial for denosumab was the Phase III randomized FREEDOM trial [25]. The FREEDOM trial had 7868 women, with mean baseline T scores LS–2.8, TH–1.9, and FN–2.2. 22% had prevalent vertebral compression fracture; 82–84% of patients completed the study. The patients received every six months denosumab

Table 5.6 Clinical development of denosumab

Study ID/study name	Trial design/study context	Study population	Comparator groups/denosumab dose	Primary endpoint	Conclusions
Phase I Trials: Pharmacokinetics, Pharmacodynamics, Safety, and Tolerability Studies					
A Single-Dose Placebo-Controlled Study of AMG 162, a Fully Human Monoclonal Antibody to RANKL, in Postmenopausal women [23]	Randomized control trial	49 Postmenopausal women	Single Dose Denosumab for 6–9 months or placebo **Denosumab Doses**: Subcutaneous 0.01, 0.03, 0.1, 0.3, 1.0 and 3.0 mg/kg	(1) Safety/tolerability. (2) Pharmacokinetics/pharmacodynamics	(1) Dose-dependent reduction of urinary NTX, within 12 h, was profound and sustained (2) No serious adverse events
Phase II Trials: Dose/Schedule Finding, Preliminary Efficacy, and Safety Studies in Postmenopausal Women with Osteoporosis or Low Bone Mass					
Denosumab in postmenopausal women with low bone mineral density [26]	Randomized control trial	412 Postmenopausal women with low bone mineral density	Denosumab for 12 months or Open-label alendronate, or placebo **Denosumab Doses**: Subcutaneous – 6, 14, or 30 mg every 3 months –14, 60, 100, or 210 mg every 6 months	(1) Percentage change from baseline in bone mineral density at the lumbar spine at 12 months	(1) Bone mineral density gain in the lumbar spine 3.0–6.7% at various doses (compared with an increase of 4.6% with alendronate and -0.8% with placebo) (2) Bone mineral density gain in the total hip of 1.9–3.6%, distal third of the radius of 0.4–1.3% at various doses (3) Rapid reduction of serum levels of c-telopeptide within three days (more pronounced than alendronate)

(continued)

Table 5.6 (continued)

Study ID/study name	Trial design/study context	Study population	Comparator groups/denosumab dose	Primary endpoint	Conclusions
Effects of Denosumab on Bone Mineral Density and Bone Turnover in Postmenopausal Women [31]	Randomized control trial **Study Context**: Prevention trial of treatment native women with a lower risk for fracture	332 postmenopausal women with lumbar spine BMD T score between −1.0 and −2.5	Denosumab for 24 months or placebo **Denosumab Doses**: Subcutaneous 60 mg every six months	(1) Percentage change from baseline in bone mineral density at the lumbar spine at 24 months	(1) Bone mineral density gain in the lumbar spine 6.5% at 12 months with denosumab (−0.6% in placebo) (2) Statistically significant greater BMD gains with denosumab than alendronate at the total hip, 1/3 distal radius, and total body (3) Significantly suppressed serum CTX, tartrate-resistant acid phosphatase 5b, P1NP when compared to placebo
Phase III Trials: Clinical Efficacy and Safety in Osteoporosis Patients Compared to Placebo, or Standard of Care					
Study ID: NCT00089791 Fracture Reduction Evaluation of Denosumab in Osteoporosis Every 6 Months (FREEDOM) [25]	Randomized control trial **Study Context**: Pivotal trial for denosumab: Fracture prevention efficacy and safety study	7868 women with osteoporosis with bone mineral density of less than −2.5 but not less than −4.0 at the lumbar spine or total hip	Denosumab for three years or placebo **Denosumab Doses**:Subcutaneous 60 mg every six months	(1) New vertebral fracture (X-ray) over study period	(1) New vertebral fracture reduced by 68% (2) Non-vertebral fracture reduced by 20% (3) Hip fracture reduced by 40% (4) No increased risk of cancer, infection, cardiovascular disease, delayed fracture healing, or hypocalcemia

(continued)

Table 5.6 (continued)

Study ID/study name	Trial design/study context	Study population	Comparator groups/denosumab dose	Primary endpoint	Conclusions
Study ID: NCT00330460 A Randomized, Double Blind Study to Compare the Efficacy of Treatment with Denosumab versus Alendronate Sodium in Postmenopausal Women with Low Bone Mineral Density (DECIDE) [27]	Randomized control trial **Study Context**: Head-to-head study of denosumab versus alendronate in postmenopausal women with low bone mass	1189 early postmenopausal women with low bone mass	Denosumab for 12 months or alendronate 70 mg weekly **Denosumab Doses**: Subcutaneous 60 mg every six months	(1) Percentage change from baseline in bone mineral density in the total hip at month 12	(1) Total hip BMD increased by 3.5% at month 12, as compared to 2.6% in patients on alendronate (Statistically Significant) (2) Statistically significant greater BMD gains with denosumab than alendronate at the lumbar spine, femoral neck, and 1/3 distal radius
Study ID: NCT00377819 A randomized study to evaluate safety and efficacy of transitioning therapy from alendronate to denosumab in postmenopausal women with low bone mineral density (STAND) [28]	Randomized control trial **Study Context**: Head-to-head study of denosumab versus alendronate in postmenopausal women with low bone mass with prior alendronate treatment	504 postmenopausal women (1) Age 55 or older. BMD T-score of −2.0 or less and −4.0 or more (2) Prior alendronate treatment for at least six months	Denosumab for 12 months or alendronate 70 mg weekly **Denosumab Doses**: Subcutaneous 60 mg every six months	(1) Percentage change from baseline in bone mineral density in the total hip at month 12	(1) Total hip BMD increased by 1.9% at month 12, as compared to 1.05% in patients on alendronate (Statistically Significant) (2) Statistically significant greater BMD gains with denosumab than alendronate at the lumbar spine, femoral neck, and 1/3 distal radius

(continued)

Table 5.6 (continued)

Study ID/study name	Trial design/study context	Study population	Comparator groups/denosumab dose	Primary endpoint	Conclusions
Study ID: NCT00980174 A Multicenter, Randomized, Double-Blind, Placebo Controlled Study to Compare the Efficacy and Safety of Denosumab Versus Placebo in Males with Low Bone Mineral Density (ADAMO) [29]	Randomized control trial **Study Context**: Placebo-controlled efficacy and safety study in men with osteoporosis	242 male patients with osteoporosis	Denosumab for 12 months or placebo, followed by one year of denosumab **Denosumab Doses**: Subcutaneous 60 mg every six months	1) Percentage change from baseline in bone mineral density at the lumbar spine at 12 months	(1) Bone mineral density gain in the lumbar spine 3.0–5.7% at 12 months (2) Bone mineral density gain in the total hip of 2.4%, femoral neck 2.1%, and 0.6% at the one-third distal radius
Extension Trials					
Study ID: NCT00523341 Extension Study to Evaluate the Long-Term Safety and Efficacy of Denosumab in the Treatment of Osteoporosis [30] FREEDOM trial Extension Study		4550 women with osteoporosis who completed the FREEDOM Study	All patients received denosumab for additional seven years Long term group-total of 10 years of denosumab Cross over group- Three years of placebo followed by three years of denosumab Denosumab dose Subcutaneous 60 mg every six months	(1) Safety Monitoring	(1) Yearly exposure adjusted participant incidence of adverse events decreased from 165.3 to 95.9 per 100 participant years over ten years (2) Serious adverse events were stable over time-11.5 and 14.4 per 100 participant years (3) One atypical femoral fracture in each group during extension (4) Seven cases of osteonecrosis of the jaw in the long-term group, six cases of osteonecrosis of the jaw in the crossover group

(continued)

Table 5.6 (continued)

Study ID/study name	Trial design/study context	Study population	Comparator groups/denosumab dose	Primary endpoint	Conclusions
Study ID: NCT00091793 Study to evaluate AMG 162 in the Prevention of Postmenopausal Osteoporosis DEFEND Extension Study		256 women with low bone density who completed the DEFEND trial	After two years in the study, all patients followed for 24 months off treatment	(1) Safety Monitoring (2) Change in bone mineral density and bone turnover markers	(1) Bone mineral density declined after treatment was discontinued in the group the received denosumab for two years (2) The previously treated denosumab group maintained higher bone mineral density than the previously treated placebo group at all sites 3) Bone turnover markers increased above baseline. CTX increased within three months, NTX increased within six months. Both turnover markers returned to baseline by month 48

subcutaneously versus placebo for 36 months. The primary endpoint in the study was vertebral compression fracture at 36 months. The secondary endpoints were non-vertebral compression fractures and hip fractures. At three years, patients who received denosumab had 68% less new vertebral fractures, 20% less non-vertebral fractures, and 40% fewer hip fractures as compared to placebo.

Two Phase III head-to-head studies compared denosumab and alendronate, with the primary endpoint being gain in hipbone mineral density. The DECIDE trial included postmenopausal women with low bone mass who have never received alendronate (n = 1189 patients) [27]. In contrast, the STAND trial included postmenopausal women with low bone mass who had received at least six months of bisphosphonates [28]. The primary endpoint for both studies was the change in total hip BMD from baseline to month twelve. Other endpoints included change in bone turnover markers and change in BMD at the femoral neck, trochanter, lumbar spine, and one-third radius at 12 months. In both studies, denosumab was superior to alendronate in increasing hip BMD at the total hip. Denosumab increased bone density in all skeletal sites, but particularly cortical bone [27, 28]. Denosumab has been studied in men, as compared to placebo, denosumab treatment leads to gain in bone density at 12 months in the spine, hips and 33% distal radius in men with low bone mineral density [29].

5.7.4 Denosumab: Safety

Overall, in the FREEDOM study, there was no significant difference in the risk of infection, except for cellulitis between placebo and treatment groups. More than half of the cellulitis was erysipelas from a site in Mexico City. There were numerically more infection-related serious adverse events in the denosumab arm (4.1%) versus the placebo group (3.5%). However, the difference was not statistically significant [25]. The concern of the possibility of increased risk from infection is because certain lymphocytes express RANK ligand. However, the role of RANKL in the immune system is unclear. Eczema occurred more commonly in the denosumab group (3.0%, vs. 1.7%).

The FREEDOM trial was continued further for ten years (Fig. 5.3). Patients who received placebo for three years received Prolia subsequently for seven years, and patients who received denosumab in the first three years continued to receive denosumab (FREEDOM extension) [30]. The main reason for FREEDOM extension was mainly to monitor for the safety of denosumab. The FREEDOM Extension results were encouraging, where the yearly rates of adverse events for patients receiving denosumab being stable or declining throughout the study (165.3 per 100 patient-years to 95.9 per 100 patient-years) over ten years. Serious adverse events, including cellulitis, were stable over time between 11.5 and 14.4 per 100 patient-years [30].

Osteonecrosis of the jaw and atypical femoral fracture were uncommon. One case of atypical femoral fracture in the long-term group (year 7 of denosumab treatment) and one case in the cross over group (year 3 of denosumab treatment). The incidence

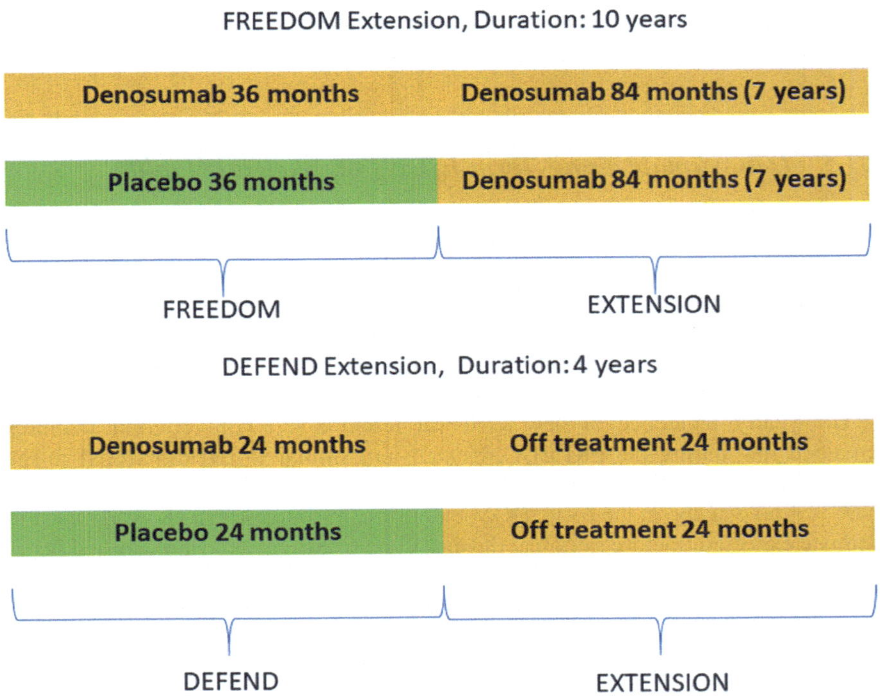

Fig. 5.3 Design of FREEDOM extension study and DEFEND extension study

of atypical femoral fractures was 0.8 per 10,000 patient-years. There were 13 cases of osteonecrosis of the jaw, 7 cases in the long-term group, and six in the crossover group. The incidence of osteonecrosis of the jaw was 5.2 per 10,000 patient-years.

At ten years in the FREEDOM Extension trial, cellulitis rates were persistently low (less than 0.1–0.2 per 100 patient-years). This finding was reassuring given the higher rates of cellulitis in patients who received denosumab as compared to patients who received placebo in the first three years of the FREEDOM study [30]. Infections declined over ten years, from 38.6 per 100 patient-years to 23.0 per 100 patient-years. Infection-related serious adverse events were stable at 1.1–2.6 per 100 patient-years. There was no increased risk of infection in the long-term treatment group and the crossover group.

A separate analysis has been done to compare adverse events of interest including malignancy, eczema, dermatitis, pancreatitis, endocarditis, and delayed fracture healing and serious adverse events of interest including infections, opportunistic infections, and cellulitis, or erysipelas between the first three years of the extension in the crossover and long-term denosumab groups relative to placebo and denosumab groups in FREEDOM. Notably, rates of adverse events incidence rates and serious adverse events incidence rates were similar or lower in the first three years of the extension in the crossover and long-term denosumab groups compared to the placebo and denosumab groups in the FREEDOM trial [32].

5.7.5 Denosumab: Long-Term Effects on Skeletal Health

With ten years of denosumab, there was an increase of bone mineral density in the spine of approximately 20%. Hip bone mineral density gained approximately 9–10% [30]. The bone density continues to increase and does not plateau as compared to teriparatide, where the BMD gain plateaus. Although denosumab mainly suppresses bone remodeling, denosumab may permit a modeling effect on the bone to proceed even though bone resorption is marked suppressed [33].

The yearly incidence of new vertebral fracture and non-vertebral fractures remained low during the FREEDOM extension, similar to rates observed in the denosumab group during the first three years of the FREEDOM study. Although there was no placebo in the FREEDOM extension, a simulation method was used to estimate expected fracture rates had the denosumab participants who enrolled in the extension study received placebo [30]. As compared to the simulated placebo group, the relative risk for new vertebral fractures and non-vertebral fractures were lower (new vertebral fractures 0.62, 95% Confidence Interval 0.47–0.80; non-vertebral fractures 0.54, 95% Confidence Interval 0.43–0.68).

In the FREEDOM extension, patients who continued to receive denosumab after receiving denosumab in the first three years had numerically lower non-vertebral fractures. The lower number of fractures suggests the possibility that non-vertebral fractures may decrease with prolonged denosumab use. On denosumab, a relationship of higher the BMD and lower the non-vertebral fracture incidence was noted [34]. This relationship plateaued at a T score of between −1.5 and −2.0. These data support the notion that treating osteoporosis to a specific T score threshold may be a reasonable target for osteoporosis treatment. Treating to a target may be feasible with denosumab because most osteoporotic women could achieve non-osteoporotic T scores after ten years of denosumab treatment. After ten years of treatment, 95% of patients had a total hip T score of more than −2.5, 81% had a total hip T score of −2.0, and 61% had a T-score of more than −1.5 [34].

Although denosumab effectively and rapidly suppresses bone remodeling, this effect is reversible after stopping of treatment [35–37]. This effect is consistent with the pathophysiology of osteoporosis and the pharmacokinetics and pharmacodynamics of monoclonal antibodies. In the DEFEND trial extension study, bone mineral density declined after denosumab treatment was discontinued in the group the received denosumab for two years (Fig. 5.3). After discontinuation of denosumab, bone turnover markers increased above baseline. CTX increased within three months, NTX increased within six months. Both turnover markers returned to baseline by month 48. Similar findings were noted in a study by Miller et al., which included a group of patients who received denosumab for 24 months. After 24 months patients either continued denosumab for an additional 24 months, discontinued therapy or discontinued treatment for 12 months then reinitiated denosumab for 12 months. Twenty-four months of denosumab treatment suppressed bone turnover markers. With stopping of denosumab at 24 months, bone turnover markers overshoot that of placebo and take two years to decrease back to baseline [35]. This correlated with a loss of BMD in the first 12 months of discontinuation of treatment. However, retreatment with denosumab led to regaining in BMD. The gain with retreatment suggests bone remained responsive with resumed denosumab treatment. Levels of bone turnover markers increased on discontinuation and decreased with retreatment [35].

This reversible effect is of consequence because of the concept of a drug holiday for osteoporosis. During a drug holiday, treatment is stopped temporarily for low-risk osteoporosis patients who have been on bisphosphonate treatment for five years to reduce the risk of side effects [38]. With the reversible effects of denosumab on bone, a drug holiday without transitioning to another agent may not be recommended. There have been reports of increased risk of fracture has after abrupt discontinuation of denosumab [39, 40]. Based on limited data, treatment with oral bisphosphonate after denosumab treatment has been suggested as a prudent strategy when discontinuing denosumab [36].

5.7.6 Denosumab in Patients with Chronic Kidney Disease

As previously described, monoclonal antibodies, including denosumab, are eliminated by reticuloendothelial macrophages. Because denosumab is not eliminated in the kidneys, no renal dosing is needed. Examining patients in the FREEDOM study stratified by renal function, Jamal et al. noted that fracture risk reduction and bone mineral density changes at all sites favored denosumab [41]. Treatment efficacy did not differ with kidney function. Notably, changes in calcium, creatinine level, and incidence of adverse events did not defer with kidney function [39]. It is essential to be aware of the increased risk of hypocalcemia in patients with chronic kidney disease. Therefore, ensuring normal calcium and vitamin D levels before administering denosumab is vital in chronic kidney disease patients.

5.8 Romosozumab

5.8.1 The Wnt Signaling Pathway and Sclerostin in Skeletal Health

The canonical Wnt-β-catenin pathway plays an essential role in skeletal health and bone remodeling [42]. When the Wnt ligand binds to a specific Frizzled family receptor and an LDL-receptor-related protein (LRP) co-receptor (LRP-5 or LRP-6 co-receptor), this activates wnt-responsive genes. This activation leads to a cascade of changes, ultimately leading to increased bone formation and decreased bone resorption.

When the Wnt-β-catenin pathway is activated, β-catenin accumulates within the cell. The accumulated β-catenin enters the nucleus, binds to the T-cell factor transcription factor leading to activation of Wnt-responsive genes. One of the effects is inducing osteoblast precursors to differentiate. Furthermore, activation of the Wnt-β-catenin pathway also leads to increased expression of osteoprotegerin. As previously described, osteoprotegerin inhibits the binding of RANK to RANKL, leading to inhibition of osteoclast differentiation [42]. The net effect of activation of the Wnt-β-catenin pathway is increased bone mass from increased osteoblasts and inhibition of osteoclasts. When the Wnt-β-catenin pathway is inhibited, a scaffolding protein called axin builds a destruction complex that phosphorylates β-catenin. Phosphorylated β-catenin is ubiquitinated and degraded by a proteasome. When β-catenin does not enter the nucleus, Wnt-responsive genes are not activated, leading to decreased bone formation and increased bone resorption [43–45].

Sclerostin is a glycoprotein encoded by the SOST gene and is produced by osteocytes. Sclerostin is an inhibitor of the Wnt-β-catenin and bone morphogenic protein signaling pathways, stimulating osteoblasts' proliferation. By binding to LRP 5/6, Sclerostin prevents Wnt from binding to LRP 5/6 and Frizzled family receptor, inhibiting the canonical Wnt signaling pathway [46]. The complex interaction between osteocytes, sclerostin, and the Wnt pathway mediates the coupling of mechanical stress on bone to the anabolic response. Sclerostin may be a crucial link between osteocyte mechanosensing and osteoblastic bone formation [47, 48].

5.8.2 Development of a Monoclonal Antibody Targeting Sclerostin to Treat Osteoporosis

Two rare genetic diseases provided insight into the possibility of pursuing sclerostin inhibition in osteoporosis treatment. Sclerosteosis is a condition where patients have homozygous mutations in the SOST gene [49]. Van Bunchem's disease is a condition where patients have a deleted region in distant regulatory elements involved in transcribing the SOST gene [50]. Both diseases have genetic abnormalities in the SOST gene leading to biologically less active sclerostin. In both conditions, the Wnt-β-catenin pathway is uninhibited, therefore leading to osteoblast hyperactivity. Osteoblast hyperactivity leads to bone overgrowth in the skull, mandible, ribs, clavicles, and long bones [49, 51]. These patients have a low risk of fracture due to their increased bone mass [49, 51].

For sclerosteosis, heterozygote carriers of the SOST mutation had low but detectable levels of sclerostin. They had high bone mass compared to healthy same-aged patients but lower than homozygous patients and were also less prone to fractures. This observation led to the initial hypothesis of circulating sclerostin's gene-dose effect, leading to different clinical phenotypes [52]. These observations were further supported by SOST knockout mice's findings having a high bone mass phenotype [53], while transgenic mice overexpressing the SOST gene having a low bone mass phenotype [54].

Several animal models have shown the effect of anti-sclerostin antibodies on bone formation. When ovariectomized rats, which develop estrogen deficiency-induced bone loss, were given anti-sclerostin antibody treatment reversed estrogen deficiency-induced bone loss. There was a significant bone mass increase with the anti-sclerostin antibody. When cynomolgus monkeys were treated with anti-sclerostin antibodies, BMD, and bone strength improved [55].

5.8.3 Romosozumab: A Monoclonal Antibody Against Sclerostin Leading to Significant Bone Growth

Romosozumab is a humanized IgG2 monoclonal antibody that targets the glycoprotein sclerostin. By inhibiting the wnt signaling pathway's inhibitor, the wnt signaling pathway is activated, leading to bone gain. With romosozumab, in contrast to anti-remodeling agents (i.e., denosumab): there is a rapid initial increase in bone formation, followed by a slighter more prolonged decrease of bone resorption. This combined bone formation and bone resorption reduction lead to significant bone gain with romosozumab treatment, with romosozumab being one of the most potent bone anabolic agents to date [56]. Inhibition of sclerostin by romosozumab predominantly seems to stimulate modeling-based bone formation at cancellous and endocortical surfaces [57]. Romosozumab is administered as a subcutaneous injection 210 mg once monthly. Romosozumab was approved in 2019 to treat osteoporosis in postmenopausal women at high risk of fracture. In the following paragraphs, we described the clinical development of romosozumab. The key findings from Phases I, II, and III trials are summarized in Table 5.7 [58–63].

Two Phase I trials of romosozumab assessed the safety of romosozumab. Both studies were placebo-controlled, randomized trials. In the first study, 72 healthy subjects were randomized to receive romosozumab subcutaneously (various doses), intravenously (various doses), or placebo [58]. Depending on the dose of romosozumab given, patients were followed for 29–85 days. The second was study investigated multiple doses of romosozumab in 48 healthy men and postmenopausal women [59]. The study involved three months of treatment followed by three months of follow-up after treatment. Both studies showed that romosozumab was well tolerated, with no deaths or significant safety findings.

Table 5.7 Clinical development of romosozumab

Study ID / Study Name	Trial Design / Study Context	Study Participants	Comparator groups / Romosozumab Dose	Primary endpoint	Conclusions
Phase I trials: pharmacokinetics, pharmacodynamics, safety, and tolerability studies					
Study ID: NCT01059435 A first-in-human study evaluating AMG 785 in healthy men and postmenopausal women [58]	Randomized control trial	72 healthy males or female (Age: 45–59)	One dose of romosozumab or placebo **Romosozumab Doses**: Subcutaneous 0.1, 0.3, 1, 3, 5 or 10 mg/kg Intravenous dose 1 or 5 mg/kg	(1) Safety/tolerability (2) Pharmacokinetics/pharmacodynamics	(1) Dose related increases in bone formation markers, with a dose related decrease in bone resorption markers leading to a large anabolic window (2) AMG 785 well tolerated, with no deaths or study discontinuations (3) One treatment-related serious adverse event of non-specific hepatitis but resolved after cessation of the drug

(continued)

Table 5.7 (continued)

Study ID / Study Name	Trial Design / Study Context	Study Participants	Comparator groups / Romosozumab Dose	Primary endpoint	Conclusions
Study ID: NCT01825785 A multiple dose study to evaluate the safety, tolerability, pharmacokinetics and pharmacodynamics of AMG 785 [59]	Randomized control trial	16 healthy males and 32 postmenopausal females (Age: 45–80 with low bone mass)	Romosozumab or placebo **Romosozumab Doses**: Subcutaneous *Postmenopausal women* – 6 doses of 1 mg/kg or 2 mg/kg (every 2 weeks) – 3 doses of 2 or 3 mg/kg (every 4 weeks) *Healthy men* – 6 doses of 1 mg/kg (every 2 weeks) – 3 doses of 3 mg/kg (every 4 weeks)	(1) Safety/tolerability (2) Pharmacokinetics/pharmacodynamics	(1) Increased bone formation marker P1NP by 66–147%, decreased bone resorption markers serum C-telopeptide by 15–50%. Bone mineral density gain in the lumbar spine by 4–7% (2) AMG 785 well tolerated, with no deaths (3) Adverse event rates balanced between groups. No significant safety findings
Phase II trials: dose/schedule finding, preliminary efficacy and safety studies in postmenopausal women with osteoporosis or low bone mass					

(continued)

Table 5.7 (continued)

Study ID / Study Name	Trial Design / Study Context	Study Participants	Comparator groups / Romosozumab Dose	Primary endpoint	Conclusions
Study ID: NCT00896532 Phase 2 Study of AMG 785 in Postmenopausal Women with Low Bone Mineral Density [60]	Randomized control trial	419 postmenopausal women (Age: 55–85 with low bone mineral density)	Romosozumab for 12 months at various doses **Romosozumab Doses**: Subcutaneous – Monthly: 70 mg, 140 mg, or 210 mg – Every 3 months: 140 mg, 210 mg	(1) Percentage change from baseline at month 12 in lumbar spine bone mineral density	(1) There were significant increases in lumbar spine bone mineral density at month 12 with all doses of romosozumab (2) Of all doses studied, romosozumab 210 mg monthly dosing had the largest bone mineral density gain of 11.3% (3) The lumbar spine bone density gain with monthly romosozumab 210 mg was more significant as compared to 0.1% decrease in patients on placebo, 4.1% increase in patients on alendronate, 7.1% in patients on teriparatide (4) Bone mineral density gain in the lumbar spine, total hip at various doses of romosozumab (5) Transitory increase in bone formation markers, with sustained decreases in bone resorption markers
Phase III trials: clinical efficacy, and safety in osteoporosis patients compared to placebo, or standard of care					

(continued)

Table 5.7 (continued)

Study ID / Study Name	Trial Design / Study Context	Study Participants	Comparator groups / Romosozumab Dose	Primary endpoint	Conclusions
Study ID: NCT01796301 An Open-label, Randomized, Teriparatide-controlled Study to Evaluate the Effect of Treatment with Romosozumab in Postmenopausal Women With Osteoporosis Previously Treated with Bisphosphonate Therapy (STRUCTURE) [61]	Randomized control trial **Study Context:** Head-to-head study of romosozumab versus teriparatide in postmenopausal women with osteoporosis previously on bisphosphonate	436 with postmenopausal osteoporosis Prior oral bisphosphonate for at least 3 years Alendronate the year before screening	Romosozumab or Teriparatide **Romosozumab Doses**: Subcutaneous 210 mg monthly	(1) Percentage change from baseline in areal BMD by dual-energy x-ray absorptiometry at the total hip through month 12	(1) Mean percentage change from baseline in the total hip areal bone mineral density was 2.6% in the romosozumab group and -0.6% in the teriparatide group
Study ID: NCT01575834 Registrational Study with AMG 785 to Treat Postmenopausal Osteoporosis (FRAME) [62]	Randomized control trial **Study Context**: Fracture prevention efficacy and safety study	7180 postmenopausal women, with osteoporosis (T-score of -2.5 to -3.5 at the total hip or femoral neck)	Romosozumab or denosumab for 12 months, followed by denosumab for 12 months **Romosozumab Doses**: Subcutaneous 210 mg monthly	(1) Incidence of vertebral fracture at 12 months and 24 months	(1) Decreased risk of vertebral fracture by 73% at 12 months, and 75%at 24 months in the romosozumab group as compared to the placebo group

(continued)

Table 5.7 (continued)

Study ID / Study Name	Trial Design / Study Context	Study Participants	Comparator groups / Romosozumab Dose	Primary endpoint	Conclusions
Study ID: NCT01631214 Study to Determine the Efficacy and Safety of Romosozumab in the Treatment of Postmenopausal Women with Osteoporosis (ARCH) [63]	Randomized control trial **Study Context**: Fracture prevention efficacy and safety study in a population with higher risk of fracture than FRAME	4093 postmenopausal women with osteoporosis and a fragility fracture	Romosozumab or alendronate for 12 months, followed by open label alendronate **Romosozumab Doses**: Subcutaneous 210 mg monthly	(1) Cumulative incidence of new vertebral fracture at 24 months (2) Cumulative incidence of clinical fracture (non-vertebral and symptomatic vertebral fracture) at the time of the primary analysis	(1) 48% lower risk of new vertebral fracture in the romosozumab to alendronate group (6.2%) as compared to alendronate-to-alendronate group (11.9%) (2) 27% lower risk of clinical fracture in the romosozumab to alendronate group (9.7%) as compared to alendronate-to-alendronate group (13.0%)

A subsequent Phase II trial evaluated the effectiveness and safety of romosozumab in postmenopausal women with low bone density [60]. In this study, 419 postmenopausal women 55–85 yearsold, with BMD T-score < −2.0 and > −3.5. Three hundred and eighty three (91%) patients were randomized to receive romosozumab monthly (doses 70, 140, 210 mg) or every 3-months (doses 140, 210 mg), placebo, or open-label comparator group (oral alendronate 70 mg weekly, or subcutaneous teriparatide 20 μg daily). In this study, the primary endpoint was percentage change from baseline of lumbar spine BMD at 12 months in patients who received romosozumab as compared to the pooled placebo group.

At 12 months, the study showed that pooled romosozumab group participants achieved a statistically significant increase in BMD at the lumbar spine, total hip, and femoral neck compared to the pooled placebo group participants at all doses and frequencies evaluated in the study. Among the assessed doses, romosozumab 210 mg administered subcutaneously monthly was associated with the highest BMD gain at 12 months with no increased incidence of adverse events. Romosozumab led to bone mineral density increases of 11.3% in the lumbar spine, 4.1% in the total hip, and 3.7% in the femoral neck. There were no significant differences in the percentage of serious adverse events between all groups. Injection site reactions (pain, hematoma, erythema reaction, discomfort, hemorrhage, or rash at the injection site) were more common with romosozumab treatment.

There was a transient rise in bone formation markers that peaked one month after romosozumab administration. After peaking at one month, bone formation markers decreased to baseline or dropped below baseline at months 2–9 depending on the romosozumab dose. Bone resorption markers decreased the most in the first week but remained below baseline up to month 12 of treatment. This suggests an anabolic window with romosozumab treatment that allows a rapid initial increase in bone formation, followed by a slighter more prolonged decrease of bone resorption as previously described. Romosozumab demonstrated non-linear pharmacokinetics.

The STRUCTURE study (**ST**udy evaluating effect of **R**omosoz**U**mab **C**ompared with **T**eriparatide in postmenopa**U**sal women with osteoporosis at high risk for fracture p**R**eviously treated with bisphosphonat**E** therapy) which was a randomized, open-label, international multicenter trial assessed the effect of a 12-month treatment course with either romosozumab or teriparatide on BMD after bisphosphonate treatment [61]. The study included 436 postmenopausal women 55–85 years old with osteoporosis (T-score <−2.5 at the lumbar spine, femoral neck or total hip, history of a vertebral fracture, or non-vertebral fracture after 50 years of age) who had taken an oral bisphosphonate for more than three years before screening and, specifically, had taken weekly alendronate one year before screening. Patients previously on bisphosphonates were randomized to receive subcutaneous romosozumab or teriparatide. The study's primary endpoint was the change of total hip BMD on DXA after treatment with 12 months of either romosozumab or teriparatide. After 12 months of therapy with romosozumab, total hip BMD increased by 2.6% versus teriparatide (decrease of 0.6%). Romosozumab led to an increase in lumbar spine bone mineral density as compared to teriparatide [61].

5.8.4 *Romosozumab: Efficacy in Treatment of Osteoporosis*

The efficacy of romosozumab was subsequently demonstrated in the FRAME study (**FRA**cture study in postmenopausal wo**M**en with ost**E**oporosis). FRAME was a multicenter, international, randomized, double-blind, placebo-controlled, parallel-group study that compared the one-year treatment of romosozumab (subcutaneous 210 mg monthly) followed by denosumab (subcutaneously 60 mg every six months) with one-year treatment with placebo followed by denosumab (subcutaneously 60 mg every six months) (Fig. 5.4) [62]. The study included 7180 postmenopausal women 55–90 years old. The study participants had a total hip or femoral neck BMD T-score of −2.5 to −3.5. Women who had a hip fracture history, any severe or more than two moderate vertebral fractures were excluded. The primary endpoints for this study were new vertebral fracture cumulative incidence at 12 and 24 months.

Twelve months of romosozumab treatment led to a 73% risk reduction of vertebral fractures compared to placebo (incidence of vertebral compression fracture in romosozumab group 0.5%, placebo group 1.8%. At 12 months, romosuzumab treatment led to a 36% reduction of clinical fractures (composite of non-vertebral fracture and symptomatic vertebral fracture) as compared to the placebo group. At 24 months, vertebral fracture incidence in the group who received romosozumab in the 1st year and denosumab in the 2nd year was 0.6%. In the group that received placebo in the 1st year and denosumab in the 2nd year, vertebral fracture incidence was 2.5%. One year of romosozumab followed by one year of denosumab was associated with 75% fracture risk vertebral fracture risk reduction and 33% reduction in clinical fractures compared to one year of placebo followed by one year of denosumab. There was no significant difference in non-vertebral fracture incidence at 12 and 24 months

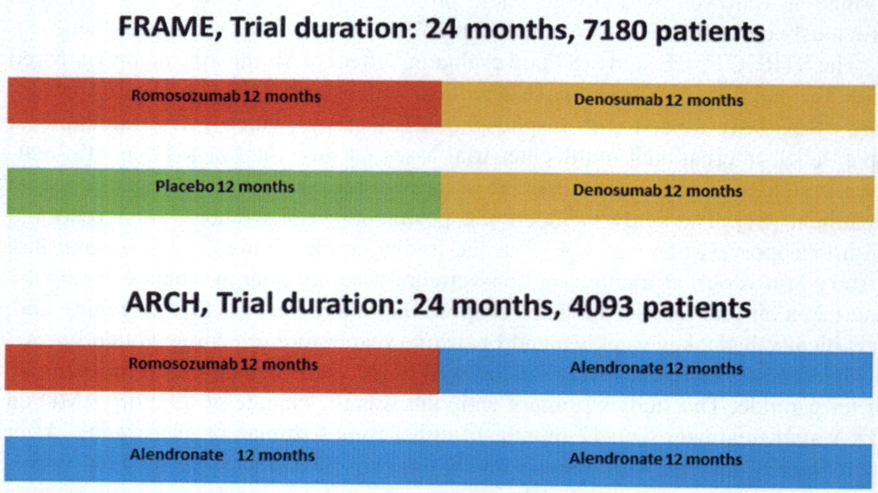

Fig. 5.4 Design of FRAME study and ARCH study

between the two groups, although there were numerically fewer fractures in the group that received romosuzumab followed by denosumab [62].

Another study compared romosozumab in the first year, followed by alendronate for the second year, with two years of alendronate therapy in higher-risk patients [64]. The ARCH study (**A**ctive-cont**R**olled Fra**C**ure Study in Postmenopausal Women with Osteoporosis at **H**igh Risk of Fracture) was a Phase III multicenter, international, randomized, double-blind, alendronate-controlled study of Romosozumab (Fig. 5.4) [63]. The ARCH study included a population of higher risk for fracture compared to the FRAME trial by having patients with osteoporosis and a history of fragility fracture. Inclusion criteria included patients with a bone mineral density T score of −2.5 or less at the total hip or femoral neck and either one or more moderate or severe vertebral fractures, or two or more mild vertebral fractures; or a bone mineral density T score of −2.0 or less at the total hip or femoral neck and either two or more moderate or severe vertebral fractures or a fracture of the proximal femur sustained 3–24 months before randomization. The study's primary outcome was cumulative incidence of new vertebral fractures at 24 months and cumulative incidence of clinical fractures (symptomatic vertebral fractures + non-vertebral fractures) at the time of primary analysis (after clinical fractures had been confirmed in 330 or more patients).

In the ARCH study, 4093 postmenopausal women were enrolled were randomized to receive either two years of alendronate (alendronate group) or one year of romosozumab followed by alendronate (romosozumab group). At 12 months, the romosozumab group showed a 37% risk reduction of vertebral compression fractures and clinical fractures as compared to the alendronate group. The benefit persisted in the 2nd year when both groups received alendronate. The romosozumab group had 48% less vertebral compression fractures, 27% less clinical fractures, 19% less non-vertebral fractures, and 18% fewer hip fractures than the alendronate group [63]. At 24 months, 6.2% of patients in the romosozumab group had new vertebral fractures compared to 11.9% in the alendronate group. After 2.7 years, 9.7% of patients in the romosozumab group developed clinical fractures, while 13.0% in the alendronate group developed clinical fractures.

5.8.5 *Romosozumab: Safety*

In both the ARCH and FRAME study, adverse events and serious adverse events were well balanced between active romosozumab groups and comparator groups [62, 63]. In the FRAME study, patients who received romosozumab had higher injection site reactions (5.3% as compared to 2.9% in placebo). Seven patients receiving romosozumab had hypersensitivity severe adverse events. Adverse events of more than 10% included joint pains, nasopharyngitis, and back pain. There was one case of osteonecrosis of the jaw after 12 months of romosozumab treatment and one case after 12 months of romosozumab, followed by one dose of denosumab. There was one case of atypical femoral fracture after three months of romosozumab administration.

In the ARCH study's first year, a higher frequency of serious cardiovascular adverse events was observed patients who received the romosozumab group (2.5% versus 1.9%). A higher cardiovascular risk was not noted in the FRAME study. This led to a boxed warning for romosozumab that it may increase the risk of heart attack, stroke and cardiovascular death. Romosozumab should not be used in patients who have had a heart attack or stroke within the previous 1 year. In the ARCH study, there was one case of osteonecrosis of the jaw and two cases of atypical femoral fractures in the group of patients who received romosozumab followed by alendronate. In patients who received alendronate for two years, there was one case of osteonecrosis of the jaw and four cases of atypical femoral fractures [62, 63].

5.9 Concluding Remarks

In this chapter, we describe the remarkable discovery and utility of monoclonal antibodies in the treatment of human disease. The specificity of monoclonal antibodies towards the target antigen allows precise targeting of pathways involved in disease processes. With osteoporosis, increased understanding of specific pathways involved in regulating bone metabolism has allowed significant treatment advances. Harnessing monoclonal antibodies' ability to target specific targets within the RANKL/RANK/OPG system and the wnt signaling pathway, denosumab and romosozumab were developed. These two new drugs have expanded the tools we have available to address osteoporosis, a global health epidemic associated with significant comorbidity and mortality.

References

1. Tsumoto K, Isozaki Y, Yagami H et al (2019) Future perspectives of therapeutic monoclonal antibodies. Immunotherapy 11:119–127
2. Lim SY, Bolster MB (2015) Current approaches to osteoporosis treatment. Curr Opin Rheumatol 27:216–224
3. Chiu ML, Goulet DR, Teplyakov A et al (2019) Antibody structure and function: the basis for engineering therapeutics. Antibodies 8:55. https://doi.org/10.3390/antib8040055
4. Albayda J, Christopher-Stine L (2014) Identifying statin-associated autoimmune necrotizing myopathy. Cleve Clin J Med 81:736–741
5. Silverstein AM (2002) The Clonal Selection Theory: what it really is and why modern challenges are misplaced. Nat Immunol 3:793–796
6. Flego M, Ascione A, Cianfriglia M et al (2013) Clinical development of monoclonal antibody-based drugs in HIV and HCV diseases. BMC Med 11:4. https://doi.org/10.1186/1741-7015-11-4
7. Köhler G, Milstein C (1975) Continuous cultures of fused cells secreting antibody of predefined specificity. Nature 256:495–497
8. Anchal S, Sushmita C, Ashima A et al (2013) Antibodies: monoclonal and polyclonal. In: Verma AS, Singh A (eds) Animal biotechnology: models in discovery and translation, 1st edn. Academic Press, Massachusetts, pp 265–287

9. Pageau SC (2009) Denosumab. MAbs 1:210–215. https://doi.org/10.4161/mabs.1.3.8592
10. Ryman JT, Meibohm B (2017) Pharmacokinetics of monoclonal antibodies. CPT Pharmacomet Syst Pharmacol 6:576–588
11. Saunders KO (2019) Conceptual approaches to modulating antibody effector functions and circulation half-life. Front Immunol 10:1296. https://doi.org/10.3389/fimmu.2019.01296
12. Lacey DL, Boyle WJ, Simonet WS et al (2012) Bench to bedside: elucidation of the OPG-RANK-RANKL pathway and the development of denosumab. Nat Rev Drug Discov 11:401–419
13. Foltz IN, Karow M, Wasserman SM (2013) Evolution and emergence of therapeutic monoclonal antibodies: what cardiologists need to know. Circulation 127:2222–2230
14. Keizer RJ, Huitema AD, Schellens JH et al (2010) Clinical pharmacokinetics of therapeutic monoclonal antibodies. Clin Pharmacokinet 49:493–507
15. Wang W, Wang EQ, Balthasar JP (2008) Monoclonal antibody pharmacokinetics and pharmacodynamics. Clin Pharmacol Ther 84:548–558
16. Langdahl B, Ferrari S, Dempster DW (2016) Bone modeling and remodeling: potential as therapeutic targets for the treatment of osteoporosis. Ther Adv Musculoskelet Dis 8:225–235
17. Clarke B (2008) Normal bone anatomy and physiology. Clin J Am Soc Nephrol 3(Suppl 3):S131–S139
18. Wallwork R, Lim SY (2020) What is osteoporosis. In: Bolster MB, Stern TA (eds) Facing osteoporosis, a guide for patients and their families, 1st edn. Massachusetts General Hospital Psychiatry Academy, Boston, pp 1–16
19. Manolagas SC (2000) Birth and death of bone cells: basic regulatory mechanisms and implications for the pathogenesis and treatment of osteoporosis. Endocr Rev 21:115–137
20. Hadjidakis DJ, Androulakis II (2006) Bone remodeling. Ann N Y Acad Sci 1092:385–396
21. Siddiqui JA, Partridge NC (2016) Physiological bone remodeling: systemic regulation and growth factor involvement. Physiology 31:233–245
22. Boyce BF, Xing L (2007) The RANKL/RANK/OPG pathway. Curr Osteoporos Rep 5:98–104
23. Bekker PJ, Holloway DL, Rasmussen AS et al (2004) A single-dose placebo-controlled study of AMG 162, a fully human monoclonal antibody to RANKL, in postmenopausal women. J Bone Miner Res 19:1059–1066
24. Bekker PJ, Holloway D, Nakanishi A et al (2001) The effect of a single dose of osteoprotegerin in postmenopausal women. J Bone Miner Res 16:348–360
25. Cummings SR, San Martin J, McClung MR et al (2009) Denosumab for prevention of fractures in postmenopausal women with osteoporosis. N Engl J Med 361:756–765
26. McClung MR, Lewiecki EM, Cohen SB et al (2006) Denosumab in postmenopausal women with low bone mineral density. N Engl J Med 354:821–831
27. Brown JP, Prince RL, Deal C et al (2009) Comparison of the effect of denosumab and alendronate on BMD and biochemical markers of bone turnover in postmenopausal women with low bone mass: a randomized, blinded, phase 3 trial. J Bone Miner Res 24:153–161
28. Kendler DL, Roux C, Benhamou CL et al (2010) Effects of denosumab on bone mineral density and bone turnover in postmenopausal women transitioning from alendronate therapy. J Bone Miner Res 25:72–81
29. Orwoll E, Teglbjærg CS, Langdahl BL et al (2012) A randomized, placebo-controlled study of the effects of denosumab for the treatment of men with low bone mineral density. J Clin Endocrinol Metab 97:3161–3169
30. Bone HG, Wagman RB, Brandi ML et al (2017) 10 years of denosumab treatment in postmenopausal women with osteoporosis: results from the phase 3 randomised FREEDOM trial and open-label extension. Lancet Diabetes Endocrinol 5:513–523
31. Bone HG, Bolognese MA, Yuen CK et al (2008) Effects of denosumab on bone mineral density and bone turnover in postmenopausal women. J Clin Endocrinol Metab 93:2149–2157
32. Watts NB, Brown JP, Papapoulos S et al (2017) Safety observations with 3 years of denosumab exposure: comparison between subjects who received denosumab during the randomized FREEDOM trial and subjects who crossed over to denosumab during the FREEDOM extension. J Bone Miner Res 32:1481–1485

33. Ominsky MS, Libanati C, Niu QT et al (2015) Sustained modeling-based bone formation during adulthood in Cynomolgus Monkeys may contribute to continuous BMD gains with denosumab. J Bone Miner Res 30:1280–1289
34. Ferrari S, Libanati C, Lin CJF et al (2019) Relationship between bone mineral density T-score and nonvertebral fracture risk over 10 years of denosumab treatment. J Bone Miner Res 34:1033–1040
35. Miller PD, Bolognese MA, Lewiecki EM et al (2008) Effect of denosumab on bone density and turnover in postmenopausal women with low bone mass after long-term continued, discontinued, and restarting of therapy: a randomized blinded phase 2 clinical trial. Bone 43:222–229
36. McClung MR (2016) Cancel the denosumab holiday. Osteoporos Int 27:1677–1682
37. Bone HG, Bolognese MA, Yuen CK et al (2011) Effects of denosumab treatment and discontinuation on bone mineral density and bone turnover markers in postmenopausal women with low bone mass. J Clin Endocrinol Metab 96:972–980
38. Adler RA, El-Hajj Fuleihan G, Bauer DC et al (2016) Managing osteoporosis in patients on long-term bisphosphonate treatment: report of a task force of the American Society for Bone and Mineral Research. J Bone Miner Res 31:16–35
39. Anastasilakis AD, Polyzos SA, Makras P et al (2017) Clinical features of 24 patients with rebound-associated vertebral fractures after denosumab discontinuation: systematic review and additional cases. J Bone Miner Res 32:1291–1296
40. Aubry-Rozier B, Gonzalez-Rodriguez E, Stoll D et al (2016) Severe spontaneous vertebral fractures after denosumab discontinuation: three case reports. Osteoporos Int 27:1923–1925
41. Jamal SA, Ljunggren O, Stehman-Breen C et al (2011) Effects of denosumab on fracture and bone mineral density by level of kidney function. J Bone Miner Res 26:1829–1835
42. Canalis E (2013) Wnt signalling in osteoporosis: mechanisms and novel therapeutic approaches. Nat Rev Endocrinol 9:575–583
43. Deal C (2009) Potential new drug targets for osteoporosis. Nat Clin Pract Rheumatol 5:20–27
44. MacDonald BT, Tamai K, He X (2009) Wnt/beta-catenin signaling: components, mechanisms, and diseases. Dev Cell 17:9–26
45. Rawadi G, Roman-Roman S (2005) Wnt signalling pathway: a new target for the treatment of osteoporosis. Expert Opin Ther Targets 9:1063–1077
46. Poole KE, van Bezooijen RL, Loveridge N et al (2005) Sclerostin is a delayed secreted product of osteocytes that inhibits bone formation. FASEB J 19:1842–1844
47. van Bezooijen RL, Svensson JP, Eefting D et al (2007) Wnt but not BMP signaling is involved in the inhibitory action of sclerostin on BMP-stimulated bone formation. J Bone Miner Res 22:19–28
48. Krause C, Korchynskyi O, de Rooij K et al (2010) Distinct modes of inhibition by sclerostin on bone morphogenetic protein and Wnt signaling pathways. J Biol Chem 285:41614–41626
49. Balemans W, Ebeling M, Patel N et al (2001) Increased bone density in sclerosteosis is due to the deficiency of a novel secreted protein (SOST). Hum Mol Genet 10:537–543
50. Loots GG, Kneissel M, Keller H et al (2005) Genomic deletion of a long-range bone enhancer misregulates sclerostin in Van Buchem disease. Genome Res 15:928–935
51. Moester MJ, Papapoulos SE, Löwik CW et al (2010) Sclerostin: current knowledge and future perspectives. Calcif Tissue Int 87:99–107
52. Yavropoulou MP, Xygonakis C, Lolou M et al (2014) The sclerostin story: from human genetics to the development of novel anabolic treatment for osteoporosis. Hormones 13:323–337
53. Li X, Ominsky MS, Niu QT et al (2008) Targeted deletion of the sclerostin gene in mice results in increased bone formation and bone strength. J Bone Miner Res 23:860–869
54. Winkler DG, Sutherland MK, Geoghegan JC et al (2003) Osteocyte control of bone formation via sclerostin, a novel BMP antagonist. EMBO J 22:6267–6276
55. Ominsky MS, Vlasseros F, Jolette J et al (2010) Two doses of sclerostin antibody in cynomolgus monkeys increases bone formation, bone mineral density, and bone strength. J Bone Miner Res 25:948–959

56. Lim SY, Bolster MB (2017) Profile of romosozumab and its potential in the management of osteoporosis. Drug Des Devel Ther 11:1221–1231
57. Ominsky MS, Niu QT, Li C et al (2014) Tissue-level mechanisms responsible for the increase in bone formation and bone volume by sclerostin antibody. J Bone Miner Res 29:1424–1430
58. Padhi D, Jang G, Stouch B et al (2011) Single-dose, placebo-controlled, randomized study of AMG 785, a sclerostin monoclonal antibody. J Bone Miner Res 26:19–26
59. Padhi D, Allison M, Kivitz AJ et al (2014) Multiple doses of sclerostin antibody romosozumab in healthy men and postmenopausal women with low bone mass: a randomized, double-blind, placebo-controlled study. J Clin Pharmacol 54:168–178
60. McClung MR, Grauer A, Boonen S et al (2014) Romosozumab in postmenopausal women with low bone mineral density. N Engl J Med 370:412–420
61. Langdahl BL, Libanati C, Crittenden DB et al (2017) Romosozumab (sclerostin monoclonal antibody) versus teriparatide in postmenopausal women with osteoporosis transitioning from oral bisphosphonate therapy: a randomised, open-label, phase 3 trial. Lancet 390:1585–1594
62. Cosman F, Crittenden DB, Adachi JD et al (2016) Romosozumab treatment in postmenopausal women with osteoporosis. N Engl J Med 375:1532–1543
63. Saag KG, Petersen J, Brandi ML et al (2017) Romosozumab or alendronate for fracture prevention in women with osteoporosis. N Engl J Med 377:1417–1427
64. Amgen (2018) Study to determine the efficacy and safety of romosozumab in the treatment of postmenopausal women with osteoporosis. In: U.S. National Library of Medicine. Available via ClinicalTrials.gov. https://clinicaltrials.gov/ct2/show/NCT01631214?term=NCT01631214&rank=1. Accessed 16 Jan 2021

Sian Yik Lim Dr. Lim is a rheumatologist at Pali Momi Bone and Joint Center, Hawaii Pacific Health. He currently runs a specialty osteoporosis clinic at Pali Momi Bone and Joint Center and has been involved in efforts to improve the quality of osteoporosis care in Hawaii. He is a clinical densitometrist certified by the International Society for Clinical Densitometry. After graduating from Osaka University School of Medicine, he subsequently trained in internal medicine and rheumatology in the United States. He completed his rheumatology fellowship at Massachusetts General Hospital and was a research fellow at Harvard Medical School. He has received the American Federation of Medical Research Resident Research Day Award and Marshal J Schiff, MD Memorial Fellow Research Award for his research. He has published more than 30 papers and abstracts about gout, osteoporosis, and septic arthritis in respected journals such as JAMA, rheumatology, and current opinions in rheumatology.

Chapter 6
Antibody Treatment and Osteoporosis: Clinical Perspective

Giacomina Brunetti, Sara Todisco, and Maria Grano

Abstract Osteoporosis is the most worldwide diffuse skeletal disease requiring new therapeutic strategies for its cure. The discovery of the pro-osteoclastogenic receptor activator of nuclear factor kappa-B ligand (RANKL) and anti-osteoblastogenic sclerostin's role is strongly changing the therapeutic approach. In this chapter, we overview the literature and data on the use of denosumab and romosozumab, antibodies against RANKL and sclerostin respectively, for osteoporosis management. Clinical trials show that denosumab long-term treatment determines a continuous augment of the bone mineral density (BMD) with few adverse effects. Most recent trials on romosozumab treatment reports bone formation increase and BMD improvement, although there are controversial reports on its adverse effects, with particular regard to cardiovascular events.

Keywords Denosumab · Romosozumab · Osteoporosis · Sclerostin · RANKL · Bone remodeling

6.1 Introduction

Osteoporosis is the most worldwide diffuse skeletal disease with an incidence of about 2 million fractures yearly in the United States [1]. World Health Organization (WHO) sets a definition of osteoporosis considering a bone mineral density (BMD) T-score more than 2.5 standard deviations below young normal reference ranges for the spine, hip or radius [2].

Hip fractures are the gravest in fact they are related to a 3-fold augment in all-cause mortality, reduced mobility, infirmity, and loss of capability perform

G. Brunetti (✉) · S. Todisco
Department of Biosciences, Biotechnologies and Biopharmaceutics, University of Bari, Piazza Giulio Cesare, 11, 70124 Bari, Italy
e-mail: giacomina.brunetti@uniba.it

M. Grano
Department of Emergency and Organ Transplantation, Section of Human Anatomy and Histology, University of Bari, Bari, Italy

© The Author(s), under exclusive license to Springer Nature Singapore Pte Ltd. 2022

A. H. Choi and B. Ben-Nissan (eds.), *Innovative Bioceramics in Translational Medicine II*, Springer Series in Biomaterials Science and Engineering 18,
https://doi.org/10.1007/978-981-16-7439-6_6

quotidian activities [3]. In 2010, about 2.7 million hip fractures have been reported in the world [4]. This value increases yearly and is predictable to rise to 4.5 to 6.3 million by 2050 [5, 6].

The female to male ratio of hip fractures is about 2:1.0 [7, 8]. The incidence of these fractures exponentially rises with advancing age. Conversely, in the UK and the US, wrist fracture occurrence ranges from approximately 400 to 800 per 100,000 women but is quite stable over several decades of elderly [8]. Women often show Colle's fracture than men (for example, a ratio of 10:1 by the age of 75) [8]. Vertebrae compression fractures are problematic to assess and frequently these can remain asymptomatic. The female to male ratio of incidence is about 2:1.

Osteoporosis is characterized by the altered bone remodeling leading to skeletal fragility and an increased fracture risk [9]. In physiological conditions, bone undergoes a continuous process of "regeneration", knows as bone remodeling, involving the coupling activities of bone resorption by osteoclasts and bone formation by osteoblasts [10–12]. Additionally, bone can be shaped through bone modelling, a process featured by the uncoupling of bone formation or resorption [10–12]. Bone modelling is required for shaping bone architecture subsequently mechanical load, and it is associated to genetic and hormonal factors. Bone modelling occurred both before puberty and in adult life [13]. Active modelling occurs in different sites, such as the distal radius, tibia, ribs, and femoral diaphysis in the elderly [14, 15].

In osteolytic bone disease, as in osteoporosis, the amount of resorbed bone by osteoclasts surpasses the quantity formed by osteoblasts with consequent skeletal architecture damage and reduced bone strength [16]. Biological studies highlighted the mechanisms regulating bone remodeling thus leading to the finding of new pharmacological targets that may help to improve bone health in osteoporosis. Particularly, the identification of receptor activator of nuclear factor kappa-B ligand (RANKL) [17] in supporting osteoclast formation, and of sclerostin [18] in inhibiting osteoblast differentiation, led to the development of two monoclonal antibodies: one anti-RANKL (denosumab) and anti-sclerostin (romosozumab) for the therapy of osteoporosis. In this chapter, we report RANKL and sclerostin roles in physiological and pathological bone remodeling, along with the current use of denosumab and romosozumab.

6.2 RANKL

The control of osteoclastogenesis is exerted by receptor activator of nuclear factor kappa-B (RANK), RANKL and osteoprotegerin [17, 19]. RANKL is produced by osteoblasts, osteocytes, bone marrow stromal cells, and activated T cells [17, 20]. RANKL works in concert with macrophage colony-stimulating factor (M-CSF) to determine the fusion of monocyte-macrophage precursors with consequent formation of mature and active osteoclasts [19]. Osteoclast apoptosis is also blocked by RANKL [17]. Consistently, RANKL knockout mice develop severe osteopetrosis [21]. RANKL levels can be modulated by glucocorticoids, Tumor necrosis factor-α

(TNF-α) and interleukins [20, 22]. Furthermore, the pro-osteoclastogenic activity of RANKL can be increased by other cytokines, such as TNF-α and LIGHT/TNFSF14 (homologous to Lymphotoxins exhibiting Inducible expression and competing with herpes simplex virus Glycoprotein D for herpes virus entry mediator [HVEM], a receptor expressed by T lymphocytes) [23–29].

The RANKL receptor is RANK, a transmembrane heterotrimer expressed by pre-osteoclasts and mature osteoclasts [17]. RANK activation leads to the transcription of genes involved in osteoclast differentiation, activity and survival [30]. Osteoprotegerin is a soluble "decoy receptor," that inhibits RANKL-RANK binding with consequent inhibition of osteoclast formation and resorption [31]. The RANKL and osteoprotegerin ratio represent thus a key issue for bone homeostasis. In detail, if the ratio is shifted toward RANKL, the likelihood of bone remodeling shifting towards osteoclastogenesis increases, and thus towards bone diseases, such as genetic disorders, post-menopausal osteoporosis, cancer-related bone loss and inflammatory disorders [32].

6.3 Denosumab

Denosumab is a fully human monoclonal IgG2 antibody, that selectively binds RANKL with high affinity, mimicking an osteoprotegerin inhibitory effect with consequent reduction of bone resorption. It is the first antibody approved by the FDA for the management of osteoporosis or high fracture risk patients [33–36].

FDA approval arose from the results of a 3-year multicenter, randomized, double-blind, placebo-controlled, phase 3 trial—the Fracture Reduction Evaluation of Denosumab (FREEDOM) study [37]. This study involved postmenopausal osteoporotic women aged 60–90 years, who were enrolled in 214 centers in North America, Europe, Latin America, and Australasia and were randomly assigned (1:1) to receive 60 mg subcutaneous Denosumab or placebo every 6 months for 3 years. It was found that Denosumab led to an increase of BMD at the lumbar spine 9.2%, hip 6.0%, femoral neck 4.8%, trochanter 7.9%, and 3.5% at the distal third radius [37]. Simultaneously, hip, new vertebral, and non-vertebral fractures were reduced by 40%, 68%, and 20%, respectively [37].

Additionally, the enhancement of volumetric BMD in the cortical and trabecular compartments of the tibia was also reported with Denosumab treatment [38]. Dual-energy X-ray absorptiometry (DXA, previously known as DEXA) scans discovered continuous increases of Trabecular Bone Score (TBS) from baseline at 12, 24, and 36 months following Denosumab management [39]. A significant reduction of the levels of the bone resorption marker CTX (crosslinked telopeptide of type 1 collagen) was measured during 3 years follow-up of Denosumab treatment [37]. Furthermore, the reduction of bone-specific alkaline phosphatase (BSAP) and type 1 collagen amino-terminal pro-peptide (P1NP) occurred after Denosumab first injection both in humans and in monkeys [37, 40–42]. Histomorphometry studies displayed a significant inhibition of bone turnover, associated with a regular bone microarchitecture and mineralization [43].

All subjects who experienced the FREEDOM trial without treatment discontinuation also participated in the open-label, 7-year extension [44]. The study also comprised of women who received 3 years of placebo and transitioned to Denosumab in the extension (crossover group). The primary outcome consisted of supervision safety, whereas the secondary outcomes were finalized to monitor BMD and new fractures. Rare cases of jaw osteonecrosis, hypocalcemia and atypical femoral fracture, were documented. Remarkably, in the long-term group, BMD augmented from the FREEDOM baseline by 21.7% at the lumbar spine, 9.0% the femoral neck, 9.2% the total hip, and 2.7% the one-third radius. In the crossover group, BMD increased from extension baseline by 16.5% at the lumbar spine, 7.1% at the femoral neck, 7.4% at the total hip, and 2.3% at the distal third radius. Denosumab treatment for up to 10 years was associated to a low occurrence of adverse events, low fracture rates, and a constant augment in BMD [44].

P1NP and CTX serum levels were reduced during the 7 years of the extension for participants in the long-term group [44]. Otherwise, in the crossover group, CTX and P1NP serum levels rapidly decreased after the initial administration of Denosumab [44]. In the crossover group, the reductions were preserved during 7 years of treatment and were comparable to the findings observed for the long-term group during the first 7 years of Denosumab treatment. The same trend was found for BSAP.

6.3.1 Bone Turnover Rebound and Post-discontinuation Effects

The therapeutic effect of Denosumab is rapidly vanished after treatment discontinuation [45]. Thus, it is probable that following Denosumab withdrawal, RANKL is promptly available and thus osteoclasts quickly differentiate and bone is resorbed. Furthermore, it is important to consider the key role of osteocytes in RANKL secretion.

In post-menopausal women, Denosumab discontinuation causes P1NP and CTX serum levels to rebound to levels of 60% and 40% above the pre-treatment levels, respectively, that are maintained for about 2 years [45]. Moreover, Denosumab discontinuation was associated with hypercalcemia in an adult subject treated with the long-term therapy [46]. Bone turnover reversibility following the discontinuation of Denosumab also ascertains a quick loss of its therapeutic effects. During the FREEDOM trial, the increased bone density achieved with the treatment disappeared over a 1-year period in patients discontinuing Denosumab [47]. This event can clarify the findings of the post-marketing period: in post-menopausal patients, spontaneous single or multiple vertebral fractures appeared during the discontinuation period [48–51]. All these findings suggest that the Denosumab action on bone cells is cytostatic and not cytotoxic.

6.4 Sclerostin

The Wnt/βcatenin pathway plays a key role in the differentiation of mesenchymal stem cells inhibiting adipogenic and chondrogenic differentiation and promoting osteoblast differentiation [52]. Wnt/βcatenin signaling also supports osteoblast and osteocyte survival and inhibits osteoclast differentiation increasing osteoprotegerin expression in osteoblasts and osteocytes [18]. Osteocytes are crucial in the regulation of canonical Wnt/βcatenin signalling, as they secrete the Wnt inhibitor sclerostin, a protein encoded by the SOST gene largely expressed by mature osteocytes but not by early osteocytes or osteoblasts [53]. In humans, the functional loss of sclerostin results in sclerosteosis [54] and van Buchem's disease [55], both showing augmented bone mass as well as resistance to fractures. Similarly,the SOST deficiency or neutralizing antibodies for sclerostin in mice reproduced the high bone mass phenotype [54–56]. Otherwise, SOST/sclerostin over-expression reduces bone mass [57–60], as demonstrated in numerous bone diseases, thus sustaining anti-sclerostin use for their management [61–69].

6.5 Romosozumab

Romosozumab, formerly known as AMG 785/CDP7851, is a humanized IgG2 monoclonal antibody with high affinity and specificity for sclerostin. It binds sclerostin and inhibits its activity leading to bone formation, an increase in BMD and in turn bone strength. The antibody has an approximate molecular weight of 149 kDa and it is produced in a mammalian cell line (Chinese hamster ovary) by recombinant DNA technology [70]. Romosozumab is currently being used for the treatment of osteoporosis.

In particular, romosozumab has been assessed for the treatment of postmenopausal osteoporosis in women in two Phase II studies and three Phase III studies. It has also been assessed in men for osteoporosis management in a single-Phase III study. A phase II study also evaluated romosozumab effect on fracture healing.

6.5.1 Phase II Studies

A total of 419 postmenopausal women with low BMD (NCT00896532) were included in this phase II/dose-finding study and were randomized to be treated with one of five subcutaneous romosozumab regimens (70, 140 or 210 mg monthly; 140 or 210 mg every 3 months), oral alendronate 70 mg weekly, subcutaneous teriparatide 20 μg daily or placebo for 12 months. At 12 months, all romosozumab dose groups had significantly increase in BMD compared to placebo at the lumbar spine, femoral neck, and total hip. The high gain was observed in the monthly subcutaneous romosozumab 210 mg group with a BMD gain of 11.3% at lumbar spine, 3.7% at femoral neck and 4.1% at total hip. All sites displayed a significantly higher BMD gain compared with the alendronate and teriparatide groups [71]. An extension study was executed in subjects who had received 24 months of romosozumab treatment or placebo; these subjects received an additional 12 months of placebo or subcutaneous denosumab 60 mg every 6 months [72]. Women treated with romosozumab followed by denosumab kept gain bone mass (BMD significantly increased by 2.5% and insignificantly at the total hip by 2.0%), while BMD declined to baseline in those receiving placebo from 24 to 36 months. A second extension study was then completed in the same group of subjects for other 12 months (months 36 to 48) where all subjects received subcutaneous romosozumab 210 mg monthly [73]. After 12 months of placebo, a second romosozumab course determined an important BMD. Interestingly, after 12 months of denosumab, a second romosozumab course further improved BMD at the lumbar spine by 2.3% and preserved BMD total hip.

In a parallel study of the Phase II trial (NCT00896532), quantitative CT was utilized to quantify the modification of bone mineral content (BMC) and volumetric BMD after 12 months of treatment with romosozumab or teriparatide (20 μg daily) or placebo at the lumbar spine and total hip. Romosozumab outclassed teriparatide either parameters or sites [74]. In another sub-study of the same Phase II trial, quantitative CT was used to measure percent change in strength at the lumbar spine showed a greater increase in strength than teriparatide by finite element analysis [75].

A total of 252 postmenopausal Japanese women with osteoporosis (NCT01992159) were enrolled in a Phase 2 trial and received subcutaneous romosozumab (70, 140, 210 mg monthly) or placebo for 12 months. All romosozumab doses determined a significant increase in BMD at the lumbar spine, femoral neck, and total hip respect to with placebo. The highest gain was found with romosozumab 210 mg monthly subcutaneously, showing an improvement of 16.9 at lumbar spine, 4.7 at total hip and 3.8% at femoral neck [76].

6.5.2 *Phase III Studies*

A total of 7180 postmenopausal women with osteoporosis were randomized to receive monthly subcutaneous romosozumab of 210 mg or placebo for 12 months followed by open-label subcutaneous denosumab of 60 mg every 6 months for an additional 12 months for both groups in the Phase III FRAME study (Fracture Study in Postmenopausal Women with Osteoporosis, NCT01575834). There was a 73% decreased risk of new vertebral fractures after 12 months of treatment with romosozumab compared to placebo in postmenopausal women with osteoporosis, but no significant treatment effect was seen in non-vertebral fractures. The risk of vertebral fracture was decreased by 75% with romosozumab compared with placebo at 24 months. At 12 months, no significant reduction in non-vertebral fractures was detected with 1.6% in the romosozumab group and 2.1% in the placebo group [77]. Interestingly in the FRAME extension, following denosumab treatment for two years, it has been observed a significant reduction in the risk of non-vertebral, clinical and vertebral fractures [78].

If BMD T-scores from FRAME were related to those of FREEDOM (Phase III trial on denosumab), the analysis revealed that 1 year of romosozumab treatment correspond to 3 and 4.5 years of denosumab treatment at the total hip and lumbar spine, respectively. Otherwise, 12 months of romosozumab followed by additional 12 months of denosumab correspond to 7 years of denosumab for both bone segments [79].

A total of 4093 postmenopausal women with osteoporosis and a previous fracture were enrolled in the Phase III trial ARCH (Active-Controlled Fracture Study in Postmenopausal Women with Osteoporosis at High Risk, NCT01631214) [80]. Patients were randomized to double-blinded monthly subcutaneous romosozumab of 210 mg or weekly oral alendronate of 70 mg for 12 months followed by a supplementary open-label alendronate for further 12 months in both groups. Interestingly, romosozumab followed by alendronate compared to alendronate alone showed a 48% reduction in risk associated with additional vertebral fractures at 24 months. Furthermore, at the time of primary analysis the romosozumab-alendronate group showed a 19% reduction in risk associated with non-vertebral fracture as well as 38% reduction in risk for hip fracture compared to the alendronate-alendronate group [80].

A total of 436 postmenopausal women with osteoporosis, a previous fracture and who had received bisphosphonate for the last three years were enrolled in the Phase III trial STRUCTURE (Study evaluating effect of Romosozumab Compared with Teriparatide in postmenopausal women with osteoporosis at this risk of fracture previously treated with bisphosphonate therapy, NCT01796301). For 12 months, they were treated with monthly subcutaneous romosozumab of 210 mg or daily subcutaneous teriparatide of 20 μg. Romosozumab increased hip BMD by 2.6%; while on the other hand, teriparatide induced a change of −0.6% in hip BMD at 12 months [81]. BMD loss at the hip consequent to teriparatide administration in this trial is consistent with other studies showing a blunted BMD variation with teriparatide in patients who were beforehand treated with anti-resorptives [82–84].

The significant improvements in BMD displayed by romosozumab seems to be a consequence of bone formation going on modeling surfaces, thus not dependent on ongoing remodeling sites. Otherwise, teriparatide seems to augment bone formation principally on remodeling surfaces and consequently would have reduced surface to work in patients previously on bisphosphonates, due to the reduced continuous remodeling [81]. Another study demonstrated that in patients who were previously on alendronate, teriparatide exhibited a mean hip BMD to be at or below baseline up to 12 months, but an improvement from baseline of 0.3% was detected at 18 months [82].

A total of 245 men with BMD T-score at femoral neck of <1.5 or a T-score at the lumbar spine of <2.5 together with a history of a fragility non-vertebral fracture or a vertebral fracture were enrolled in the Phase III BRIDGE trial (NCT02186171). After 12 months of treatment with monthly subcutaneous romosozumab of 210 mg or placebo, romosozumab showed statistically significant improvements in BMD at the lumbar spine and total hip comparing to placebo in men with osteoporosis [85].

An interesting fact when comparing the different randomized trials for osteoporosis treatment, it is possible to observe the use of romosozumab followed by an antiresorptive agent or teriparatide in combination with denosumab establishes a quick and strong gain in BMD for patients who at very high risk of fractures. Furthermore, it is fundamental to note that normally there is a correct sequence for administrating osteoanabolic agent and antiresorptive agent in order to achieve a gain in BMD. Interestingly, this will not happen when using romosozumab.

The STRUCTURE and DATA-switch studies have demonstrated that in osteoporotic patients, the switch from alendronate or denosumab to teriparatide caused transient bone loss [86, 87]. However, in the STRUCTURE study and the Phase II studies by Kendler et al. [73, 81], this effect was not observed for patients switching from denosumab or alendronate to romosozumab as they continued to gain bone mass. These observations and findings are consistent with the theory that romosozumab works primarily on modeling bone surfaces whereas teriparatide works mainly on remodeling surfaces, which are educed following treatment with an anti-resorptive drugs, as denosumab or a bisphosphonate [88]. Furthermore, quantification of changes in P1NP (bone formation marker) and CTX (bone resorption marker) shows that romosozumab has a peak onset of 2 weeks after administration. The effect on CTX is preserved during the therapy, whereas a reduction in the concentration of P1NP from baseline after 9 months was observed and the concentration declines thereafter. Upon termination of therapy, an increase in the CTX levels above baseline levels was observed within 3 months. After therapy discontinuation, the levels of CTX, P1NP, and BMD returned to baseline levels within 12 months [70].

6.5.3 Hip Fractures

A total of 332 patients in a phase II study were enrolled to evaluate romosozumab improvement for the fracture-healing-related clinical and radiographic outcomes. A total of 243 were randomized for the treatment with different doses of romosozumab (70 mg, n = 60; 140 mg, n = 93; and 210 mg, n = 90), while 89 received a placebo. The difference in the mean timed "Up & Go" (TUG) score represented the first end point. Furthermore, the score on the Radiographic Union Scale for Hip (RUSH) and the time to radiographic evidence of healing were evaluated as additional end points. However, no parameter among those chosen showed statistically significant changes [89].

6.5.4 Indication, Dosing and Administration

Romosozumab has been approved by FDA for a maximum use of 12 months in postmenopausal women with a high risk for osteoporotic fractures [70]. It is offered in ready-to-use syringes for subcutaneous injection, each with 105 mg romosozumab in 1.17 mL of solution. The therapeutic dose is 210 mg and should be administered once monthly via injection of 2 syringes at different sites. It is divided for optimal absorption as the concentration requires a higher volume respect to recommended for subcutaneous injection.

6.5.5 Safety and Tolerability

In clinical trials, Romosozumab has been relatively well tolerated and the most frequent adverse reactions (≥5%) described were headache and arthralgia [90].

Given that romosozumab works by decreasing bone resorption and increasing bone formation, hypocalcemia was observed in 0 to 3% of patients in phase II trials [77, 80, 81, 85]. In contrast, hypercalcemia was recorded in <1% of patients treated with romosozumab compared to 10% of patients treated with teriparatide in the STRUCTURE trial [81].

Injection site reactions (often mild in severity) has been recorded in 4.4 to 8.0% of participants during Phase III studies [77, 80, 81, 85]. In the STRUCTURE trial, one severe injection site reaction was recorded resulting in the postponement of treatment [81]. Hypersensitivity reactions that have been observed consist of erythema multiforme, dermatitis, angioedema, urticaria and rash [90].

In the FRAME study, two cases with osteonecrosis of the jaw and one case of atypical femoral fracture were recorded in the romosozumab group [75]. In the ARCH study, two atypical femoral fractures and one osteonecrosis of the jaw occurred in the romosozumab population; however, all three cases were found during the alendronate period and a major case numbers have been discovered in the alendronate group [90].

Disturbingly, serious cardiovascular events took place more frequently for patients (50 of 2040 patients or about 2.5%) treated with romosozumab than compared to patients (38 of 2014 patients or around 1.9%) treated with alendronate in the ARCH Phase III trial. In detail, 16 patients (0.8%) in the romosozumab group and 6 patients (0.3%) in the alendronate groups suffered from cardiac ischemic events. Moreover, 16 patients (0.8%) in the romosozumab group and 7 patients (0.3%) suffered from cerebrovascular events. Unfortunately, these discrepancies could not be elucidated by baseline cardiovascular risk or simultaneous use of cardiovascular drugs [80]. Additionally, a large number of cardiovascular adverse events were also depicted in the romosozumab group (8 patients (4.9%)) compared to placebo (2 patients (2.5%)) in the Phase III BRIDGE trial. In the romosozumab group, these events consisted of cardiac ischemia in three patients; cerebrovascular events in three patients; cardiovascular death in one patient; and heart failure in one patient. However, it is important to note that in this study the patients in the romosozumab group started with a major number of cardiovascular risk factors at baseline compared to the placebo group (77.3 vs. 72.0%) [85]. The higher number of cardiovascular events has not been illustrated in other studies; in the FRAME study for instance, incidence of serious cardiovascular event during the first 12 months of the treatment was 1.2% (44 patients) for the romosozumab group compared to 1.1% (41 patients) for the placebo group. At 24 months, both groups transitioned to denosumab and the number of patients increased, and the incidence for the romosozumab to denosumab was 2.3% (82 patients) and placebo to denosumab was 2.2% (79 patients) [77]. Despite ample analysis, there were no elucidations found for this dissimilarity with the ARCH trial.

Numerous researchers tried to explain the cause of cardiovascular events associated to the use of romosozumab. In detail, Asadipooya and Weinstock reported that both cardiovascular and bone remodeling share different pathways and markers, including the Wnt pathway, thus the use of romosozumab may cause endothelial dysfunction by promoting endothelial inflammation as well as the secretion of reactive oxygen species, lipid accumulation by smooth muscle cells and macrophages, and vascular wall calcification [91]. Additionally, it has been reported that sclerostin is upregulated in calcified vessels [92]. Previously, we also demonstrated the pro-angiogenic effect of sclerostin [93].

There was a theoretical apprehension for a possible pro-malignancy effect with anti-sclerostin treatment as in different tissues cellular proliferation is regulated by the Wnt signaling; however, in a lifetime study of rats treated with romosozumab, it has no effect on tumor incidence or in any clinical trials [94]. Conversely, osteosarcoma induction has been showed in analogous rat toxicity studies using the teriparatide and abaloparatide (PTH agonists), which has led to the approval for these drug treatments for a maximum of 2 years [95].

6.6 Concluding Remarks

The WHO has acknowledged osteoporosis as a public health emergency because the high number of subjects who manifested the disease as well as the morbidity and mortality linked to fractures. Anti-resorptive treatments with bisphosphonates or denosumab represent the gold standard for therapy. However, it has recently been recommended to treat patients at highest fracture risk using an anabolic agent, followed by an antiresorptive agent because osteoanabolic agents seemed to increase BMD and decrease fracture risk more quickly and effectively than antiresorptive agents [96], as previously reported.

Until now, the results obtained stimulated the use of romosozumab for osteoporosis management. Since its first approval in January 2019 in Japan for the treatment of osteoporosis, romosozumab treatment has been permitted by 37 nations within 12 months. Clinical trials have revealed that 1 year of romosozumab treatment is better than compared to all the other standard therapies in improving BMD and decreasing fracture risk. This crucial increase in BMD is maintained during ongoing treatments that utilizes antiresorptive regime.

More importantly, future studies are required to define whether or not romosozumab resulted in cardiovascular risk to patients.

References

1. Burge R, Dawson Hughes B, Solomon DH et al (2007) Incidence and economic burden of osteoporosis-related fractures in the United States, 2005–2025. J Bone Miner Res 22:465–475
2. Kanis JA (1994) Assessment of fracture risk and its application to screening for postmenopausal osteoporosis: synopsis of a WHO report. Osteoporosis Int 4: 368–381
3. Panula J, Pihlajamäki H, Mattila VM et al (2011) Mortality and cause of death in hip fracture patients aged 65 or older-a population-based study. BMC Musculoskel Disord 12:105
4. Odén A, McCloskey EV, Johansson H et al (2013) Assessing the impact of osteoporosis on the burden of hip fractures. Calcif Tissue Int 92:42–49
5. Gullberg B, Johnell O, Kanis JA (1997) World-wide projections for hip fracture. Osteoporos Int 7:407–413
6. Cooper C, Campion G, Melton LJ (1992) Hip fractures in the elderly: a world-wide projection. Osteoporos Int 2:285–289
7. Eliffors I, Allander E, Kanis JAS (1994) The variable incidence of hip fracture in southern Europe: The Medos Study. Osteoporosis Int 4:253–261
8. DeLaet CEDH, Pols HAP (2000) Fractures in the elderly: epidemiology and demography. Bailliere's Clin Endocrinol Metab 14:171–179
9. NIH (2001) Consensus Development Panel. Osteoporosis prevention, diagnosis, and therapy. JAMA 285:785–795
10. Brunetti G, Di Benedetto A, Mori G (2014) Bone remodeling. In: Albanese C, Faletti C (eds) Imaging of prosthetic joints—a combined radiological and clinical perspective. Springer, Milan, pp 27–37
11. Kylmaoja E, Nakamura M, Tuukkanen J (2016) Osteoclasts and remodeling based bone formation. Curr Stem Cell Res Ther 11:626–633
12. Seeman E (2009) Bone modeling and remodeling. Crit Rev Eukaryot Gene Expr 19:219–233
13. Hattner R, Epker BN, Frost HM (1965) Suggested sequential mode of control of changes in cell behaviour in adult bone remodelling. Nature 206:489–490
14. Epker BN, Frost HM (1966) Periosteal appositional bone growth from age two to age seventy in man. A tetracycline evaluation. Anat Rec 154:573–577
15. Ruff CB, Hayes WC (1982) Subperiosteal expansion and cortical remodeling of the human femur and tibia with aging. Science 217:945–948
16. Raisz LG (2005) Pathogenesis of osteoporosis: concepts, conflicts, and prospects. J Clin Invest 115: 3318–3325
17. Lacey DL, Boyle WJ, Simonet WS et al (2012) Bench to bedside: elucidation of the OPG-RANK-RANKL pathway and the development of denosumab. Nat Rev Drug Discov 11:401–419
18. Delgado-Calle J, Sato AY, Bellido T (2017) Role and mechanism of action of sclerostin in bone. Bone 96:29–37
19. Lacey DL, Timms E, Tan HL et al (1998) Osteoprotegerin ligand is a cytokine that regulates osteoclast differentiation and activation. Cell 93:165–176
20. Mori G, D'Amelio P, Faccio R et al (2013) The interplay between the bone and the immune system. Clin Dev Immunol 2013:720504. https://doi.org/10.1155/2013/720504

21. Kong YY, Yoshida H, Sarosi I et al (1999) OPGL is a key regulator of osteoclastogenesis, lymphocyte development and lymph-node organogenesis. Nature 397:315–323
22. Ventura A, Brunetti G, Colucci S et al (2013) Glucocorticoid-induced osteoporosis in children with 21-hydroxylase deficiency. Biomed Res Int 2013:250462. https://doi.org/10.1155/2013/250462
23. Lam J, Takeshita S, Barker JE et al (2000) TNF-alpha induces osteoclastogenesis by direct stimulation of macrophages exposed to permissive levels of RANK ligand. J Clin Invest 106:1481–1488
24. Brunetti G, Faienza MF, Colaianni G et al (2018) Impairment of bone remodeling in LIGHT/TNFSF14-deficient mice. J Bone Miner Res 33:704–719
25. Brunetti G, Rizzi R, Storlino G et al (2018) LIGHT/TNFSF14 as a new biomarker of bone disease in multiple myeloma patients experiencing therapeutic regimens. Front Immunol 9:2459
26. Brunetti G, Belisario DC, Bortolotti S et al (2020) LIGHT/TNFSF14 promotes osteolytic bone metastases in non-small cell lung cancer patients. J Bone Miner Res 35:671–680
27. Brunetti G, Storlino G, Oranger A et al (2020) LIGHT/TNFSF14 regulates estrogen deficiency-induced bone loss. J Pathol 250:440–451
28. Cafiero C, Gigante M, Brunetti G et al (2018) Inflammation induces osteoclast differentiation from peripheral mononuclear cells in chronic kidney disease patients: crosstalk between the immune and bone systems. Nephrol Dial Transplant 33:65–75
29. Brunetti G, Rizzi R, Oranger A et al (2014) LIGHT/TNFSF14 increases osteoclastogenesis and decreases osteoblastogenesis in multiple myeloma-bone disease. Oncotarget 5:12950–12967
30. Hsu H, Lacey DL, Dunstan CR et al (1999) Tumor necrosis factor receptor family member RANK mediates osteoclast differentiation and activation induced by osteoprotegerin ligand. Proc Natl Acad Sci US A 96:3540–3545
31. Simonet WS, Lacey DL, Dunstan CR et al (1997) Osteoprotegerin: a novel secreted protein involved in the regulation of bone density. Cell 89:309–319
32. Brunetti G, D'Amato G, Chiarito M et al (2019) An update on the role of RANKL-RANK/osteoprotegerin and WNT-ß-catenin signaling pathways in pediatric diseases. World J Pediatr 15:4–11
33. Ellis GK, Bone HG, Chlebowski R et al (2008) Randomized trial of denosumab in patients receiving adjuvant aromatase inhibitors for nonmetastatic breast cancer. J Clin Oncol 26:4875–4882
34. Amgen (2016) Prolia [Denosumab]. Thousand Oaks, CA; Inc
35. Stopeck AT, Lipton A, Body JJ et al (2010) Denosumab compared with zoledronic acid for the treatment of bone metastases in patients with advanced breast cancer: a randomized, double-blind study. J Clin Oncol 28:5132–5139
36. Fizazi K, Carducci M, Smith M et al (2011) Denosumab versus zoledronic acid for treatment of bone metastases in men with castration resistant prostate cancer: a randomised, double-blind study. Lancet 377:813–822
37. Cummings SR, San Martin J, McClung MR et al (2009) Denosumab for prevention of fractures in postmenopausal women with osteoporosis. N Engl J Med 361:756–765
38. Zebaze RM, Libanati C, Austin M et al (2014) Differing effects of denosumab and alendronate on cortical and trabecular bone. Bone 59:173–179
39. McClung MR, Lippuner K, Brandi ML et al (2017) Effect of denosumab on trabecular bone score in postmenopausal women with osteoporosis. Osteoporos Int 28:2967–2973
40. McClung MR, Lewiecki EM, Cohen SB et al (2006) Denosumab in postmenopausal women with low bone mineral density. N Engl J Med 354:821–831
41. Bone HG, Chapurlat R, Brandi ML et al (2013) The effect of three or six years of denosumab exposure in women with postmenopausal osteoporosis: results from the FREEDOM extension. J Clin Endocrinol Metab 98:4483–4492
42. Ominsky MS, Libanati C, Niu QT et al (2015) Sustained modeling-based bone formation during adulthood in cynomolgus monkeys may contribute to continuous BMD gains with denosumab. J Bone Miner Res 30:1280–1289

43. Reid IR, Miller PD, Brown JP et al (2010) Effects of denosumab on bone histomorphometry: the FREEDOM and STAND studies. J Bone Miner Res 25:2256–2265
44. Bone HG, Wagman RB, Brandi ML et al (2017) 10 years of denosumab treatment in postmenopausal women with osteoporosis: results from the phase 3 randomised FREEDOM trial and open-label extension. Lancet Diabetes Endocrinol 5:513–523
45. Bone HG, Bolognese MA, Yuen CK et al (2011) Effects of denosumab treatment and discontinuation on bone mineral density and bone turnover markers in postmenopausal women with low bone mass. J Clin Endocrinol Metab 96:972–980
46. Koldkjær Sølling AS, Harsløf T, Kaal A et al (2016) Hypercalcemia after discontinuation of long-term denosumab treatment. Osteoporos Int 27:2383–2386
47. Wang HD, Boyce AM, Tsai JY et al (2014) Effects of denosumab treatment and discontinuation on human growth plates. J Clin Endocrinol Metab 99:891–897
48. Brown JP, Roux C, Torring O et al (2013) Discontinuation of denosumab and associated fracture incidence: analysis from the Fracture Reduction Evaluation of Denosumab in Osteoporosis Every 6 Months (FREEDOM) trial. J Bone Miner Res 28:746–752
49. Aubry-Rozier B, Gonzalez-Rodriguez E, Stoll D et al (2016) Severe spontaneous vertebral fractures after denosumab discontinuation: three case reports. Osteoporos Int J 27:1923–1925
50. Anastasilakis AD, Polyzos SA, Makras P et al (2017) Clinical features of 24 patients with rebound-associated vertebral fractures after denosumab discontinuation: systematic review and additional cases. J Bone Miner Res 32:1291–1296
51. Lamy O, Gonzalez-Rodriguez E, Stoll D et al (2017) Severe rebound-associated vertebral fractures after denosumab discontinuation: nine clinical cases report. J Clin Endocrinol Metab 102:354–358
52. Balemans W, Ebeling M, Patel N et al (2001) Increased bone density in sclerosteosis is due to the deficiency of a novel secreted protein (SOST). Hum Mol Genet 10:537–543
53. Balemans W, Van Den Ende J, Freire Paes-Alves A et al (1999) Localization of the gene for sclerosteosis to the van Buchem disease-gene region on chromosome 17q12-q21. Am J Hum Genet 64:1661–1669
54. Baron R, Kneissel M (2013) WNT signaling in bone homeostasis and disease: from human mutations to treatments. Nat Med 19:179–192
55. Poole KE, Van Bezooijen RL, Loveridge N et al (2005) Sclerostin is a delayed secreted product of osteocytes that inhibits bone formation. FASEB J 19:1842–1844
56. Li X, Ominsky MS, Niu QT et al (2008) Targeted deletion of the sclerostin gene in mice results in increased bone formation and bone strength. J Bone Miner Res 23:860–869
57. Tu X, Rhee Y, Condon KW et al (2012) Sost downregulation and local Wnt signaling are required for the osteogenic response to mechanical loading. Bone 50:209–217
58. Rhee Y, Allen MR, Condon K et al (2011) PTH receptor signaling in osteocytes governs periosteal bone formation and intra-cortical remodeling. J Bone Miner Res 26:1035–1046
59. Niziolek PJ, MacDonald BT, Kedlaya R et al (2015) High bone mass-causing mutant lrp5 receptors are resistant to endogenous inhibitors in vivo. J Bone Miner Res 30:1822–1830
60. Kramer I, Loots GG, Studer A et al (2010) Parathyroid hormone (PTH)- induced bone gain is blunted in SOST overexpressing and deficient mice. J Bone Miner Res 25:178–189
61. Faienza MF, Ventura A, Delvecchio M et al (2017) High Sclerostin and Dickkopf-1 (DKK-1) Serum Levels in Children and Adolescents With Type 1 Diabetes Mellitus. J Clin Endocrinol Metab 102:1174–1181
62. Tsentidis C, Gourgiotis D, Kossiva L et al (2017) Increased levels of Dickkopf-1 are indicative of Wnt/β-catenin downregulation and lower osteoblast signaling in children and adolescents with type 1 diabetes mellitus, contributing to lower bone mineral density. Osteoporos Int 28:945–953
63. Giordano P, Brunetti G, Lassandro G et al (2016) High serum sclerostin levels in children with haemophilia A. Br J Haematol 172:293–295
64. Colucci S, Brunetti G, Oranger A et al (2011) Myeloma cells suppress osteoblasts through sclerostin secretion. Blood Cancer J 1:27
65. Terpos E, Christoulas D, Katodritou E et al (2012) Elevated circulating sclerostin correlates with advanced disease features and abnormal bone remodeling in symptomatic myeloma: reduction post-bortezomib monotherapy. Int J Cancer 131:1466–1471

66. Eda H, Santo L, Wein MN et al (2016) Regulation of sclerostin expression in multiple myeloma by Dkk-1: a potential therapeutic strategy for myeloma bone disease. J Bone Miner Res 31:1225–1234
67. McDonald MM, Reagan MR, Youlten SE et al (2017) Inhibiting the osteocyte specific protein sclerostin increases bone mass and fracture resistance in multiple myeloma. Blood 129:3452–3464
68. Delgado-Calle J, Anderson J, Cregor MD et al (2017) Genetic deletion of Sost or pharmacological inhibition of sclerostin prevent multiple myeloma-induced bone disease without affecting tumor growth. Leukemia 31:2686–2694
69. Brunetti G, Grugni G, Piacente L et al (2018) Analysis of circulating mediators of bone remodeling in prader-willi syndrome. Calcif Tissue Int 102:635–643
70. Amgen (2019) Product monograph including patient medication information; Evenity. Canada Inc. 1–38. https://pdf.hres.ca/dpdpm/00051819.PDF. Accessed 31 Dec 2020
71. McClung MR, Grauer A, Boonen S et al (2014) Romosozumab in postmenopausal women with low bone mineral density. N Engl J Med 370:412–420
72. Mcclung MR, Brown JP, Diez-perez A et al (2018) Effects of 24 months of treatment with romosozumab followed by 12 months of denosumab or placebo in postmenopausal women with low bone mineral density: a randomized, double-blind, Phase 2, parallel group study. J Bone Miner Res 33:1397–1406
73. Kendler DL, Bone HG, Massari F et al (2019) Bone mineral density gains with a second 12-month course of romosozumab therapy following placebo or denosumab. Osteoporos Int 30:2437–2448
74. Keaveny TM, Crittenden DB, Bolognese MA et al (2017) Greater gains in spine and hip strength for romosozumab compared with teriparatide in postmenopausal women with low bone mass. J Bone Miner Res 32:1956–1962
75. Graeff C, Campbell GM, Pena J et al (2015) Administration of romosozumab improves vertebral trabecular and cortical bone as assessed with quantitative computed tomography and finite element analysis. Bone 81:364–369
76. Ishibashi H, Crittenden DB, Miyauchi A et al (2017) Romosozumab increases bone mineral density in postmenopausal Japanese women with osteoporosis: a Phase 2 study. Bone 103:209–215
77. Cosman F, Crittenden DB, Adachi JD et al (2016) Romosozumab treatment in postmenopausal women with osteoporosis. N Engl J Med 375:1532–1543
78. Lewiecki EM, Dinavahi RV, Lazaretti-castro M et al (2019) One year of romosozumab followed by two years of denosumab maintains fracture risk reductions: results of the FRAME extension study. J Bone Miner Res 34:419–428
79. Cosman F, Crittenden DB, Ferrari S et al (2018) FRAME study: the foundation effect of building bone with 1 year of romosozumab leads to continued lower fracture risk after transition to denosumab. J Bone Miner Res 33:1219–1226
80. Saag KG, Petersen J, Brandi ML et al (2017) Romosozumab or alendronate for fracture prevention in women with osteoporosis. N Engl J Med 377:1417–1427
81. Langdahl BL, Libanati C, Crittenden DB et al (2017) Romosozumab (sclerostin monoclonal antibody) versus teriparatide in postmenopausal women with osteoporosis transitioning from oral bisphosphonate therapy: a randomised, open-label, Phase 3 trial. Lancet 390:1585–1594
82. Ettinger B, Martin SJ, Crans G et al (2004) Differential effects of teriparatide on BMD after treatment with raloxifene or alendronate. J Bone Miner Res 19:745–751
83. Miller PD, Delmas PD, Lindsay R et al (2008) Early responsiveness of women with osteoporosis to teriparatide after therapy with alendronate or risedronate. J Clin Endocrinol Metab 93:3785–3793
84. Boonen S, Marin F, Obermayer-Pietsch B et al (2008) Effects of previous antiresorptive therapy on the bone mineral density response to two years of teriparatide treatment in postmenopausal women with osteoporosis. J Clin Endocrinol Metab 93:852–860
85. Lewiecki EM, Blicharski T, Goemaere S et al (2018) A Phase III randomized placebo-controlled trial to evaluate efficacy and safety of romosozumab in men with osteoporosis. J Clin Endocrinol Metab 103:3183–3193

86. Kanis JA, Harvey NC, McCloskey E et al (2020) Algorithm for the management of patients at low, high and very high risk of osteoporotic fractures. Osteoporos Int 31:1–12
87. McClung MR (2018) Romosozumab for the treatment of osteoporosis. Osteoporos Sarcopenia 4:11–15
88. Leder BZ, Tsai JN, Uihlein AV et al (2015) Denosumab and teriparatide transitions in postmenopausal osteoporosis (the DATA-Switch study): extension of a randomised controlled trial. Lancet 386:1147–1155
89. Schemitsch EH, Miclau T, Karachalios T et al (2020) A randomized, placebo-controlled study of romosozumab for the treatment of hip fractures. J Bone Joint Surg Am 102:693–702
90. Amgen (2019) Biologics License Application for Romosozumab, Inc. http://www.fda.gov/media/121255/download. Accessed 31 Dec 2020
91. Asadipooya K, Weinstock A (2019) Cardiovascular outcomes of romosozumab and protective role of alendronate. Arterioscler Thromb Vasc Biol 39:1343–1350
92. De Maré A, Maudsley S, Azmi A et al (2019) Sclerostin as regulatory molecule in vascular media calcification and the bone-vascular axis. Toxins 11:428. https://doi.org/10.3390/toxins11070428
93. Oranger A, Brunetti G, Colaianni G et al (2017) Sclerostin stimulates angiogenesis in human endothelial cells. Bone 101:26–36
94. Chouinard L, Felx M, Mellal N et al (2016) Carcinogenicity risk assessment of romosozumab: A review of scientific weight-of-evidence and findings in a rat lifetime pharmacology study. Regul Toxicol Pharmacol 81:212–222
95. Vahle JL, Sato M, Long GG et al (2002) Skeletal changes in rats given daily subcutaneous injections of recombinant human parathyroid hormone (1–34) for 2 years and relevance to human safety. Toxicol Pathol 30:312–321
96. Shoback D, Rosen CJ, Black DM et al (2020) Pharmacological management of osteoporosis in postmenopausal women: an endocrine society guideline update. J Clin Endocrinol Metab 105:dgaa048. https://doi.org/10.1210/clinem/dgaa048

Giacomina Brunetti Dr. Brunetti is Assistant Professor of Histology, at University of Bari. She carried out research on the morphology, biology and physiology of bone tissue cells, using cytological techniques applied to cell culture systems and mouse models. During these years, she has studied the techniques for the culture of osteoclasts in vitro and the mechanisms that regulate the differentiation of bone cells, in particular in multiple myeloma, solid tumors, in periodontal disease, in psoriatic arthritis and in pediatric endocrine diseases. In addition, her studies also focused on the apoptosis of bone cells. She is also carried out studies on the characteristics and potential of the differentiation of stem cells obtained from dental pulp, in order to obtain cell populations to be used in bone division surgery in the dental and orthopedic field. She is taking part in studies evaluating the effect of microgravity on bone cells. She is also studying the role of LIGHT in physiological and pathological bone remodeling.

Maria Grano Dr. Grano is full Professor of Histology and Embryology, at School of Medicine, University of Bari, Italy. She has been studying bone metabolism since the beginning of her career. Early research has involved morphological and functional studies on osteoclasts and osteoblasts. Subsequent studies have allowed the development of a human cell model for the study of osteoclasts. Since 2000 she has been coordinating research projects carried out as part of a programme of Space Biomedicine aimed at the study of osteoporosis and bone functions in microgravity. In recent years she also focused on the understanding the mechanisms that regulate osteolytic phenomena in various diseases in the field of Oncology, Dentistry and Rheumatology. In the last years she has been particularly dedicated to the study of the interaction between bone and muscle adipose tissue and the experiments conducted in this field have led her to the important discovery of the anabolic effect, on bone mass, of a molecule called Irisin that is produced by muscle during physical exercise.

Chapter 7
Fabrication of Fully Artificial Carbonate Apatite Bone Substitutes

Kanji Tsuru, Michito Maruta, and Kunio Ishikawa

Abstract Elder people tend to be bedridden by accidents accompanied by bone fractures. Therefore, a solution that maintains the quality of life for patients by providing early recovery of wounded or damaged hard tissues supported by high-functional bone substitutes would be ideal. In this chapter, we will describe the successful fabrication of novel bone substitutes that is similar to natural bone from the perspective of inorganic component, crystallinity, and bioresorbability. These properties are essential if natural bones damaged through falls are fully replaced by artificially synthesized bone substitutes. We will also discuss the concept, synthesis route, and the introduction of interconnected pores into the structures of bone substitutes.

Keywords Bone substitutes · Carbonate apatite · Carbonation · Phosphatization · Porous scaffolds

7.1 Introduction

Bone graft is used in the reconstruction of hard tissue damaged or destroyed due to disease or injury [1, 2]. Amongst the possible treatment options, the utilization of autografts (grafts that are harvested from the patient) is the first priority. However, there are many disadvantages such as intervention of a healthy site, limited bone morphology and amount of collectable bone, risk of infection at the bone harvest site, and prolonged treatment periods [3, 4]. Therefore, the fabrication of a superior functional bone substitute identical to natural bone is an objective worth pursuing.

The main constituents of typical bone substitutes (in the forms of granules and paste) commercially available in Japan are briefly summarized in Table 7.1. Bone

---

K. Tsuru (✉) · M. Maruta
Department of Dental Engineering, Fukuoka Dental College, Sawara-ku, Fukuoka 814-0193, Japan
e-mail: tsuruk@college.fdcnet.ac.jp

K. Ishikawa
Faculty of Dental Science, Department of Biomaterials, Kyushu University, Higashi-ku, Fukuoka 812-8582, Japan

© The Author(s), under exclusive license to Springer Nature Singapore Pte Ltd. 2022

A. H. Choi and B. Ben-Nissan (eds.), *Innovative Bioceramics in Translational Medicine II*, Springer Series in Biomaterials Science and Engineering 18,
https://doi.org/10.1007/978-981-16-7439-6_7

Table 7.1 Main constituent of typical bone substitutes commercially available in Japan

Type	Main Constituent	Chemical Formula
Granules	Hydroxyapatite (HAp)	$Ca_{10}(PO_4)_6(OH)_2$
	β-tricalcium phosphate (β-TCP)	$Ca_3(PO_4)_2$
	Biphasic of HAp and β-TCP	
	HAp of bovine origin	
Paste	Tetracalcium phosphate (TTCP)	$Ca_4(PO_4)_2O$
	α-tricalcium phosphate (α-TCP)	$Ca_3(PO_4)_2$
	Dicalcium phosphate anhydrate (DCPA)	$CaHPO_4$
	Dicalcium phosphate dihydrate (DCPD)	$CaHPO_4 \cdot 2H_2O$

cements in the form of a paste possess a number of advantages such as the ability to establish connection with the surrounding bone structure, applicable to bone defect with complicated shape, injectable into bone defect. However, they are limited in providing a macro-porous structure that allows newly formed bone tissue to penetrate into the bone substitute once it is set [5–7].

On the other hand, macro pore, which affects bone ingrowth, can be designed and incorporated into granular bone substitutes [8, 9]. The important point to remember is that powders are not suitable for use as bone substitutes. This is due to the fact that the small dimensions of the powders are incorporated into the macrophages because of phagocytosis irrespective of the biocompatibility of the material. Consequently, inflammation is induced. For this reason, bone cement having insufficiently setting can lead to inflammatory response [10].

As shown in Table 7.1, nearly all granular bone substitutes consists of hydroxyapatite (HAp) [11, 12], β-tricalcium phosphate (β-TCP) [12, 13] or their biphasic component [14–16], while bone substitutes in the form of a paste [5–7, 17–19] consists of tetracalcium phosphate (TTCP), α-tricalcium phosphate (α-TCP), dicalcium phosphate anhydrate (DCPA), dicalcium phosphate dihydrate (DCPD).

The solubility of Ca^{2+} in various calcium phosphates versus pH, calculated from the solubility product of each calcium phosphate is shown in Fig. 7.1 [20]. HAp is the most thermodynamically stable phase and β-TCP is the second most stable phase in neutral pH. Granular bone substitutes are essential for cells as scaffolds for creating new bony tissues and hence scaffolds should be designed based on calcium phosphates such as HAp and β-TCP that has a comparatively low solubility (thermodynamically stable phase) under a physiological condition. On the other hand, bone substitutes in the form of a paste are designed from calcium phosphates such as TTCP, α-TCP, DCPA and DCPD that has comparatively higher solubility (thermodynamically metastable phase) under a physiological condition because initial dissolution of the powdery raw material is necessary for determining their setting property.

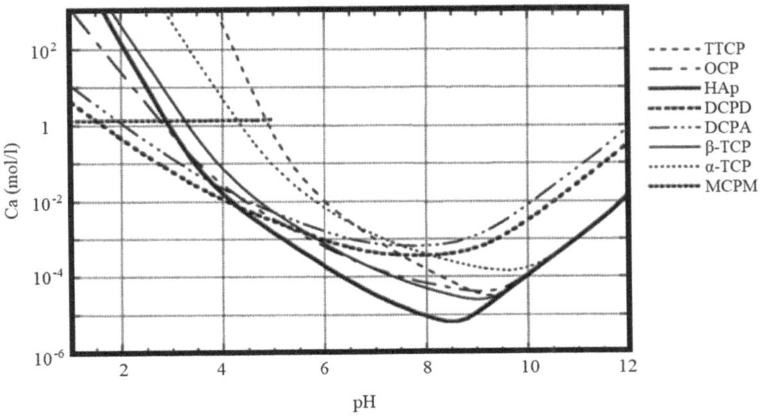

Fig. 7.1 Solubility of Ca^{2+} in various calcium phosphates versus pH calculated from the solubility product of each calcium phosphate. Adapted from [20]

The characteristics of HAp are different from β-TCP when it comes to granular bone substitutes. HAp shows superior osteoconductivity but its bioresorbability is much lower compared to β-TCP. Despite β-TCP displays bioresorbability, the bone-bonding ability is slightly lower than HAp. The rate of recovery to natural bone decreases as the size of bone defect increases when using β-TCP bone substitutes [21]. Biphasic bone substitute composed of HAp and β-TCP demonstrate a behavior that is an amalgamation of HAp and β-TCP [14–16].

Recently, new types of bone substitutes based on the concept of hybrid material were developed and commercialized in Japan: HAp-collagen [22, 23] and octacalcium phosphate (OCP)-collagen [24]. On the other hand, we gained an understanding into the inorganic component of natural bone, which contains around 8% of carbonate ions. Therefore, the fabrications of artificial bone substitute based on carbonate ions containing hydroxyapatite (CO_3Ap) would be ideal [20, 25]. In this chapter, we will describe the concept, synthesis route, and the introduction of interconnected pores into the structures of bone substitutes.

7.2 Fabrication of Bone Substitutes Through Chemical Reaction Without Sintering

Generally, HAp or β-TCP bone substitutes are fabricated using sintering approach. However, this approach cannot be used in the manufacture of CO_3Ap bone substitutes because of its low thermal stability and hence decomposition will take place during firing [26, 27]. Although a study by Doi et al. reported the fabrication of CO_3Ap bone substitutes by sintering CO_3Ap powder at 600 °C [28], the final product was different to natural bone in terms of mechanical strength and crystallinity.

On the other hand, fabrication of calcium phosphate bone substitutes without sintering has been reported and the use of organic substances such as collagen is one of strategy employed. As mentioned previously, inorganic-organic hybrid bone substitute consisting of HAp-collagen [22, 23] and octacalcium phosphate (OCP)-collagen [24] have been commercialized in Japan. Another strategy is a technique that utilizes phase transformation of precursor through chemical reaction. A typical example is known as coralline HAp [29–32].

Natural coral consists of calcite or aragonite that is thermodynamically metaphase under ambient condition and thus it possess the characteristic of being moderately dissolvable under an aqueous solution. The immersion of coral consisting calcium carbonate into an aqueous solution containing phosphate ions, the solution reaches saturation with respect to carbonate apatite due to the appropriate dissolution of coral. As a result, the precipitation of carbonate apatite on the coral is observed.

Through a series of dissolution-precipitation reaction, the phase of calcium carbonate gradually transforms into carbonate apatite while maintaining the macroscopic morphology of coral. Since the interconnected porous morphology of coral is similar to that of human cancellous bone, coral must be a suitable precursor for the fabrication of artificial bone substitutes. In effect, coralline apatite is used for clinical application [29–31]. However, coral is a natural resource so that its utilization will lead to environmental destruction if they are not harvested in a sustainable manner. In addition, there are concerns related to the time and cost required for the purification of natural coral as it contains impurity. Therefore, the application of a fully artificial precursor instead of using natural coral is anticipated for the synthesis of bone substitute via chemical reaction without sintering.

7.3 Precursor Block Utilized for Fabrication of CO_3Ap Bone Substitutes Through Chemical Reaction

As described above, the preparation of a fully artificial precursor block is a key component in the production of CO_3Ap bone substitutes through chemical reaction without sintering. There are a number of properties essential for precursor blocks. Firstly, precursor block must have anti-washout property when immersed in the surrounding solution because the precursor block gradually transforms into carbonate apatite block while at the same time maintaining its macroscopic morphology. Over again, the use of powders as bone substitutes is not suitable as it gives rise to an inflammation response once implanted. Secondly, the precursor block must have at least one constituent ion of carbonate apatite, such as calcium ion (Ca^{2+}), phosphate ion (PO_4^{3-}) or carbonate ion (CO_3^{2-}). Furthermore, precursor block must be relatively dissolvable (thermodynamically metastable phase [20]) in aqueous solution as described previously. Calcium carbonate ($CaCO_3$), calcium sulfate ($CaSO_4$), α-TCP ($Ca_3(PO_4)_2$), DCPA ($CaHPO_4$), and DCPD ($CaHPO_4 \cdot 2H_2O$) for instance are candidates ideal as precursor blocks as they satisfy all of the requirements. Ions missing or unavailable during the constructing CO_3Ap can be supplied from surrounding aqueous solution during the treatment of bone defects. Phosphate salt solution is used in case of precursor containing $CaCO_3$ or $CaSO_4$, while carbonate salt solution is employed in case of precursor involving calcium phosphates. In the following section, we will describe the process in the production of precursor blocks.

7.4 Fabrication of Calcite Precursor Blocks

Calcite block is one of the candidates that can be used as a precursor because it is the most thermodynamically stable phase among $CaCO_3$ polymorphs. The process of obtaining calcite is relatively easy and this can be achieved through the carbonation of calcium hydroxide ($Ca(OH)_2$). Furthermore, microporous calcite is ideal as the precursor for the fabrication of CO_3Ap bone substitutes as the presence of micropores allows the diffusion of PO_4^{3-} ions into the interior of the precursor block easily followed by the phase transformation from calcite to CO_3Ap after immersion in an aqueous solution. The following are some examples of the various routes used in the fabrication of microporous calcite precursor blocks before compositional transformation to CO_3Ap bone substitutes.

7.4.1 Calcite Precursor Blocks Derived from $Ca(OH)_2$ Compact

7.4.1.1 Carbonation of $Ca(OH)_2$ Compact by Heating Under CO_2 Atmosphere

$Ca(OH)_2$ compact can be prepared using a stainless steel mold and an oil pressure press machine. The resultant $Ca(OH)_2$ compacts are placed in electronic tubular furnace and heated under CO_2 atmosphere by the carbonation of $Ca(OH)_2$ compacts as shown in the following equation (Eq. 1):

$$Ca(OH)_2 + CO_2 \rightarrow CaCO_3 + H_2O \tag{1}$$

Phase composition of the compact prepared at various molding pressure followed by firing at different temperature under CO_2 atmosphere is summarized in Table 7.2 [33]. It was discovered that pure calcite could be obtained under a very specific set of conditions (0.2 MPa and 600 °C). Trace quantities of CaO are formed in the synthesized precursor block when carbonated at temperatures above 700 °C. The reason is that calcite decomposes to CaO as shown in the following equation (Eq. 2) and the diffusion of CO_2 is not sufficient to penetrate the compact [33]:

Table 7.2 Phase composition (relative amount (wt%) in parentheses) of specimens compacted at various molding pressure (MPa) followed by firing at different temperature under CO_2 atmosphere [33]

Molding Pressure (MPa)	Heat Treatment Temperature (°C)				
	200	400	600	700	800
0.2	$Ca(OH)_2$/Calcite (99.0/1.0)	$Ca(OH)_2$/Calcite (92.0/8.0)	Calcite (100)	Calcite/CaO (100/trace)	Calcite/CaO (100/trace)
0.5	$Ca(OH)_2$/Calcite (99.0/1.0)	$Ca(OH)_2$/Calcite (91.0/9.0)	Calcite/$Ca(OH)_2$ (100/trace)	Calcite/CaO (100/trace)	Calcite/CaO (100/trace)
1.0	$Ca(OH)_2$/Calcite (98.0/2.0)	$Ca(OH)_2$/Calcite (90.5/9.5)	Calcite/$Ca(OH)_2$ (100/trace)	Calcite/CaO (100/trace)	Calcite/CaO (100/trace)
2.0	$Ca(OH)_2$/Calcite (96.5/3.5)	$Ca(OH)_2$/Calcite (90.0/10.0)	Calcite/$Ca(OH)_2$ (100/trace)	Calcite/CaO (100/trace)	Calcite/CaO (100/trace)

$$CaCO_3 \rightarrow CaO + CO_2 \tag{2}$$

It is a well-known fact that CaO is sensitive to moisture and will transform to $Ca(OH)_2$ upon contact. This leads to the destruction of the CaO structure, resulting in a reduction of mechanical strength. A restriction in CO_2 diffusion also affects the carbonation of $Ca(OH)_2$. For instance, no traces of $Ca(OH)_2$ was found if the $Ca(OH)_2$ compact was produced using 0.2 MPa of pressure, and on the other hand, trace quantities of unreacted $Ca(OH)_2$ was found in compacts produced with pressures greater than 0.2 MPa. It can be concluded that with this approach, pure calcite can be obtained under a very specific set of conditions (0.2 MPa and 600 °C) and the obtained calcite has a higher crystallinity due to the way it is sintered.

7.4.1.2 Carbonation of $Ca(OH)_2$ Compact via Exposure to CO_2 Gas at Room Temperature

An approach where $Ca(OH)_2$ compact is exposed CO_2 atmosphere at room temperature is also investigated to fabricate microporous calcite precursor blocks. The important point with this method is that the carbonation process is affected by the humidity. Calcite formation from $Ca(OH)_2$ compact (the specimen is 3 mm in height and 6 mm in diameter and compacted using 20 MPa of pressure) after exposure to CO_2 gas under different humidity has been reported (Fig. 7.2) [34]. $Ca(OH)_2$ compact can be transformed into pure calcite under 100% relative humidity whereas no compositional changes were observed even after 2 weeks when a $Ca(OH)_2$ compact was exposed to CO_2 under 0% relative humidity. It is discovered that the carbonation of $Ca(OH)_2$ compact can take place with relative ease at room temperature and under the high humidity.

As described above, carbonation of $Ca(OH)_2$ compact can take place effectively by dry-heating $Ca(OH)_2$ compact under CO_2 atmosphere as shown in Eq. 1, while

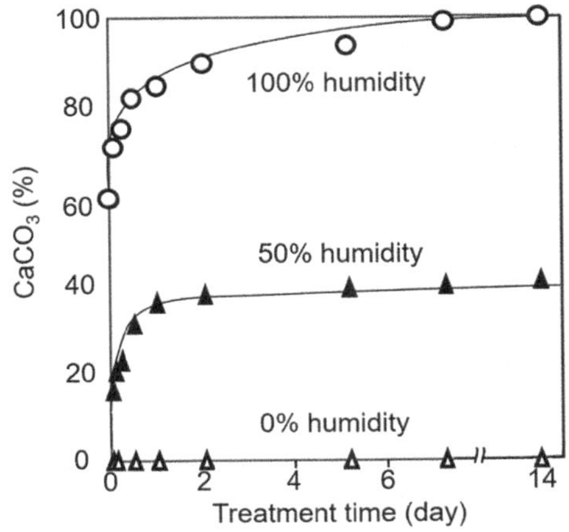

Fig. 7.2 Calcite formation from $Ca(OH)_2$ compact by exposed to CO_2 gas under different humidity [34]

the reaction will not take place if the $Ca(OH)_2$ compact is exposed to a dry-CO_2 atmosphere at room temperature. Under a dry-CO_2 condition, direct reaction of $Ca(OH)_2$ with CO_2 can occur at high temperature while it hardly occurs in room temperature. On the other hand, $Ca(OH)_2$ compact can transform into pure calcite block if $Ca(OH)_2$ compact is exposed to CO_2 under 100% relative humidity even if is the transformation is performed at room temperature. This means that the presence of water or moisture is needed for the carbonation of $Ca(OH)_2$ compact at low temperature. In case of carbonation of $Ca(OH)_2$ compact at room temperature, both $Ca(OH)_2$ and CO_2 will dissolve in water to be Ca^{2+} and CO_3^{2-}, respectively according to Eqs. 3 and 4. Consequently, calcite is precipitated as shown in Eq. 5.

$$Ca(OH)_2 \rightarrow Ca^{2+} + 2OH^- \tag{3}$$

$$CO_2 + H_2O \rightarrow CO_3^{2-} + 2H^+ \tag{4}$$

$$Ca^{2+} + CO_3^{2-} \rightarrow CaCO_3 \tag{5}$$

Mechanical strength of the resultant calcite block increases with increasing molding pressure used in the manufacture of $Ca(OH)_2$ compact [35, 36]. However, carbonation of $Ca(OH)_2$ does not proceed efficiently due to the narrow intergranular space which makes the diffusion of CO_3^{2-} ions difficult [36]. The low crystallinity of the obtained calcite block is advantageous during the compositional transformation to CO_3Ap bone substitutes.

7.4.2 *Calcite Precursor Blocks Derived from Gypsum Hardened Blocks*

Gypsum has self-setting ability so that a precursor block with arbitrary configuration can be fabricated. Adding gypsum hemihydrate powder to water will result in the formation of gypsum dihydrate according to Eq. 6. Anhydrous gypsum is obtained through the heating of gypsum dihydrate (Eq. 7).

$$CaSO_4 \cdot \frac{1}{2}H_2O + \frac{3}{2}H_2O \rightarrow CaSO_4 \cdot 2H_2O \tag{6}$$

$$CaSO_4 \cdot 2H_2O \rightarrow CaSO_4 + 2H_2O \tag{7}$$

Although the solubility of both gypsum dihydrate and gypsum anhydrate are low [37], their partial dissolution in aqueous solution will lead to the release of Ca^{2+} and SO_4^{2-} ions according to Eqs. 8 and 9. If the solution already contains CO_3^{2-} as well as Ca^{2+}, it will become supersaturated with calcium carbonate and calcite will precipitate (Eq. 10) as the solubility of calcite is very low ($K_{sp} = 10^{-8.4}$ at 20 °C) [38].

$$CaSO_4 \cdot 2H_2O \rightarrow Ca^{2+} + SO_4^{2-} + 2H_2O \tag{8}$$

$$CaSO_4 \rightarrow Ca^{2+} + SO_4^{2-} \tag{9}$$

$$Ca^{2+} + CO_3^{2-} \rightarrow CaCO_3 \tag{10}$$

Immersion of $CaSO_4 \cdot 2H_2O$ block or $CaSO_4$ block in Na_2CO_3 solution will result in compositional transformation to $CaCO_3$ block through dissolution-precipitation reaction [39]. As a byproduct, Na_2SO_4 is dissolved in the solution due to its high solubility as shown in Eqs. 11 and 12:

$$CaSO_4 \cdot 2H_2O + Na_2CO_3 \rightarrow CaCO_3 + Na_2SO_4 + H_2O \tag{11}$$

$$CaSO_4 + Na_2CO_3 \rightarrow CaCO_3 + Na_2SO_4 \tag{12}$$

7.5 Fabrication of Precursor Blocks Consisting Chemical Composition Other Than Calcite

Once again, precursors must be in blocks or granules crushed from blocks and must possess at least one constituent ion of carbonate apatite. Furthermore, precursor must have a property of being relatively dissolvable in an aqueous solution as previously described.

α-TCP block satisfy the requirements set for use as precursor blocks. The α-TCP block can be fabricated using a sintering approach. First, α-TCP compact is prepared using a stainless steel mold and an oil-pressure press machine. The resultant α-TCP compacts are placed in electronic furnace and heated above the α,β-transition temperature (1180 °C) in order to obtain α-TCP phase. After heating for a few hours, it must be quenched to room temperature in order to avoid the formation of β-TCP phase. In the manufacture of porous body that is prepared by the burning out of polymer template, the use of α-TCP precursor block has an added advantage due to the fact that it remains stable at high temperature. The effects of sintering temperature on physical and compositional properties of the α-TCP foam precursor were reported in a study by Udoh et al. [40]. Details on the fabrication of ceramics foam are described later with respect to the manufacture of porous body.

Similar to α-TCP, DCPD block can also be utilized as precursor block given it met all the requirements. DCPD block can be synthesized using the setting reaction of DCPD or brushite cement [17, 41]. Both β-TCP and monocalcium phosphate monohydrate (MCPM) powders dissolve in water to supply H^+ as well as Ca^{2+} and PO_4^{3-} ions according to Eqs. 13 and 14, and thus supersaturation with respect to DCPD is reached under acidic conditions since DCPD is most stable at pH 2-4.5 (Fig. 7.1). As a result, DCPD crystals are precipitated and interlock with each other during setting (Eq. 15).

Dissolution:

$$\text{MCPM}: Ca(H_2PO_4)\cdot H_2O \rightarrow Ca^{2+} + 2H^+ + 2HPO_4^{2-} + H_2O \tag{13}$$

$$\beta-\text{TCP}: Ca_3(PO_4)_2 + 2H^+ \rightarrow 3Ca^{2+} + 2HPO_4^{2-} \tag{14}$$

Precipitation:

$$Ca(H_2PO_4)\cdot H_2O + Ca_3(PO_4)_2 + 7H_2O \rightarrow 4Ca^{2+} + 4HPO_4^{2-} + 8H_2O \tag{15}$$

$$4Ca^{2+} + 4HPO_4^{2-} + 8H_2O \rightarrow 4CaHPO_4\cdot 2H_2O[\text{DCPD}]$$

Mixture powder of β-TCP and MCPM were mixed with methanol at a Ca/P molar ratio of 1.0. The methanol was allowed to evaporate at room temperature, and the mixture was placed into a split plastic steel mold 6 mm in diameter and 3 mm in height. Water was added dropwise until a water to powder weight ratio of 0.001 was reached. The set specimen was kept at 100% humidity for 24 h prior to testing. Porosity of the obtained DCPD block is approximately 37% and the pore structure (Fig. 7.3) must be favorable for the precursor block before compositional transformation to CO_3Ap bone substitutes due to the fact that the pore structure helps to diffuse CO_3^{2-} ions (essential for carbonation) into the interior structure of the precursor block [42].

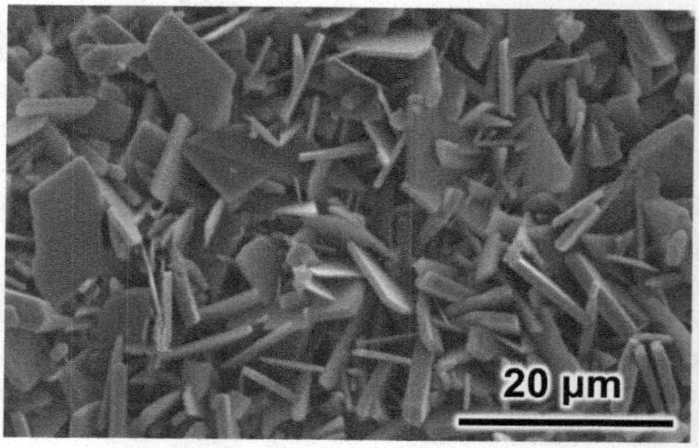

Fig. 7.3 SEM micrograph of a set specimen prepared by the reaction of β-TCP-MCPM mixture powder with water [42]

A set gypsum is also useful as the precursor block. Recently, an arbitrary shaped structure consisting of gypsum can be fabricated by 3D printer so that it can be utilized for the precursor of CO_3Ap bone substitutes [43, 44]. As a final product, SO_4^{2-} contamination in CO_3Ap bone substitutes should be avoided. Consequently, carbonation of the set gypsum precursor is recommended before the compositional transformation to CO_3Ap [39].

7.6 Fabrication of CO_3Ap Bone Substitutes Through Compositional Transformation of Precursors

Human hard tissue does not consist of calcium carbonate but calcium phosphate instead; in particular, approximately 8% CO_3^{2-} are contained in hydroxyapatite [45]. Therefore, compositional transformation of precursor to CO_3Ap is ideal for the fabrication of artificial bone substitutes. The following are some examples used in the fabrication of CO_3Ap bone substitutes through compositional transformation of precursor block.

7.6.1 Phosphatization of Calcite Precursor Block

Compositional transformation of calcite to CO_3Ap is achieved by immersing into a phosphate solution [46]. Calcite is partially dissolved in the phosphate solution to supply Ca^{2+} and CO_3^{2-} according to Eq. 16. Consequently, the surrounding solution would be supersaturated relating to CO_3Ap, which is thermodynamically the most stable phase under that condition. Thus, CO_3Ap is precipitated by consuming PO_4^{3-} ions from the phosphate solution (Eq. 17).

Calcite

$$CaCO_3 \rightarrow Ca^{2+} + CO_3^{2-} \quad (16)$$

CO_3Ap

$$Ca^{2+} + CO_3^{2-} + PO_4^{3-} + OH^- \rightarrow Ca_{10-a}(PO_4)_b(CO_3)_{6-c}(OH)_{2-c} \quad (17)$$

The compositional transformation of calcite to CO_3Ap is influenced by porosity and the crystallinity of calcite precursor block [47–49]. Calcite precursor block with high porosity and low crystallinity is preferable for the compositional transformation to CO_3Ap. This is due to the fact that the high porosity assists in the diffusion of PO_4^{3-} to the interior structures of the precursor block, while the low crystallinity affects the reactivity on phosphatization of the calcite precursor block. Compositional transformation of calcite to CO_3Ap can be achieved by heating at 60 °C for 2 weeks (Fig. 7.4) [49]. Carbonate levels of the obtained CO_3Ap block (8–10 mass%) is similar to that of natural bone (6–8 mass%). In fact, CO_3Ap granules fabricated using this approach are approved for clinical use in the dental clinic by the Pharmaceuticals and Medical Devices Agency (PMDA), Japan and commercialized as Cytrans® Granules in 2017.

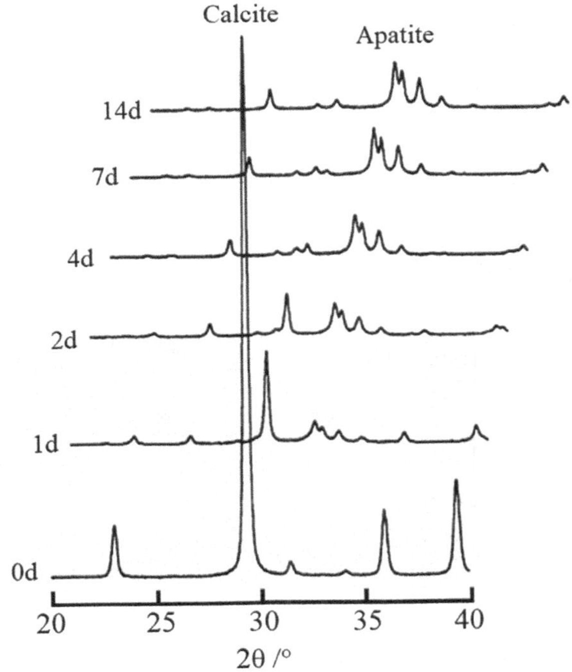

Fig. 7.4 Changes in X-ray diffraction (XRD) patterns after treatment in 1 mol·dm^{-3} Na_2HPO_4 solutions at 60 °C for various periods up to 14 days [49]

7.6.2 *Carbonation of α-TCP Precursor Block or DCPD Precursor Block*

Compositional transformation of α-TCP or DCPD to CO_3Ap can be achieved through immersion in carbonate solution. α-TCP or DCPD partially dissolves to provide Ca^{2+} and PO_4^{3-} in the carbonate solution according to Eq. 18 or Eq. 19, respectively. Ca^{2+} and PO_4^{3-} dissolved from α-TCP or DCPD could be precipitated as CO_3Ap by reacting with CO_3^{2-} ions in the carbonate solution (Eq. 17).

α-TCP

$$Ca_3(PO_4)_2 \rightarrow 3Ca^{2+} + 2PO_4^{3-} \tag{18}$$

DCPD

$$CaHPO_4 \cdot 2H_2O \rightarrow Ca^{2+} + H^+ + PO_4^{3-} + 2H_2O \tag{19}$$

In the fabrication of CO_3Ap block using α-TCP precursor block, hydrothermal treatment using a temperature of more than 100 °C is needed for compositional transformation [50–52]. The carbonate contents of the resultant CO_3Ap tends to decrease with increasing hydrothermal temperature. On the other hand, the hydrothermal treatment is not necessary for the CO_3Ap block fabrication from DCPD precursor block. DCPD will be converted to CO_3Ap within 3 days after immersion in $NaHCO_3$ or Na_2CO_3 solution at 80 °C [42, 53].

Despite the fact that the solubility of DCPD in the carbonate solution is higher than the other precursors such as calcite or α-TCP, it will not be dissolved to such as extent that the shape will be altered or vanished during the reaction and thus the macroscopic structure can be maintained (Fig. 7.5) [42]. In comparison to CO_3Ap fabricated using other techniques, the maximum carbonate contents of the resultant CO_3Ap is relatively higher.

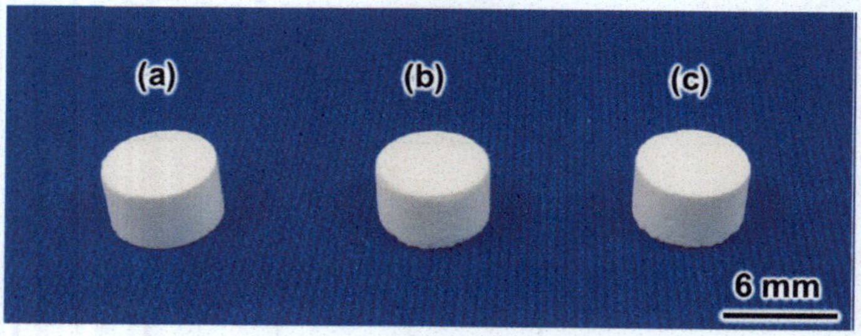

Fig. 7.5 Photographs of set samples. **a** Set sample made with the setting reaction of the β-tricalcium phosphate and monocalcium phosphate monohydrate mixture, and samples obtained by immersion in **b** 2 M $NaHCO_3$ and **c** 2 M Na_2CO_3 solutions at 80 °C for 14 days [42]

7.6.3 *Phosphatization and Carbonation of Gypsum Precursor Blocks*

Compositional transformation of gypsum to CO_3Ap is achieved through immersion in mixture solution of phosphate and carbonate. The gypsum is partially dissolved to supply Ca^{2+} and SO_4^{2-} into phosphate solution (Eq. 20). Ca^{2+} dissolved from gypsum would be precipitated as CO_3Ap by reacting with PO_4^{3-} and CO_3^{2-} from the mixture solution of phosphate and carbonate (Eq. 17).

Gypsum

$$CaSO_4 \rightarrow Ca^{2+} + SO_4^{2-} \tag{20}$$

The advantage of utilizing gypsum as the precursor is its self-setting ability while the compositional transformation of gypsum to CO_3Ap will require the use of hydrothermal treatment (Fig. 7.6) [54]. Mechanical property of block decreases after the compositional transformation from gypsum to CO_3Ap. The increase in porosity (60–70%) of the resultant CO_3Ap blocks is the reason behind the decrease

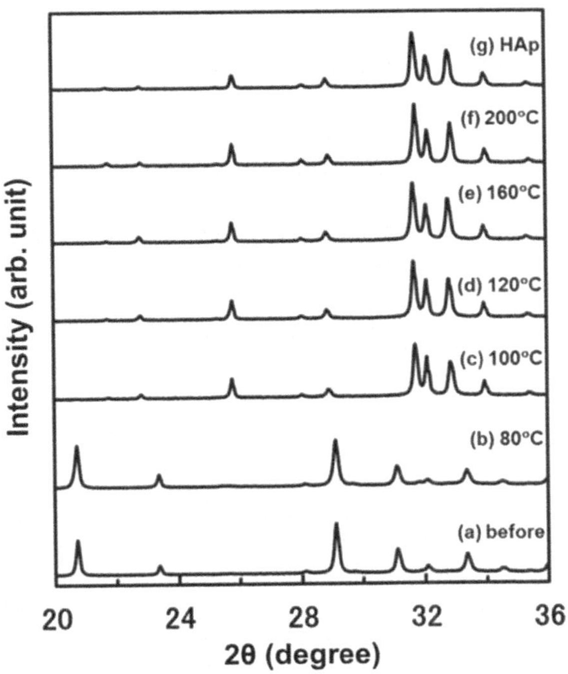

Fig. 7.6 Powder XRD patterns of set gypsum before treatment **a** and after treatment at 80 °C **b**, 100 °C **c**, 120 °C **d**, 160 °C **e**, and 200 °C **f** for 48 h. XRD pattern of commercial HAp **g** is also shown for comparison [54]

in mechanical strength. Both calcium and carbonate are used to fabricate CO_3Ap blocks if calcite is used as a precursor. On the other hand, only calcium is used to fabricate CO_3Ap blocks if set gypsum is used as a precursor. Therefore, CO_3Ap block possess higher porosity than set gypsum even if phosphate and carbonate ions were available in the solution during its fabrication [54].

7.7 Fabrication of Porous CO_3Ap Bone Substitutes and Its Efficacy on New Bone Formation

Macro-pores, which are valuable for bony tissues ingrowth, are incorporated to almost all commercially available bone substitutes [55]. In particular, interconnected macropore of an appropriate size can permit the penetration of cells followed by the new vascular formation and bony tissues [8, 56, 57]. Porous CO_3Ap bone substitute must be fabricated via compositional transformation if macro-pores are to be introduced into the porous precursor block.

7.7.1 Fabrication of Porous Calcite Precursor Block Using Microfiber as a Porogen

Organic fiber as a porogen is useful for creating pore structure in ceramics material. $Ca(OH)_2$ powder containing 10 wt% chopped nylon fiber (approximately 110 μm in diameter and 3 mm in length) is pressed at 150 MPa by isostatic pressing machine, and subsequently sintered to burn out the fiber and to carbonate the $Ca(OH)_2$ under the stream of 1:2 O_2–CO_2 in order to fabricate porous calcite block. Sintering at 770 °C results in complete combustion of the fiber and complete carbonation of $Ca(OH)_2$ to calcite without forming CaO that results in the destruction of structure as described previously. The size of macro-porosities within the porous calcite block is slightly smaller (approximately 100 μm) than the diameter of porogen (110 μm). The efficacy regarding bone tissue penetration is confirmed by in vivo test using the bone defect in femur of 12-weeks-old rat [58].

Self-setting ability of gypsum is advantageous during the introduction of porogen into the hardened body. Moreover, complete combustion of microfibers can be accomplished using a lower temperature without causing the thermal decomposition of the precursor since the thermal stability of gypsum is higher than that of $Ca(OH)_2$. Macro-porous CO_3Ap bone substitute is fabricated by compositional transformation of set gypsum precursor containing different size of microfibers ranging from 30 μm to 205 μm in diameter (Fig. 7.7) [59]. The optimal pore size of the CO_3Ap bone substitute is estimated to be approximately 85 μm, which is the most ideal for bone formation to form inside the pores of the bone substitute as shown using calvarial defects of rabbits during the early stage of recovery [59].

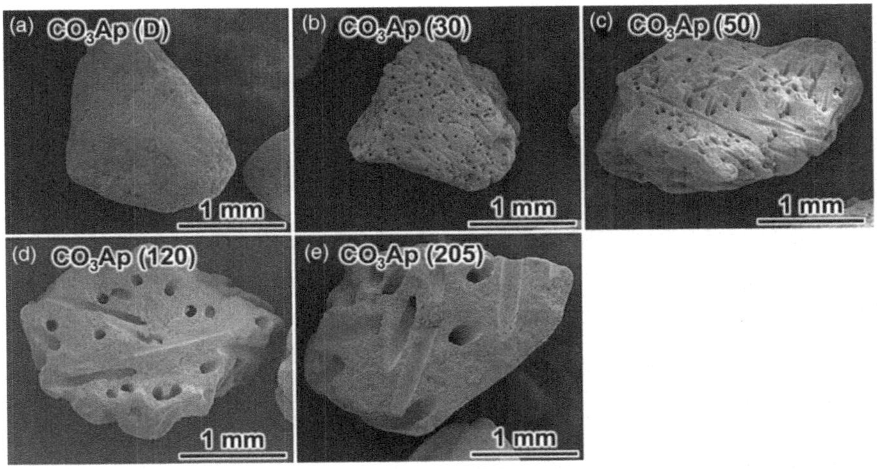

Fig. 7.7 Typical SEM images of CO_3Ap granules fabricated without microfiber **a** and with microfibers of various diameters: **b** 30 μm; **c** 50 μm; **d** 120 μm; and **e** 205 μm [59]

7.7.2 Fabrication of Interconnected Porous CO_3Ap Foam Similar to Cancellous Bone

The so-called ceramic foam method or polyurethane foam replica method is unique in that it produces precursor block with interconnected porous structure similar to cancellous bone [60–63]. In this method, polyurethane foam which has cancellous bone-like fully-interconnected pores is used as a template. Polyurethane foam is immersed in ceramic slurry so that the ceramic powder would coat the polyurethane foam strut. This is following by a sintering process in which the polyurethane foam is completely disintegrated and at the same time, the ceramic powder is sintered to produce the ceramic foam [60]. For the fabrication of CO_3Ap foam, calcite precursor construct was first prepared by coating polyurethane foam with a $Ca(OH)_2$ slurry and sintered at a temperature of 700 to 800 °C under CO_2+O_2 atmosphere. The sintered slurry is then immersed in a solution containing 1 mol/L Na_2HPO_4 at 60 °C for two weeks. The calcite foam is transformed to CO_3Ap based on the dissolution-precipitation reaction while maintaining its fully interconnected porous structure (Fig. 7.8) [48].

α-TCP precursor foam can also be manufactured using the same technique [50–53]. Although the CO_3Ap foam is similar to the fully interconnecting porous structure and inorganic composition of bone, it suffered from poor mechanical property and brittleness. Bone is known as a composite of CO_3Ap and collagen and this demonstrates excellent mechanical property including limited elasticity. Reinforcing CO_3Ap with organic material is essential during the synthesis of CO_3Ap foam intended to match as close as possible the structure and composition of cancellous bone. PLGA or gelatin is used as reinforcement and added to CO_3Ap foam through

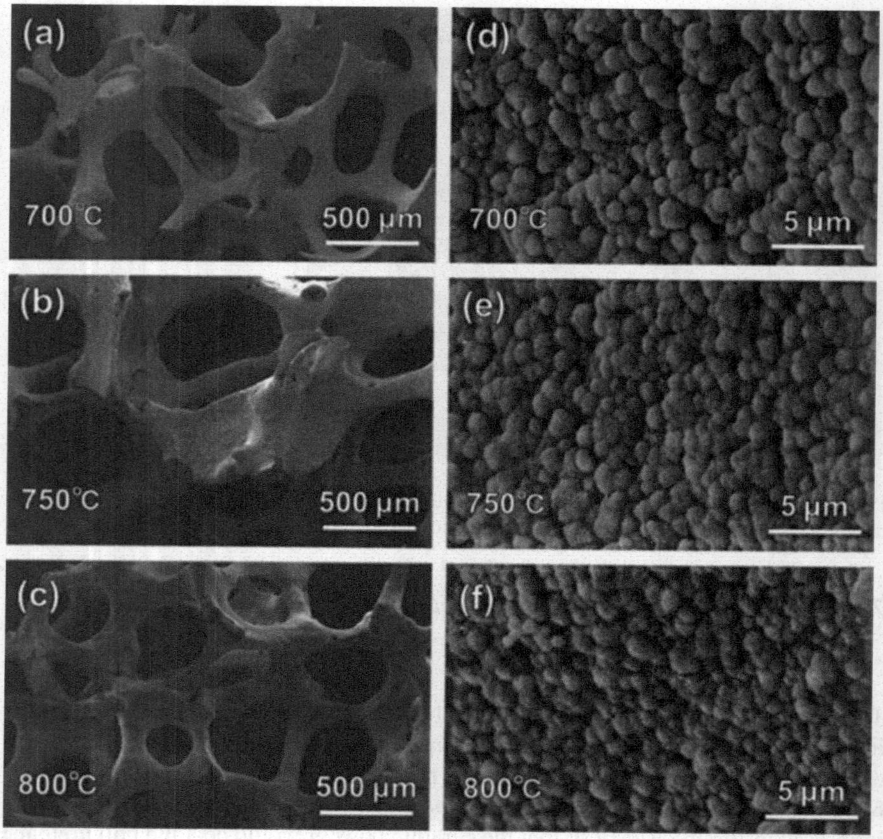

Fig. 7.8 SEM images of foams sintered at 700 °C (**a**, **d**), 750 °C (**b**, **e**), and 800 °C (**c**, **f**) after treatment in Na_2HPO_4 [48]

immersion and vacuum infiltration and consequently, the mechanical property of the resultant CO_3Ap foam can be improved using reinforcement [64–66].

7.7.3 Fabrication of Interconnected Porous CO_3Ap Bone Substitutes by Granular Bridging Method

The geometric packing of spheres is capitalized as a mean to fabricate interconnected porous structures. For instance, hexagonal close-packed structures are constructed by piling up spheres with 74% of the space being occupied by spheres. This also resulted in an interconnected porous structure with 26% porosity. Accordingly, a fully interconnected porous body can be obtained if a close-packed assembly of spheres is prepared.

The setting reaction of brushite cement [17, 41] can be explained by the bridging actions of spheres or granules [67–69]. More specifically, α-TCP spheres are exposed to the MCPM-H_3PO_4 solution in order to obtain the bridging between the spheres. DCPD crystals are precipitated and interlocked with each other at the interface between spheres (Fig. 7.9) [69]. This method can also be applied for the bridging of calcite granules [67, 68]. The application of a small external force of approximately 0.4 MPa must be loaded during the setting reaction if this technique is used to produce a set porous body. In addition, H_3PO_4 is also applied in this approach and the increase in its concentration results in an increase in the amount of DCPD precipitated. The amount of DCPD precipitated relates to the compressive strength of the obtained porous body (Fig. 7.10) [68]. Once the amount of precipitated DCPD reached 3.8 mass%, the precipitated DCPD bridged the calcite granules with one another, and thus, porous calcite was fabricated. The mechanical strength was not increased even when the amount of the precipitated DCPD is more than 5.2 mass%. This is due to the fact that the compressive strength between bridged DCPD reached a maximum level. The porous body derived from α-TCP spheres or calcite granules can be employed as a precursor for the fabrication of interconnected porous CO_3Ap bone substitutes (Fig. 7.11). In the event of α-TCP spheres, another route for bridging the spheres is available. First, porous body consisting of α-TCP spheres are exposed to steam to cause the bridging spheres via a setting reaction of α-TCP with water that is utilized for the production of apatite cement [19] (Eqs. 21 and 22).

Dissolution:

$$[\alpha - \mathrm{TCP}]\, 3Ca_3(PO_4)_2 + 6H_2O \rightarrow 9Ca^{2+} + 6HPO_4^{2-} + 6OH^- \tag{21}$$

Precipitation:

$$9Ca^{2+} + 6HPO_4^{2-} + 6OH^- \rightarrow Ca_9(HPO_4)(PO_4)_5(OH) + 5H_2O\, [\mathrm{cd} - \mathrm{HAp}] \tag{22}$$

Fig. 7.9 SEM images at interface between two α-TCP spheres after exposure to a MCPM-H_3PO_4 solution at 37 °C for 10 min. **g** low magnification; **h** high magnification [69]

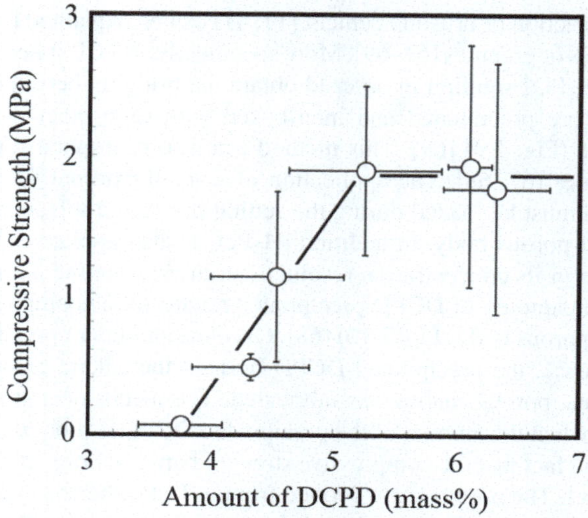

Fig. 7.10 Relationship between the compressive strength of interconnected porous calcite formed by the setting reaction of calcite granules with acidic calcium phosphate solution and the amount of formed DCPD [68]

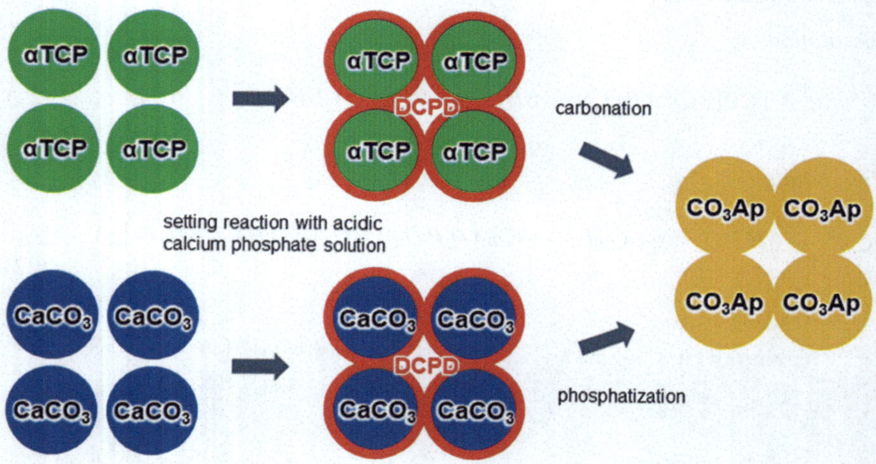

Fig. 7.11 The schematic illustration of porous CO_3Ap fabrication through the bridging of granules followed by compositional transformation

As a result, interconnected porous calcium deficient hydroxyapatite (cd-HAp) is formed (Eq. 22). The resultant porous body constituting cd-HAp can be phase-transformed into α-TCP by sintering at 1300 °C. Finally, the compositional transformation of α-TCP to CO_3Ap is achieved through carbonation (Fig. 7.12) [70].

Fig. 7.12 Photographs of **a** α-TCP spheres; **b** after exposure to steam; **c** after heating at 1300 °C for 6 h; and **d** after immersion in Na-CO_3-PO_4 solution at 80 °C for 7 days [70]

More recently, it was reported that granules can be bridged using calcite-polymethyl methacrylate (PMMA) mixture granules [71]. The calcite-PMMA granules can amalgamate in the mold through the fusion of PMMA, resulting in the formation of porous blocks composed of interconnected calcite-PMMA granules. This is followed by the thermal destruction of PMMA and the resultant interconnected calcite granules are partially sintered. Finally, the composition of the interconnected granules is converted from calcite to CO_3Ap in a dissolution-precipitation reaction.

7.7.4 Fabrication of CO_3Ap Honeycomb Scaffolds by Injection Molding

The incorporation of a porous structure is important during the fabrication of scaffolds, as it will affect cell mobility, proliferation, differentiation, calcification, eventually bone formation. As a result, ceramic honeycomb is interesting and ceramics with honeycomb structure can be obtained by injection molding using honeycomb metal extrusion die. In the fabrication of CO_3Ap honeycomb, a mixture of $Ca(OH)_2$ and a wax-based binder is first extruded through the honeycomb extrusion die. The obtained body is heated to remove the organic binder and for the carbonation to take place at a temperature of 450 °C in a mixed O_2-CO_2 atmosphere. This results in a $CaCO_3$ honeycomb. Once the $CaCO_3$ honeycomb is immersed in a solution containing 1 mol/L Na_3PO_4 solution at 80 °C for 7 days, its composition will change from $CaCO_3$ to CO_3Ap, while maintaining the structure of the original honeycomb [72]. Currently, CO_3Ap honeycomb with different macropore/channel sizes (100, 200 and 300 μm) can be fabricated (Fig. 7.13) [73]. Mechanical property of the CO_3Ap honeycomb is higher than that of porous CO_3Ap fabricated by the other methods. The tissue response and bone regeneration are described in the next section [72–75].

7.7.5 *Construction of 3-D CO_3Ap with Arbitrary Shaped Structure Using 3-D Printer*

Recently, 3-D printer has become popular and it makes 3D design easy even if the structure is complex. Despite the amount of material that can be used in this method is limited, gypsum is one material that is available for use in 3D printers. Since gypsum can be used as precursors for CO_3Ap, the construction of 3-D CO_3Ap with arbitrary shaped structure using 3-D printers are promising [43, 44].

7.8 In Vitro and in Vivo Evaluations of CO_3Ap Bone Substitutes

In vitro evaluation of CO_3Ap bone substitute is conducted using human bone marrow cells and CO_3Ap disc were prepared through the carbonation of $Ca(OH)_2$ compact followed by compositional transformation [76]. The CO_3Ap disc has the capacity to promote osteoblastic differentiation of human bone marrow cells earlier than sintered HAp disc as a control.

In vivo animal studies have been performed on CO_3Ap block or granule derived from calcite granule [20], DCPD block [77], and gypsum granule [78]. For instance, when CO_3Ap granules (Cytrans®) are implanted into a rat's cranial bone defect, the granules are gradually resorbed and replaced to new bone. From the histological examination after 24 weeks post-implantation, osteoclastic resorption can be observed along with new bone (Fig. 7.14) [20]. A similar resorption and new bone formation on CO_3Ap can be observed irrespective to the type of precursor used in the CO_3Ap blocks or granules fabrication [77, 78].

Once interconnected porous structure is incorporated into CO_3Ap bone substitutes, not only material resorption but new bone formation will also become earlier

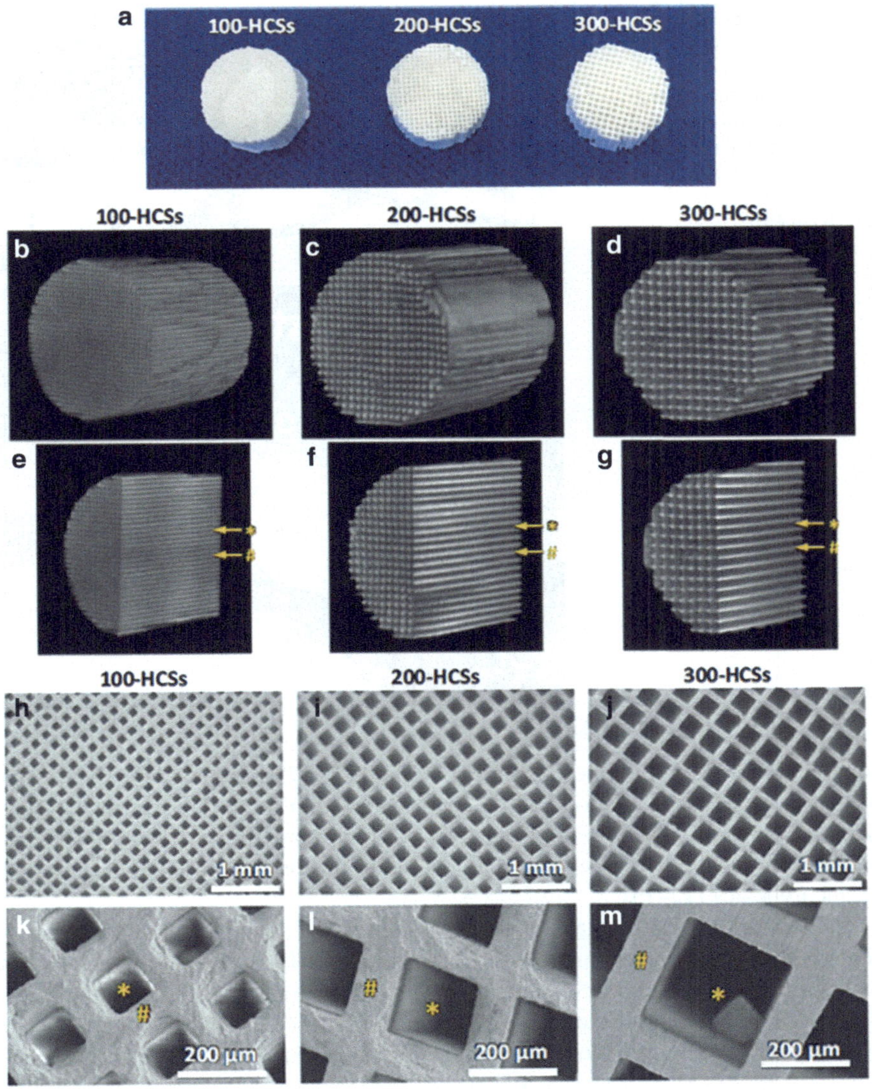

Fig. 7.13 **a** The morphology of CO_3Ap honeycomb scaffolds (HCSs). **b–g** μ-CT images; (H-J) stereomicroscopy images; and **k–m** SEM images of 100-HCSs, 200-HCSs, and 300-HCSs. "#" and "*" indicate HCS strut and channel, respectively [73]

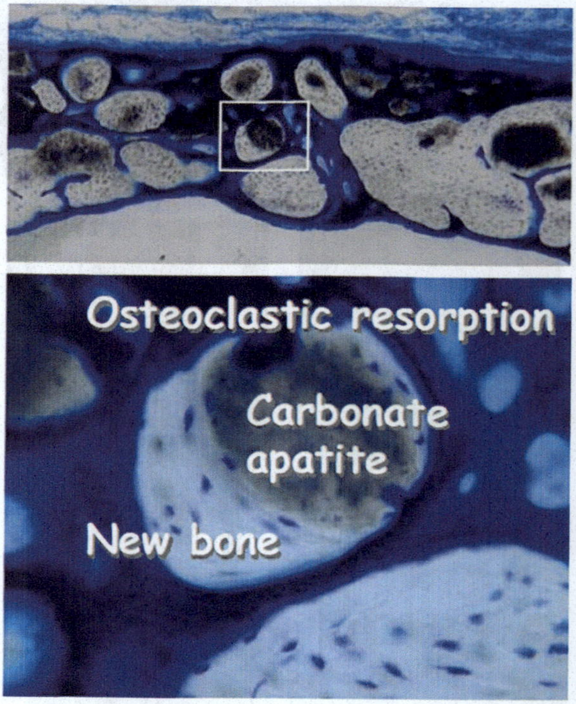

Fig. 7.14 Histological pictures of carbonate apatite granules 24 weeks post-implantation in a bone defect made to the cranial bone of rats. After 24 weeks, most of the carbonate apatite granules were replaced by new bone [20]

compared to those of non-porous one [59, 70, 72–75, 79]. Interestingly, new bone formation and vascularization can be confirmed within the pores even if the pore size is less than 100 μm in diameter [59]. More specifically, CO_3Ap honeycomb provides an ideal environment for the generation of bone marrow-like tissues and megakaryocytes [72–75].

In 2018, a study by Ishikawa et al. performed physical and histological comparisons between various commercial bone substitutes [80]. CO_3Ap (Cytrans®) has high specific surface area (SSA) and low crystalline size due to the fabrication using dissolution-precipitation reaction at low temperature whereas HAp (Neobone®) and β-TCP (Cerasorb®) have low SSA and high crystalline size due to the fabrication through sintering at high temperature. The results revealed that Cytrans® is more similar to natural bone in terms of chemical composition, SSA, and crystalline size than Neobone® and Cerasorb®. Compared to Cerasorb®, Cytrans® displayed limited dissolution under physiological condition and higher dissolution under weak acidity mimicking Howship's lacunae. It is the result of the relationship between pH and stability of each calcium phosphate (Fig. 7.1).

Furthermore, Cytrans® elicits the greatest amount of new bone formation among three bone substitute materials. Similarity to natural bone may become a prerequisite in designing artificial bone substitutes that shows osteoconductivity and resorbability like natural bone. Compositional and histological comparison of apatite bone substitutes with different carbonate content was performed [81, 82]. Carbonate contents of Cytrans® and Bio-Oss® are approximately 12 and 5.5 mass%, respectively. Faster and greater new bone formation is observed in Cytrans® then compared to Bio-Oss®. In addition, resorption of Cytrans® is higher than that of Bio-Oss® of up to 12 weeks although both materials are resorbed over a period. Human clinical trial using Cytrans® was also reported in a number of studies [83, 84].

7.9 Basic Researches on CO_3Ap Bone Substitutes

An in vitro co-culture study using osteocyte-like cells and bone marrow cells as well as in vivo implantations of hydroxyapatite with different carbonate content into rat femurs was carried out by Nakamura et al. [85]. Higher carbonate content shows greater osteoclastogenesis than hydroxyapatite containing lower carbonate content under both in vitro and in vivo conditions. The results indicate that carbonate content of CO_3Ap affect osteoclastogenesis in the vicinity of the implanted CO_3Ap bone substitutes.

Recently, Kawashita et al. investigated the adsorption behavior of fibronectin, cell adhesion molecule, on hydroxyapatite with different carbonate content [86]. The carbonate content of CO_3Ap influences the adsorption behaviour of fibronectin and the study suggests that its specific adsorption contributes to the high osteoconductivity of CO_3Ap.

As discussed earlier, paste-type apatite cement has the advantage of not only setting at the implanted site but also the resorbability of the set cement. However, the reported results are controversial regarding the resorbability, for instance, the apatite cement replaced by new bone tissue [87–94]. We hypothesized that the replacement of apatite cement by new bone might be related to the carbonation of implanted apatite cement under the body environment. Moreover, CO_3Ap is detected in set apatite cement if the cement powders are mixed with water under CO_2 atmosphere. The set cement containing CO_3Ap shows higher in vivo resorbability than the set cement without CO_3Ap prepared under N_2 atmosphere [95]. Consequently, new bone is expected to replace the set cement is it is composed of CO_3Ap and not of HAp. Based on this idea, we developed CO_3Ap-forming cement consisting of $CaCO_3$ and DCPA [96]. In this cement, $CaCO_3$ and DCPA will dissolve and provide Ca^{2+}, CO_3^{2-} and PO_4^{3-} ions followed by the precipitation of the most thermodynamically stable phase, CO_3Ap. Vaterite is the key among $CaCO_3$ polymorphs and not calcite. Since vaterite has higher solubility than calcite, the cement fully-transforms to pure CO_3Ap without residual raw material [38, 97].

References

1. Moore WR, Graves SE, Bain GI (2001) Synthetic bone graft substitutes. ANZ J Surg 71:354–361
2. Damien CJ, Parsons JR (1991) Bone graft and bone graft substitutes: a review of current technology and applications. J Appl Biomater 2:187–208
3. Khan SN, Cammisa FP Jr, Sandhu HS et al (2005) The biology of bone grafting. J Am Acad Orthop Surg 13:77–86
4. Younger EM, Chapman MW (1989) Morbidity at bone graft donor sites. J Orthop Trauma 3:192–195
5. Hughes E, Yanni T, Jamshidi P et al (2015) Inorganic cements for biomedical application: calcium phosphate, calcium sulphate and calcium silicate. Adv Appl Ceram 114:65–76
6. Ishikawa K (2011) Bioactive ceramics: cements. In: Ducheyne P, Heary KE, Hutmacher DW et al (eds) Comprehensive biomaterials, vol 1. Elsevier, San Diego, p 267–283
7. Dorozhkin SV (2008) Calcium orthophosphate cements for biomedical application. J Mater Sci 43:3028–3057
8. Karageorgiou V, Kaplan D (2005) Porosity of 3D biomaterial scaffolds and osteogenesis. Biomaterials 26:5474–5491
9. Ioku K (2001) Microstructure designing of ceramics artificial bone. J Soc Inorg Mater Jpn 8:153–159
10. Miyamoto Y, Ishikawa K, Takechi M et al (1999) Histological and compositional evaluations of three types of calcium phosphate cements when implanted in subcutaneous tissue immediately after mixing. J Biomed Mater Res 48:36–42
11. Aoki H (1994) Medical applications of hydroxyapatite. Ishiyaku Euro America, St. Louis
12. Bucholz RW, Carlton A, Holmes RE (1987) Hydroxyapatite and tricalcium phosphate bone graft substitutes. Orthop Clin North Am 18:323–334
13. Galois L, Mainard D, Delagoutte JP (2002) Beta-tricalcium phosphate ceramic as a bone substitute in orthopaedic surgery. Int Orthop 26:109–115
14. Bouler JM, LeGeros RZ, Daculsi G (2000) Biphasic calcium phosphates: influence of three synthesis parameters on the HA/beta-TCP ratio. J Biomed Mater Res 51:680–684
15. Gauthier O, Bouler JM, Aguado E et al (1998) Macroporous biphasic calcium phosphate ceramics: influence of macropore diameter and macroporosity percentage on bone ingrowth. Biomaterials 19:133–139
16. Yamada S, Heymann D, Bouler JM et al (1997) Osteoclastic resorption of calcium phosphate ceramics with different hydroxyapatite/beta-tricalcium phosphate ratios. Biomaterials 18:1037–1041
17. Lemaitre J, Mirtchi AA, Mortier A (1987) Calcium phosphate cements for medical uses: state of the art and perspectives of development. Sil Ind Ceram Sci Technol 52:141–146
18. Brown WE, Chow LC (1986) A new calcium phosphate, water-setting cement. In: Brown PW (ed) Cement research progress. The American Ceramics Society, Westerville, p 351–379
19. Monma H, Kanazawa T (1976) The hydration of α-tricalcium phosphate. Yogyo-Kyokai-Shi 84:209–213
20. Ishikawa K (2010) Bone substitute fabrication based on dissolution-precipitation reactions. Materials 3:1138–1155
21. Shinomiya K, Ishizuki M, Morioka H (2012) A phase III randomized controlled trial of self-organized hydroxyapatite/collagen composite versus beta tricalcium phosphate as bone substitute for treatment of osseous defect. Seikei-geka 63:921–926
22. Kikuchi M (2020) Developments of calcium phosphate-based bone regenerating materials utilizing interfacial interactions between inorganic-organic substances. J Ceram Soc Japan 128:547–554
23. Kikuchi M, Itoh S, Ichinose S et al (2001) Self-organization mechanism in a bone-like hydroxyapatite/collagen nanocomposite synthesized in vitro and its biological reaction in vivo. Biomaterials 22:1705–1711

24. Suzuki O (2013) Octacalcium phosphate (OCP)-based bone substitute materials. Jpn Dent Sci Rev 49:58–71
25. Ishikawa K (2019) Carbonate apatite bone replacement: learn from the bone. J Ceram Soc Japan 127:595–601
26. Landi E, Tampieri A, Celotti G et al (2004) Influence of synthesis and sintering parameters on the characteristics of carbonate apatite. Biomaterials 25:1763–1770
27. Rau JV, Cesaro SN, Ferro D et al (2004) FTIR study of carbonate loss from carbonated apatites in the wide temperature range. J Biomed Mater Res B Appl Biomater 71:441–447
28. Doi Y, Koda T, Wakamatsu N et al (1993) Influence of carbonate on sintering of apatites. J Dent Res 72:1279–1284
29. Wasielewski RC, Sheridan KC, Lubbers MA (2008) Coralline hydroxyapatite in complex acetabular reconstruction. Orthopedics 31:367
30. Irwin RB, Bernhard M, Biddinger A (2001) Coralline hydroxyapatite as bone substitute in orthopedic oncology. Am J Orthop 30:544–550
31. Thalgott JS, Fritts K, Giuffre JM et al (1999) Anterior interbody fusion of the cervical spine with coralline hydroxyapatite. Spine 24:1295–1299
32. Roy D, Linnehan S (1974) Hydroxyapatite formed from coral skeletal carbonate by hydrothermal exchange. Nature 247:220–222
33. Otsu A, Tsuru K, Maruta M et al (2012) Fabrication of microporous calcite block from calcium hydroxide compact under carbon dioxide atmosphere at high temperature. Dent Mater J 31:593–600
34. Koga N, Tsuru K, Takahashi I et al (2015) Effects of humidity on calcite block fabrication using calcium hydroxide compact. Ceram Int 41:9482–9487
35. Lin X, Matsuya S, Nakagawa M et al (2008) Effect of molding pressure on fabrication of low-crystalline calcite block. J Mater Sci Mater Med 19:479–484
36. Matsuya S, Lin X, Udoh K et al (2007) Fabrication of porous low crystalline calcite block by carbonation of calcium hydroxide compact. J Mater Sci Mater Med 18:1361–1367
37. Partridge EP, White AH (1929) The solubility of calcium sulfate from 0 to 200 °C. J Am Chem Soc 51:360–370
38. Plummer LN, Busenberg E (1982) The solubilities of calcite, aragonite and vaterite in CO_2-H_2O solutions between 0 and 90 °C, and evaluation of the aqueous model for the system $CaCO_3$-CO_2-H_2O. Geochim Cosmochim Acta 46:1011–1040
39. Ishikawa K, Kawachi G, Tsuru K et al (2017) Fabrication of calcite blocks from gypsum blocks by compositional transformation based on dissolution-precipitation reactions in sodium carbonate solution. Mater Sci Eng C Mater Biol Appl 72:389–393
40. Udoh K, Munar ML, Maruta M et al (2010) Effects of sintering temperature on physical and compositional properties of alpha-tricalcium phosphate foam. Dent Mater J 29:154–159
41. Mirtchi AA, Lemaitre J, Terao N (1989) Calcium phosphate cements: study of the beta-tricalcium phosphate—monocalcium phosphate system. Biomaterials 10:475–480
42. Tsuru K, Yoshimoto A, Kanazawa M et al (2017) Fabrication of carbonate apatite block through a dissolution-precipitation reaction using calcium hydrogen phosphate dihydrate block as a precursor. Materials 10:374
43. Lowmunkong R, Sohmura T, Suzuki Y et al (2009) Fabrication of freeform bone-filling calcium phosphate ceramics by gypsum 3D printing method. J Biomed Mater Res B Appl Biomater 90:531–539
44. Lowmunkong R, Sohmura T, Takahashi J et al (2007) Transformation of 3DP gypsum model to HA by treating in ammonium phosphate solution. J Biomed Mater Res B Appl Biomater 80:386–393
45. LeGeros RZ (1991) Calcium phosphates in oral biology and medicine. Monogr Oral Sci 15:1–201
46. Lee Y, Hahm YM, Matsuya S et al (2007) Characterization of macroporous carbonate-substituted hydroxyapatite bodies prepared in different phosphate solutions. J Mater Sci 42:7843–7849

47. Sunouchi K, Tsuru K, Maruta M et al (2012) Fabrication of solid and hollow carbonate apatite microspheres as bone substitutes using calcite microspheres as a precursor. Dent Mater J 31:549–557
48. Maruta M, Matsuya S, Nakamura S et al (2011) Fabrication of low-crystalline carbonate apatite foam bone replacement based on phase transformation of calcite foam. Dent Mater J 30:14–20
49. Ishikawa K, Matsuya S, Lin X et al (2010) Fabrication of low crystalline B-type carbonate apatite block from low crystalline calcite block. J Ceram Soc Jpn 118:341–344
50. Sugiura Y, Tsuru K, Ishikawa K (2016) Fabrication of carbonate apatite foam based on the setting reaction of α-tricalcium phosphate foam granules. Ceram Int 42:204–210
51. Takeuchi A, Munar ML, Wakae H et al (2009) Effect of temperature on crystallinity of carbonate apatite foam prepared from alpha-tricalcium phosphate by hydrothermal treatment. Biomed Mater Eng 19:205–211
52. Wakae H, Takeuchi A, Udoh K et al (2008) Fabrication of macroporous carbonate apatite foam by hydrothermal conversion of alpha-tricalcium phosphate in carbonate solutions. J Biomed Mater Res A 87:957–963
53. Sugiura Y, Tsuru K, Ishikawa K (2017) Fabrication of arbitrarily shaped carbonate apatite foam based on the interlocking process of dicalcium hydrogen phosphate dihydrate. J Mater Sci Mater Med 28:122
54. Nomura S, Tsuru K, Maruta M et al (2014) Fabrication of carbonate apatite blocks from set gypsum based on dissolution-precipitation reaction in phosphate-carbonate mixed solution. Dent Mater J 33:166–172
55. Dorozhkin SV (2010) Bioceramics of calcium orthophosphates. Biomaterials 31:1465–1485
56. Pilliar RM, Filiaggi MJ, Wells JD et al (2001) Porous calcium polyphosphate scaffolds for bone substitute applications—*in vitro* characterization. Biomaterials 22:963–972
57. Kasai T, Ishikawa K, Suzuki K et al (2000) Initial evaluation of a ceramic form as a reconstructive material for bone defects. Dent Mater J 19:381–388
58. Ishikawa K, Tram NX, Tsuru K et al (2015) Fabrication of porous calcite using chopped nylon fiber and its evaluation using rats. J Mater Sci Mater Med 26:94
59. Akita K, Fukuda N, Kamada K et al (2020) Fabrication of porous carbonate apatite granules using microfiber and its histological evaluations in rabbit calvarial bone defects. J Biomed Mater Res A 108:709–721
60. Munar ML, Udoh K, Ishikawa K et al (2006) Effects of sintering temperature over 1,300 degrees C on the physical and compositional properties of porous hydroxyapatite foam. Dent Mater J 25:51–58
61. Jayasinghe S, Edirisinghe M (2002) A Novel Method of Forming Open Cell Ceramic Foam. J Porous Mater 9:265–273
62. Milosevski M, Bossert J, Milosevski D et al (1999) Preparation and properties of dense and porous calcium phosphate. Ceram Int 25:693–696
63. Powell SJ, Evans JRG (1995) The structure of ceramic foams prepared from polyurethane-ceramic suspensions. Mater Manuf Process 10:757–771
64. Hara K, Fujisawa K, Nagai H et al (2016) Fabrication and physical evaluation of gelatin-coated carbonate apatite foam. Materials 9:711
65. Munar GM, Munar ML, Tsuru K et al (2014) Effects of PLGA reinforcement methods on the mechanical property of carbonate apatite foam. Biomed Mater Eng 24:1817–1825
66. Munar GM, Munar ML, Tsuru K et al (2013) Influence of PLGA concentrations on structural and mechanical properties of carbonate apatite foam. Dent Mater J 32:608–614
67. Ishikawa K, Koga N, Tsuru K et al (2016) Fabrication of interconnected porous calcite by bridging calcite granules with dicalcium phosphate dihydrate and their histological evaluation. J Biomed Mater Res A 104:652–658
68. Koga N, Ishikawa K, Tsuru K et al (2016) Effects of acidic calcium phosphate concentration on mechanical strength of porous calcite fabricated by bridging with dicalcium phosphate dihydrate. Ceram Int 42:7912–7917
69. Kien PT, Ishikawa K, Tsuru K (2015) Setting reaction of α-TCP spheres and an acidic calcium phosphate solution for the fabrication of fully interconnected macroporous calcium phosphate. Ceram Int 41:13525–13531

70. Ishikawa K, Arifta TI, Hayashi K et al (2019) Fabrication and evaluation of interconnected porous carbonate apatite from alpha tricalcium phosphate spheres. J Biomed Mater Res B Appl Biomater 107:269–277
71. Putri TS, Hayashi K, Ishikawa K (2020) Fabrication of three-dimensional interconnected porous blocks composed of robust carbonate apatite frameworks. Ceram Int 46:20045–20049
72. Ishikawa K, Munar ML, Tsuru K et al (2019) Fabrication of carbonate apatite honeycomb and its tissue response. J Biomed Mater Res A 107:1014–1020
73. Hayashi K, Munar ML, Ishikawa K (2020) Effects of macropore size in carbonate apatite honeycomb scaffolds on bone regeneration. Mater Sci Eng C Mater Biol Appl 111:110848
74. Hayashi K, Kishida R, Tsuchiya A et al (2019) Carbonate apatite micro-honeycombed blocks generate bone marrow-like tissues as well as bone. Adv Biosyst 3:e1900140
75. Hayashi K, Munar ML, Ishikawa K (2019) Carbonate apatite granules with uniformly sized pores that arrange regularly and penetrate straight through granules in one direction for bone regeneration. Ceram Int 45:15429–15434
76. Nagai H, Kobayashi-Fujioka M, Fujisawa K et al (2015) Effects of low crystalline carbonate apatite on proliferation and osteoblastic differentiation of human bone marrow cells. J Mater Sci Mater Med 26:99
77. Kanazawa M, Tsuru K, Fukuda N et al (2017) Evaluation of carbonate apatite blocks fabricated from dicalcium phosphate dihydrate blocks for reconstruction of rabbit femoral and tibial defects. J Mater Sci Mater Med 28:85
78. Ayukawa Y, Suzuki Y, Tsuru K et al (2015) Histological comparison in rats between carbonate apatite fabricated from gypsum and sintered hydroxyapatite on bone remodeling. Biomed Res Int 2015:579541
79. Sakemi Y, Hayashi K, Tsuchiya A et al (2019) Fabrication and histological evaluation of porous carbonate apatite block from gypsum block containing spherical phenol resin as a porogen. Materials 12:3997
80. Ishikawa K, Miyamoto Y, Tsuchiya A et al (2018) Physical and histological comparison of hydroxyapatite, carbonate apatite, and β-tricalcium phosphate bone substitutes. Materials 11:1993
81. Mano T, Akita K, Fukuda N et al (2020) Histological comparison of three apatitic bone substitutes with different carbonate contents in alveolar bone defects in a beagle mandible with simultaneous implant installation. J Biomed Mater Res B Appl Biomater 108:1450–1459
82. Fujisawa K, Akita K, Fukuda N et al (2018) Compositional and histological comparison of carbonate apatite fabricated by dissolution-precipitation reaction and Bio-Oss®. J Mater Sci Mater Med 29:121
83. Kudoh K, Fukuda N, Kasugai S et al (2019) Maxillary sinus floor augmentation using low-crystalline carbonate apatite granules with simultaneous implant installation: first-in-human clinical trial. J Oral Maxillofac Surg 77:985.e1–985.e11
84. Nakagawa T, Kudoh K, Fukuda N et al (2019) Application of low-crystalline carbonate apatite granules in 2-stage sinus floor augmentation: a prospective clinical trial and histomorphometric evaluation. J Periodontal Implant Sci 49:382–396
85. Nakamura M, Hiratai R, Hentunen T et al (2016) Hydroxyapatite with high carbonate substitutions promotes osteoclast resorption through osteocyte-like cells. ACS Biomater Sci Eng 2:259–267
86. Wang Y, Tsuru K, Ishikawa K et al (2021) Fibronectin adsorption on carbonate-containing hydroxyapatite. Ceram Int. https://doi.org/10.1016/j.ceramint.2021.01.017
87. Schneider G, Blechschmidt K, Linde D et al (2010) Bone regeneration with glass ceramic implants and calcium phosphate cements in a rabbit cranial defect model. J Mater Sci Mater Med 21:2853–2859
88. Ishikawa K (2008) Calcium phosphate cement. In: Kokubo T (ed) Bioceramics and their clinical applications. Woodhead Publishing Series in Biomaterials, Woodhead Publishing, London, p 438–463
89. Cancian DC, Hochuli-Vieira E, Marcantonio RA et al (2004) Utilization of autogenous bone, bioactive glasses, and calcium phosphate cement in surgical mandibular bone defects in Cebus apella monkeys. Int J Oral Maxillofac Implants 19:73–79

90. Gosain AK, Riordan PA, Song L et al (2004) A 1-year study of osteoinduction in hydroxyapatite-derived biomaterials in an adult sheep model: part II. Bioengineering implants to optimize bone replacement in reconstruction of cranial defects. Plast Reconstr Surg 114:1155–1163; discussion 1164-1165
91. Friedman CD, Costantino PD, Takagi S et al (1998) BoneSource hydroxyapatite cement: a novel biomaterial for craniofacial skeletal tissue engineering and reconstruction. J Biomed Mater Res 43:428–432
92. Miyamoto Y, Ishikawa K, Takechi M et al (1997) Tissue response to fast-setting calcium phosphate cement in bone. J Biomed Mater Res 37:457–464
93. Shindo ML, Costantino PD, Friedman CD et al (1993) Facial skeletal augmentation using hydroxyapatite cement. Arch Otolaryngol Head Neck Surg 119:185–190
94. Costantino PD, Friedman CD, Jones K et al (1992) Experimental hydroxyapatite cement cranioplasty. Plast Reconstr Surg 90:174–185
95. Cahyanto A, Tsuru K, Ishikawa K (2018) Effect of setting atmosphere on apatite cement resorption: An in vitro and in vivo study. J Mech Behav Biomed Mater 88:463–469
96. Cahyanto A, Maruta M, Tsuru K et al (2015) Fabrication of bone cement that fully transforms to carbonate apatite. Dent Mater J 34:394–401
97. de Leeuw NH, Parker SC (1998) Surface structure and morphology of calcium carbonate polymorphs calcite, aragonite, and vaterite: an atomistic approach. J Phys Chem B 102:2914–2922

Kanji Tsuru Dr. Tsuru is professor at Fukuoka Dental College, Section of Bioengineering, Department of Dental Engineering, Fukuoka, Japan. He completed his Ph.D. degree on the fabrication and characterization of inorganic-organic hybrid biomaterials under the supervision of Prof. Akiyoshi Osaka in Graduate School of Natural Science and Technology, Okayama University. He started to work in Okayama University as a research associate in the Department of Bioscience and Biotechnology from 1999. Since 2008, he worked as an associate professor in the Faculty of Dental Science, Kyushu University. He focused on the development of carbonate apatite bone substitutes under the leadership of Prof. Kunio Ishikawa till 2017. To date, he has authored more than 150 papers on inorganic-organic hybrids, sol-gel synthesis, surface modification, calcium phosphate bone substitutes, functionalization of dental materials etc. He majors in biomaterials based on the inorganic chemistry.

Michito Maruta Dr. Maruta is an associate professor at Fukuoka Dental College, Section of Bioengineering, Department of Dental Engineering, Fukuoka, Japan. He took his Ph.D. degree in Fabrication of low-crystalline carbonate apatite foam bone replacement based on phase transformation of calcite foam under Prof. Seiji Nakamura and Kunio Ishikawa's supervision in Graduate School of Dental Science, Kyushu University. He focused on the calcium phosphate-based bone filling materials based on his experience as a dentist who belonged to oral & maxillofacial surgery. He is currently working with Professor Kanji Tsuru to develop new dental materials and artificial bone filling materials.

Kunio Ishikawa Dr. Ishikawa is a Chairman and Professor of Department of Biomaterials, Faculty of Dental Science, Kyushu University, Fukuoka, Japan. He completed his Ph.D. degree at Faculty of Engineering, Osaka University. After he worked at Toray Ltd from 1986-1988, he became an assistant professor at Department of Dental Materials, Faculty of Dentistry, Tokushima University in 1988. He became an associate professor at Department of Dental Materials, Faculty of Dentistry, Okayama University in 1997. He became a Chairman and Professor at Kyushu University in 2001. To date, he has authored more than 250 papers on calcium phosphate bone substitutes, apatite cement, surface treatment of dental implant etc. Carbonate apatite bone substitute he invented with Prof Kanji Tsuru was commercialized as Cytrans® in 2017. He majors in biomaterials based on the inorganic chemistry.

Chapter 8
Smart Bioceramics for Orthopedic Applications

Fatma Nur Depboylu, Petek Korkusuz, Evren Yasa, and Feza Korkusuz

Abstract Smart bioceramics are mostly used to replace or reconstruct bone and joints. They can induce bone formation in hard tissues, and they can be used to increase the success of treating bone nonunion and fracture healing. A delayed union or prolonged healing can occur in certain bone fractures such as in distal tibia, scaphoid of the wrist, and talus of the ankle and in some cases can lead to nonunion fractures. Smart bioceramics can be used to prevent nonunion fractures and enhance the bone regeneration process. The addition of certain elements such as magnesium, zinc, strontium, and boron may enhance the osteoconductive property of bioceramics. They can be used to promote spinal fusion and/or assist implant integration in osteoporotic bones. Surfaces of nano-bioceramics improve osteointegration by increasing the areas available for osteoblast attachment, proliferation, differentiation and extracellular matrix formation. Nano-bioceramics are also biocompatible and may induce bone formation. Smart bioceramics could be osteoinductive when produced by combining them with active signaling molecules and/or cells. The context of this chapter reviews the recent trends on smart bioceramics.

Keywords Bone graft · Tissue engineering scaffold · Nanobioceramics · Fracture healing · Bioactive glass · Hydroxyapatite · Calcium phosphate · Mesenchymal stem cells (MSCs) · Osteoinductive · Osteoconductive · Biocomposites

F. N. Depboylu
Department of Bioengineering, Hacettepe University Institute of Science and Technology, Beytepe, 06800 Ankara, Turkey

P. Korkusuz
Faculty of Medicine, Department of Histology and Embryology, Hacettepe University, Sihhiye, 06100 Ankara, Turkey

E. Yasa
Department of Mechanical Engineering, Eskişehir Osmangazi University, 26480 Eskisehir, Turkey

F. Korkusuz (✉)
Faculty of Medicine, Department of Sports Medicine, Hacettepe University, Sihhiye, 06100 Ankara, Turkey

© The Author(s), under exclusive license to Springer Nature Singapore Pte Ltd. 2022
A. H. Choi and B. Ben-Nissan (eds.), *Innovative Bioceramics in Translational Medicine II*, Springer Series in Biomaterials Science and Engineering 18,
https://doi.org/10.1007/978-981-16-7439-6_8

8.1 Introduction

Bones are composed of organic and inorganic extracellular matrix (ECM) that is continuously renewed by cells regulated by signaling molecules (Fig. 8.1). This renewal process is a part of tissue homeostasis that is referred to as "bone remodeling". The unique physiochemical reactions between the organic and inorganic compounds of bone provide an incomparable mechanical strength that is rarely disrupted by trauma or diseases [1].

The combination of cells and organic ECM, which is mostly type 1 collagen, proteoglycans and glycoproteins and inorganic part as calcium-deficient carbonated hydroxyapatite (CHAp) nanocrystals, form the bone tissue structure. The notable feature of the apatite structure is its ability to hold different ions in its three ionic subgroups (calcium, phosphate, and hydroxyl). These apatite nanocrystals develop in the areas of regular mineralization of collagen molecules in bone [2].

An adult human skeleton has over 200 bones with various forms and sizes. Their main function is to protect internal organs, establish attachments and lever arm for skeletal muscles to enable motion and movement, store minerals and produce all blood cells including stem cells. All bones have, however, a common hierarchical structure, which ranges from nanocrystals to lamellae, lacunae and osteons at the micron-level [1]. Embryonic (woven) bones are differentiated to adult spongy or

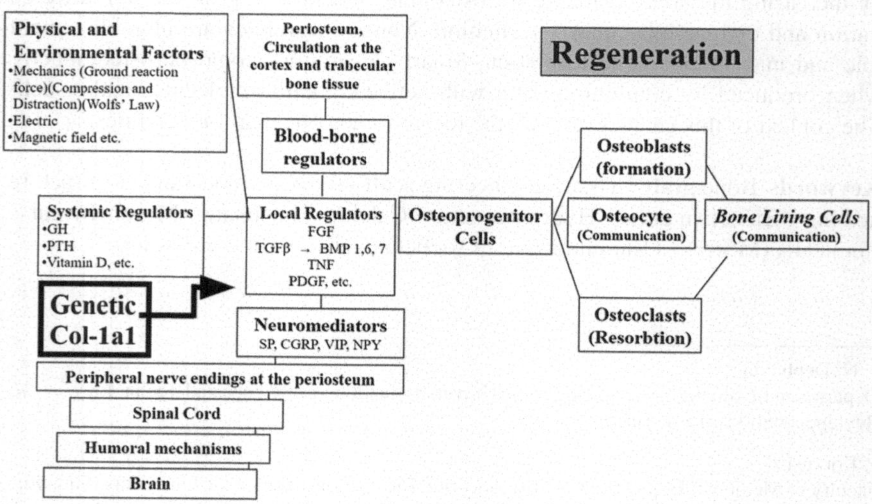

Fig. 8.1 Systems that regulate bone and regeneration are presented and these include physical and environmental factors; systemic regulators; blood-borne regulators; periosteum; brain; humoral mechanism; spinal cord and peripheral nerve endings at the periosteum. (**GH**: growth hormone; **PTH**: Parathyroid Hormone; **Genetic Col-1a1**: collagen type 1 α1 as protein coding gene; **FGF**: fibroblast growth factor; **TGF-β**: transforming growth factor-β; **BMP**: bone morphogenetic protein; **TNF**: tumor necrosis factor; **PDGF**: platelet-derived growth factor; **SP**: substance P; **CGRP**: calcitonin gene-related peptide; **VIP**: Vasoactive intestinal polypeptide; **NPY**: neuropeptide)

compact bones by intramembranous or endochondral ossification. The osteon that is made by Haversian canal and lamellae is the functional metabolic unit of the compact bone. Spongy bone consists of anastomosing trabeculae with irregular lamellae. Periosteum is the irregular connective tissue layer that surrounds the compact bone. Spongy bone is lined by endosteum that consists of bone lining cells at its interface with marrow cavity. The porosity of bone tissue enables metabolic exchange during, mineral accumulation, and pH equilibrium [3]. In addition, porosity makes a major contribution in traditional bone repair and regeneration. If the porosity of the bone tissue is under 20%, this type of bone is named as cortical/compact bone. If the porosity is higher than that, it is named as trabecular/cancellous/spongy bone [1, 4]. The mechanical strength is correlated with porosity: the lower the porosity, the higher mechanical strength.

Bioceramics, which have biomimetic features to the natural bone structure, can be naturally derived or synthetically produced. Bioceramics have many different classifications including oxides, phosphates, carbonates, nitrides, carbides, carbons and glasses for repair/reconstructions of defects or damaged tissue sites [2]. For clinical applications, these bioceramics can be in solid, powder and granule forms. While the solid form is often preferred for load-bearing applications such as in joint replacements, powder and granule forms are applied as bone grafts, coating on metal implants, bone cement and porous scaffolds. Baino et al. [5] stated that, three subgroups of bioceramics are described as inert (alumina, zirconia, sintered hydroxyapatite (HAp)), bioactive (HAp, bioglass, glass-ceramics) and biodegradable (α- and β-tri-calcium phosphates (TCP), calcium sulphates). Once an inert bioceramics are implanted, a protective fiber capsule 1–3 μm thick is formed around the implant. Although there is no serious foreign body reaction against bioceramics, interaction and bonding do not occur within the environment [6].

The main purpose of bioceramic material selection (in particular for coatings and surface modifications of implants) is to provide stabilization between the metallic implant and its surrounding connective tissue. Bioactive and biodegradable ceramics are able to fulfil this purpose with steady bonds until degradation and/or tissue replacement [7]. Although HAp belongs to the subgroup of bioactive material, it shows slow degradation with a rate of 1–3% annually. On the other hand, other calcium-based bioceramics such TCP and calcium sulfate possess faster degradation rates (35 to 40% annually for TCP and 100% annually for calcium sulfate) are preferred for bone regeneration in non-load bearing sites [8].

Tissue integration may change depending on the degradation rate of bioceramics. Recent investigations have focused on composites containing bioactive and biodegradable ceramics [9–11]. These composites established strong bonds with the tissue, and they were classified as second-generation bioceramics in orthopedic, maxillofacial and dental applications. Tissue engineering has recently changed directions from bone repair to bone regeneration by combining cells and active signaling molecules with bioceramics. Accordingly, the research and development into third generation bioceramics have gained speed and they were categorized as innovative or smart bioceramics. Smart bioceramics promote osteogenesis by inducing cell response. The porosity of bioactive and biodegradable bioceramics has become vital in the encapsulation of bioactive substances such as growth factors, hormones, peptides, which stimulate osteoprogenitor cells for osteogenesis [12]. Those bioactive substance or cell containing ceramics may have osteoinductive properties.

8.2 Implant-Cell Interactions

Various implant designs, implant materials and surface modifications have accelerated the osteointegration over the years. Bone is a vascular, dynamic and living tissue having a relatively constant remodeling rate until old ages. Implant material, cell type, bone type, cytokines and growth factors act in synchronization to regenerate bone. The type of material used plays a role in determining the regeneration potential of bone surrounding the implant, and ultimately the osseointegration process.

Shortly after surgical implantation, the growth of regional blood vessels allows the local mesenchymal stem cells (MSCs) recruitment that initiates stronger osseointegration at the implant and surgical site. The proliferation of MSCs lead to the development of osteoprogenitor cells that is then differentiated into osteoblasts, which are involved in the construction of the bone matrix and essential for implant integration [13, 14]. The osteocytes are mature bone cells residing in their lacunae one by one. They are interconnected with each other by their cytoplasmic processes located in the canaliculus between the lacunae through the gap junctions.

During regeneration, the primary focus is the formation of bone cells that will be attached to the implant surface. Determination of regenerative cell types is achieved by the protein that binds to the surface when the implant surface and body fluid comes into close contact. The process begins immediately after the implant is exposed to bone tissue after implantation. Unique proteins contain certain molecules that specifically attach to precise receptors on the cells. Therefore, it is necessary to regulate cell penetration through surface modifications that develop binding structures [15]. Consequently, surface modification of implants and specifically smart bioceramics is of clinical significance [16, 17].

Biological molecules such as ECM components, engineered peptides, and growth factors may activate osteogenic cells and accelerate bone tissue development during the initial phases of implantation once they are freshly introduced to the surface textures of bioceramic implants. In addition to acting as a scaffold for cells, ECM facilitates to their adhesion and promotes the regulation of attachment, proliferation, differentiation and migration. In addition, growth factors are bioactive substances that control several mechanisms involved in tissue regeneration. They promote angiogenesis for vessel growing. These growth factors, especially in the presence of osteoporosis, improve the surface property of the bioceramic and/or accelerate the healing if the natural renewable potential of bone tissues around the bioceramic is dysfunctional. Considering the significance of stem-progenitor cells, ECM, and growth factors in the mechanism of osteointegration, it may be beneficial to create a biochemically tunable bioceramics surface that combines these three components [13]. Cells interact with extracellular bio-signals, which trigger the cytokines and are

Table 8.1 Advantages and disadvantages of nano and micro-sized bioceramics

Bioceramics	Advantages	Disadvantages
Nano (< 50 nm)	• Enhanced life quality and shorter rehabilitation • Increased surface area • Higher cell attachment • Less integration problems • Faster tissue regeneration • Higher mechanical properties • Used in the treatment of bone infections, drug/gene delivery	• Osteolysis (particle disease) • Aseptic implant loosening due to the release of particles • Inducing allergic or inflammatory reaction • Blood coagulation
Micro (350 to 1000 nm)	• Cell aggregation • Vascularization • Ideal internal growth • Minimized stress shielding	• Lower mechanical hardness than nanoceramics • More prone to crack propagation • Less cytocompatibility

1–10 nm in size. These bio-signals are growth factors, peptides, and proteins. Recent smart bioceramic research has therefore concentrated on combining ECM-based peptides and/or proteins on nano-scale surfaces to fulfill biomimetic approaches through nanotechnology [18, 19]. The advantage of nano-bioceramics over micro-bioceramics is their increased surface area that allows cell and tissue attachment and their minimized tissue reaction problems (Table 8.1).

Biological response to bioceramics follows a familiar cascade with bone fracture healing. The cascade starts with growth of hematoma, infection, initiation of vascularity, osteoclastic resorption and ossification [20]. Interfacial bonds are formed between bone and bioceramic substitutes with the surrounding ECM as degradation of ceramic proceeds. The initial cell reaction towards the bioceramic begins with the osteoblasts and blood cells. Vessel forming endothelial cells appear first after the completion of osteoclastic resorption given that the restoration of blood circulation is essential for bone regeneration. Mesenchymal stem cells stimulate and/or transform to osteoblast directly as intramembranous ossification advances [21]. This type of ossification for rigid fractures is the main regeneration mechanism although endochondral ossification contributed to nearly all intramembranous ossifications (Fig. 8.2) [22]. Monocytes that stimulate expression and secretion of the cytokines and proteases interact with the bioceramic particles to trigger bone regeneration and remodeling (Fig. 8.3) [23].

The implantation of calcium phosphate ceramics additionally deploys the macrophages, which are responsible for the infiltration and the secretion of H^{+}. Inflammatory cells promote the secretion of Interleukin-1 (IL-1) and Interleukin-6 (IL-6), osteoprotegerin (OPG), TNF-α, and other growth factors [8]. Macrophages are able to support segregation of cytokines, fibroblast, platelet-derived and insulin-like growth factors. TGF-β superfamily regulated by macrophages comprises BMPs and vascular endothelial growth factor (VEGF) as shown in Fig. 8.4. These trigger angiogenesis and new bone formation [24]. Osteoprogenitor cells appear as osteoclastic

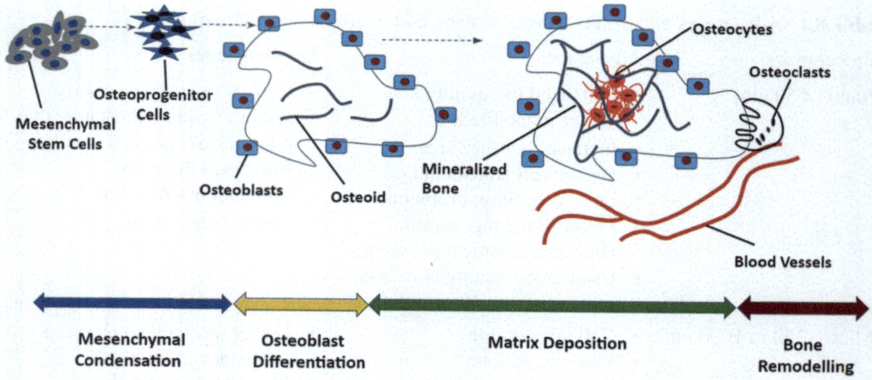

Fig. 8.2 Intramembranous ossification

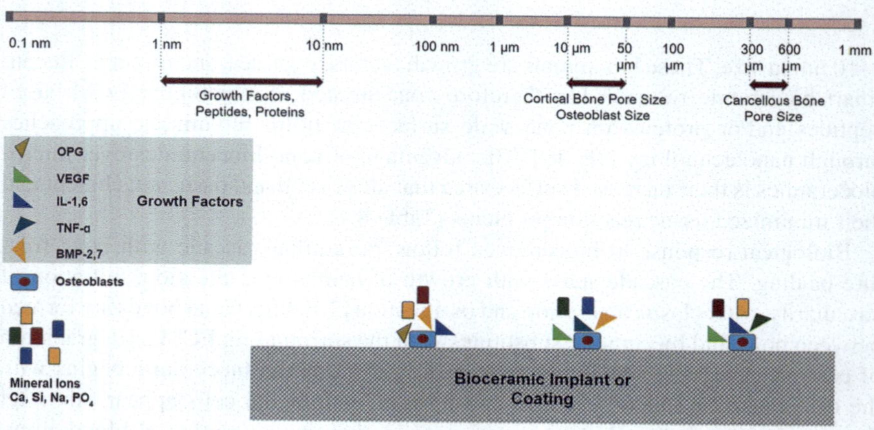

Fig. 8.3 The bar includes that growth factors, peptides, and proteins are between 1–10 nm in size; cortical bone pore size and osteoblast size are between 10–50 μm while cancellous bone pore size is between 300–600 μm. Bone regeneration on the surfaces of bioceramics implant or coating with growth factors (**OPG**: osteoprotegerin; **VEGF**: vascular endothelial growth factor; **IL**: Interleukin 1, 6; **BMP**: bone morphogenetic protein 2. 7; **TNF**: tumor necrosis factor-α), osteoblasts and mineral ions (**Ca**: calcium; **Si**: silicon; **Na**: sodium; $\mathbf{PO_4}$: Phosphate)

resorption of necrotic bone occurs. Since the osteoclasts signal the osteoblasts to trigger bone formation, constant regeneration with or without bioceramics continues during this bone-remodeling cascade by forming cutting cones (Fig. 8.5) [25].

After bioceramic implantation, the interaction rapidly begins between the ECM proteins and surfaces of the bioceramic. Proteins generate biological signals according to surface topology and chemical features of the implant. Thus, the biological response is taken from the cells against the surface features. Biological response to bioceramics provides cell adhesion stability and triggers cell proliferation and

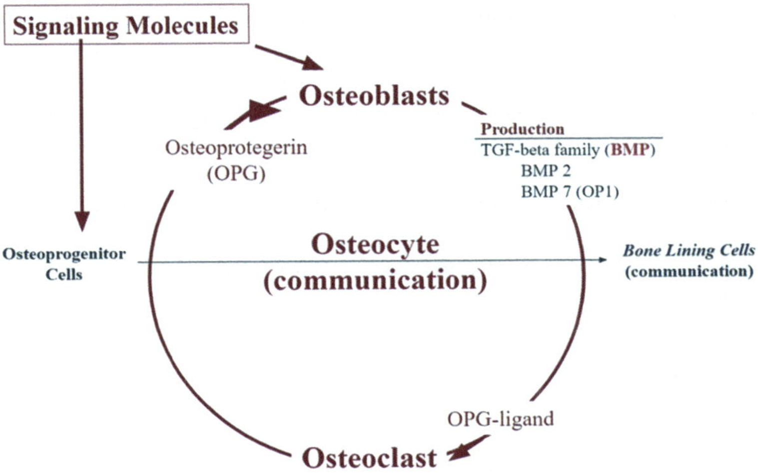

Fig. 8.4 The effect of signaling molecules on cell communication. Signaling molecules trigger angiogenesis and new bone formation. Osteoprogenitor cells appear as osteoclastic resorption of bone happens. The osteoclasts signal the osteoblasts to induce bone formation. **TGF-β**: transforming growth factor-β; **BMP**: bone morphogenetic protein 2, 7; BMP 7 also called **OP1**: osteogenic protein 1, **OPG**: osteoprotegerin

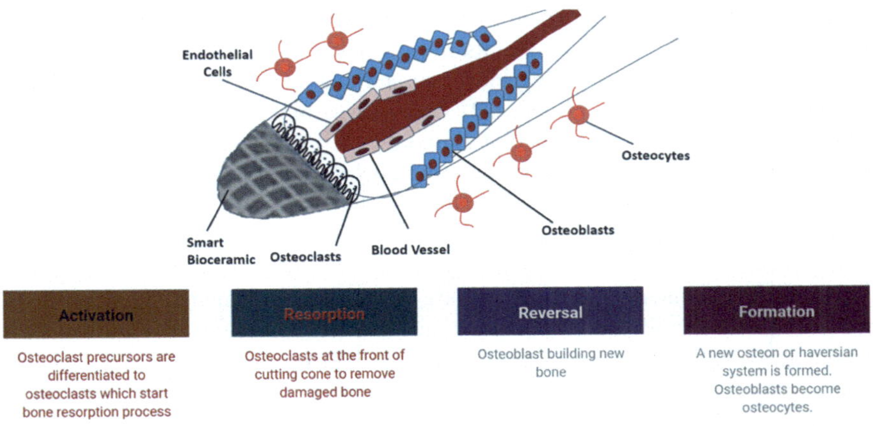

Fig. 8.5 Constant regeneration continues during the bone-remodeling cascade by forming cutting cones. Bone remodeling cascade: activation, resorption, reversal and formation

differentiation [26]. Another effective factor for interaction is the wettability, which encourages surface protein and cell adhesions [27]. Wettability is considered as the hydrophobic and hydrophilic behavior of the surfaces. Cells may react negatively for both superhydrophobic and hydrophilic surfaces. The balance, therefore, is so critical to provide an adequate wettability on the surface of the biomaterial without the interference of rejection [28]. The wettability is related to the contact angle of the material, which influences cell proliferation [27]. The contact angle is used to determine if a surface is hydrophobic or hydrophilic, and for a hydrophobic surface, the contact angle is between 90° and 150°. The contact angle for superhydrophobic surface is over the 150°. For hydrophilic surface, the contact angle is between 10° and 90°, while for superhydrophilic surface is less than 10° [28]. Menzies and Jones investigated and tested the contact angle effects on the biocompatibility of an implant [29]. They discovered that the smaller contact angle gives better wettability and promotes cell adhesion.

8.3 Bioceramics

8.3.1 Calcium Phosphates

It is well recognized that bone consists of nanosized platelets of crystals of biological HAp grown in close contact with an organic matrix rich in collagen fibers. Bone apatite can be better defined as calcium-deficient carbonate apatite with the formula $(Ca, Mg, Na)_{10}(PO_4, HPO_4, CO_3)_6(OH)_2$ [30]. Therefore, the application of calcium phosphate as the essential element of the inorganic ECM is recommended due to both cell culture and promising results from in vivo testing [30, 31]. Benefits and drawbacks of the most promising bioactive and degradable bioceramics available in clinical use are shown in Table 8.2. When smart bioceramics are designed for orthopedic applications, the choice of materials, their substitutes and combinations play important roles to meet criteria presented below in Table 8.3.

Hydroxyapatite (HAp), whose formula is given as $(Ca)_{10}(PO_4)_6(OH)_2$, is more stable and less toxic compared to other calcium phosphates as its Ca/P ratio is 1.67, which can affect its dissolution property [32]. It is primarily used in hard tissue replacements of human bone as scaffolds and coatings on dental implants [33]. Using HAp is generally preferred as a composite with biopolymer reinforcements or combined with other bioceramics such as TCP/HAp as biphasic calcium phosphate to achieve improvements in mechanical properties and to increase interfacial bonding between scaffold and tissue [34, 35]. Such properties can also be observed in macroporous biphasic calcium phosphate [36].

β-Tri-calcium phosphate is clinically used as a bone graft supply for orthopedic and dental applications [37]. β-TCP is also preferred for the treatment of maxillofacial disorders due to its outstanding bioresorbability, corrosion-resistance, and high interfacial binding ability. The sintering temperature of β-TCP is 1100 °C. If the temperature is higher than 1100°C, β-TCP transforms to α-TCP, which displays less stability and high solubility [38]. The β-TCP comprises α, α'and β' phases in its structure and has a Ca/P ratio of 1.5. Lower Ca/P ratio enhances solubility and

Table 8.2 Benefits and drawbacks of bioactive and biodegradable ceramics

Bioactive and Biodegradable Ceramics	Orthopedic Applications	Advantages	Disadvantages
Hydroxyapatite	• Coating material on load-bearing implants • Bone graft or fillers	• Excellent biocompatibility • Biodegradable • Osteoconductive and osteoinductive • Non-toxicity and non-inflammatory	• Poor tensile strength and low fracture toughness • Brittleness • Inelasticity
β-Tricalcium Phosphate	• Bone graft or fillers • Scaffold	• Rapid degradation rate compared to hydroxyapatite • Adjustable resorption and bone remodelling	• Low fatigue resistance • Faster absorption than newly formed tissue • Although it is osteoconductive, it is not osteoinductive
Bioglass	• Coating material on load-bearing implants • Used as bone cement • Used to repair bone defects	• Osteoconductive and osteoinductive • Adsorption of protein • Providing cell differentiation and proliferation	• Poor mechanical properties • Unsuitable as load-bearing implant material due to its brittleness
Glass-Ceramic	• Apatite-Wollastonite (A-W) enables in some compression load-bearing applications	• Scratch and abrasion resistant • Controlled and designed crystal phases enables the combination of high bioactivity and mechanical properties • A-W has the highest mechanical strength • Osteoconductive and osteoinductive bioglass can be derived from glass-ceramics with modifications	• Unsuitable as load-bearing implant material due to its brittleness

acidity of the environment compared to high Ca/P ratio. β-TCP has greater degradable compared with HAp [39]. Although β-TCP is accepted as bioresorbable, it does not fully replace bone when used as a bone graft. Furthermore, the amount of bone formed is lower than the reabsorption rate of β-TCP [34]. β-TCP is bioactive and biodegradable, but it cannot be regarded as osteoinductive. Moreover, it cannot be defined as a smart bioceramic unless it is combined with osteoinductive substances (Table 8.2).

Table 8.3 Features and advantages of smart bioceramics

	Features	Advantages
1	Biodegradable	• Chemical dissolution • Macrophagic phagocytosis of bioceramics
2	3-D Porosity	• Hierarchical model from macro to nano-level • Interconnectivity for cell migration
3	Bioactivity	• Sustainable ion releasing until degradation • Osseointegration • Osteoconductivity
4	Biomimicry of the bone	• Physio-chemical structure
5	Osteoinductivity	• Acceleration of new bone formation and regeneration
6	Non-cytotoxicity	• Biocompatible

Dicalcium phosphate dihydrate (DCPD) is preferred in certain bone regeneration applications due to its high biodegradability under the physiological conditions when compared to HAp. DCPD are categorized as osteoinductive material due to rapid ossification after in vitro and in vivo degradation. Thus, it can be said that DCPD and dicalcium phosphate are suitable options for use as bone cement. Dicalcium phosphate/HAp biphasic ceramics also have both osteocondutive and osteoinductive properties and can procure controllable biodegradability [34].

A study by Shuai et al. examined the biodegradable activity, mechanical features, and biochemical performances of TCP/HAp porous scaffolds via selective laser melting (SLS) rapid prototyping (RP) method [40]. A number of TCP/HAp ratios were examined: 0/100, 10/90, 30/70, 50/50, 70/30, and 100/0. Their results showed scaffolds with TCP/HAp ratios of 30/70 and 50/50 displayed better biological performance as demonstrated by the adherence and proliferation of osteoblast-like cell (MG63). After soaking the scaffolds in simulated body fluid for 7 days, it was revealed that the amount of apatite precipitated on the surfaces of scaffolds increased as the content of TCP increases. The scaffold with a TCP/HAp ratio of 30/70 revealed the formation and uniform distribution of cavities on its surface. The results of mechanical testing demonstrated that both the fracture toughness and compressive strength increased with increasing content of TCP from 0 to 30 and the scaffold with the TCP/HAp ratio of 30/70 exhibited the optimum fracture toughness and compressive strength. Schmidleithner et al. indicated that a pure β-TCP could be manufactured with 99.5% relative density using digital light processing (DLP) RP method [41]. Digital light processing also contributed to high cell compatibility and differentiation of cultured Murine pre-osteoblastic (MC3T3-E1) cells on the β-TCP scaffolds.

8.3.2 Bioglass and Glass-Ceramics

Bioglass is the silica-based amorphous structure, which possesses non-uniform distribution of atoms due to rapid cooling of the ceramic. It consists of a stable interface and stimulates angiogenesis and tissue regeneration during degradation [42, 43]. Bioglass can be characterized by the creation of a network that interconnects different silica tetrahedron network via the –Si–O–Si– bridging oxygen bond. This network can be disrupted by network modifiers such as Na and Ca creating non-bridging oxygen bonds (–Si-O–NBO–). Na_2O, K_2O, CaO, MgO can support the formation of an apatite layer on the surfaces of the bioglass that supports ossification [44]. Silica-based bioceramics that present an improvement to the 45S5 bioglass with a composition of 45 wt% SiO_2, 24.5 wt% Na_2O, 24.5 wt% CaO, and 6 wt% P_2O_5 have been used in clinical practice since 1985. They trigger the secretion of growth factors and support neovascularization [45]. Borate and phosphate bioglasses were used for muscle and ligament replacements after the developments of silica-based bioglass [18, 46, 47]. Bioglasses are applied as bone cements at defects because of their bioactivity and the ability to induce bone. They are generally preferred as coating material instead of load-bearing applications due to their weak physical characteristics. The apatite film covered around the glass acts as a barrier against metal ion release [48]. A study by Jones has revealed that the potential of bone tissue regeneration and cell response depend on the ionic degradation products of bioglass and glass-ceramics and HAp layer formation [49].

Bioglass can be manufactured using either the melt-derived approach or the sol-gel technique [5]. Silica increases the bioactivity of the glass. The interaction levels, type of bond as well as the time needed for its formation governs the rate of bioactivity of the bioglass. The rapid bond formation with soft and hard tissues can be achieved when the silica content in the bioactive glasses obtained from the melt is between 42 to 53%. Bonding occurs in about 2 to 4 weeks in hard tissue but there is no soft tissue interaction if the silica content in the bioglass is between 54 to 60%. Glass behaves immobile and does not interact with tissues as the silica content increases above 60% [44, 50].

Bioglass with a silica content of up to 90% can still achieve bioactivity and bonding due to the formation of an HAp layer if it is manufactured using the sol-gel method [5, 51]. Sol-gel manufacturing method can produce porous structures at nano scale (from 1 to 30 nm) known as meso-porous bioglasses such as 58S (with a composition of 58 wt% SiO_2, 38 wt% CaO, and 4 wt% P_2O_5) and 70S30C (70 mol% silica and 30 mol% CaO). The features of meso-porous structures such as lattice dimensions, area and volume are effective in regulating the adsorption and releasing activities of biomolecules such as proteins, growth factors, peptides and drugs. Thus, it is possible to obtain greater degradation rate and rapid generation of the apatite film on the glass surfaces [52]. Bioglass produced from sol-gel technique exhibits higher chemical and biological properties than compared to those synthesized using the melt-derived method. The reason for this is the Si-OH surface layer and dissolved ions on the surface, which increase chemical reactivity. As mentioned previously, this value cannot exceed 60% in the melt-derived bioglass, but reaches 90% in sol-gel-derived bioglass [42].

Ojansivu et al. stated fiber-reinforced composite implants (S53P4) are capable of achieving similar results comparable to titanium implants [53]. Nevertheless, the formation of osteogenic medium extracts on the composite led to enhance pH of body environment and triggered cell differentiation from hMSC to osteoblasts due to the release of calcium and phosphate.

Depending on the thermal treatment parameters, glass-ceramics are composed of fine grains with controllable size and contains a small fraction of residual glass found at the grain boundary [5, 44]. These ceramics are known as polycrystalline solid and one advantage is that a combination of special properties such as enhanced mechanical properties, bioactivity, and workability can be achieved through the design and control of the formation of crystal phases and crystallization [44, 51].

Thermal and mechanical characteristics of glass-ceramics such as scratch and abrasion resistance permit its use in orthopedic applications [54]. Apatite wollastonite (A-W) glass ceramics, which were developed from apatite matrix reinforced with wollastonite, are utilized in compression load-bearing applications since it has the highest crack resistance and Young's modulus amongst bioglasses and glass-ceramics [55]. A study by Shi et al. has demonstrated the biosafety and bone regeneration capability of bioglasses [56]. Glass-ceramics, however, are not the ideal for load-bearing applications due to their brittleness. A-W glass-ceramics cannot be regarded as smart bioceramics if their biological activities are taken into consideration. They display osteoconductive property due to crystalline phases, reduction in chemical reactivity, and osteogenesis [42].

Yuan et al. reported that bone formation on the surface of implant occurred with typical intramembranous ossification inside of the pores instead of crystal layers when they implanted bioglass with osteoconductive lattice microarchitecture into the muscles of dogs [57] (Fig. 8.2). The study presented that the osteoconductive biomaterials can be converted to osteoinductive by optimization of the porosity, geometry, and chemical surface and addition of growth factors and osteogenic cells. Thus, glass-ceramic structures can also be classified as smart bioceramics through appropriate modifications.

8.3.3 Biocomposites

8.3.3.1 Calcium Phosphate-Polymer Based Composites

Composites produced from calcium phosphate (CaP) and reinforced with polymer matrix shows promise as the key biomaterial for orthopedic applications. Composites can support tissue regeneration due to bioactive CaP and the biodegradable polymer matrix [58]. While hydroxyapatite exhibits brittleness and is more prone to crack generation, especially in load bearing applications, HAp-polymer (polylactic acid

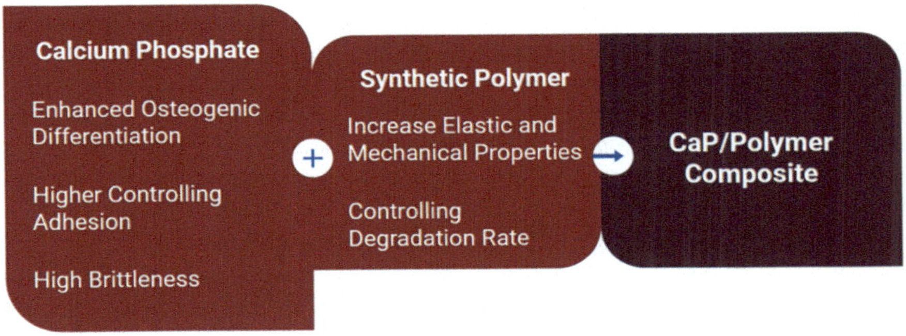

Fig. 8.6 Contribution of calcium phosphate and synthetic polymer to CaP/Polymer composite in terms of mechanical and biocompatible properties

(PLA), poly (lactic-co-glycolic acid) (PLGA), poly-Ɛcaprolactone (PCL)) composites show improved ductility and deformity due to the flexibility of polymers. In addition, the composites will have a greater lifespan as there is a reduction in the possibility for cracks to form. Combination of CaP phase into biodegradable polymer matrices, for instance by combining them with PLA, PCL and the copolymer PLGA, may improve the Young's Modulus and mechanical performance (Fig. 8.6) [59]. These composites help to control the adhesion of stem cells, which promote osteogenesis activities [60]. Acidic features of biodegradable polymers however may affect their tissue response negatively. Alkali ceramics neutralize the acidic impact of the polymers and, for this reason, their bioactivity is enhanced [44].

Chang et al. investigated the bone regeneration efficacy when collagen-enhanced particulate biphasic calcium phosphate (TCP/HAp/Col) and dicalcium phosphate and HAp (DCP/HAp) were grafted into critical bony defects of dogs. The composite bone graft composed of 12% bovine-derived collagen and 88% ceramic, in the form of HAp and β-TCP biphasic ceramic (60% HAp and 40% β-TCP) [34]. The long-term osteoconductivity and controllable biodegradation feature were provided by the HAp/TCP biphasic ceramic for bone growth. Collagen also played an osteoinductive role at the composite. Cell migration and differentiation were promoted with the porosity and interconnectivity for bone graft application of composite. When TCP/HAp/Col and DCP/HAp grafts were compared for bone defect repair, DCP/HAp exhibited a profound impact for osteoblastic bone formation and compression strength compared to TCP/HAp/Col composite.

8.3.3.2 Calcium Phosphate-Metal Based Composites

The CaP/metal matrix composites demonstrate enhanced mechanical characteristics such as improved tensile strength, yield strength and Young's modulus. Magnesium-based CaP composites have been considered as a promising candidate due to its favorable biodegradation and physical characteristics [9]. The main reason for utilizing magnesium is based on its degradation. Li et al. reported the detection of magnesium, calcium and phosphorus ions at different time intervals after magnesium alloy samples were immersed in simulated body fluid (SBF) solution [61]. During the degradation tests, products bond and form a protective layer on the surface of the metal. An increase in elasticity and yield strength was observed during the mechanical testing. After the magnesium and calcium phosphate bonds degraded, a decrease in mechanical properties was detected. Therefore, these ions are considered as the degradation product of the metal/ceramic composite. Thus, it has been proposed that magnesium/ceramic composites could be ideal for orthopedic applications. Phosphate-based biodegradable and bioactive ceramics are also desirable as they provide high corrosion resistance and appropriate mechanical properties in orthopedic implants. Examples are calcium polyphosphate grains (CPP), HAp and TCP [62].

Calcium polyphosphate grains are bone-like polymer ceramic oxides with good biocompatibility similar to other phosphate-based ceramics. Its exact composition is $Ca(PO_3)_2$. A number of calcium phosphate-based ceramics were used as reinforcement in magnesium matrix composites. The degradation rate of magnesium is extremely high and in the process releases a significant quantity of hydrogen gas, which delays the healing of fracture bone tissue. The incorporation of calcium phosphate allows the degradation to be controlled at a given rate in vivo and in vitro [9]. Magnesium has a biodegradable and osteoinductive characteristics and thus contributes to the osteogenic differentiation of MSCs [63]. Also, the bioactive ceramics are resistant to corrosion and can control the rate of biodegradation. The osteoinductivity and controlled biodegradability as the crucial properties in smart materials are the biggest reasons why magnesium/ceramic composites show promise as candidate for smart composites. In addition, reinforcing titanium with HAp and TCP will improve the mechanical strength of apatite by eliminating the brittle nature of bioceramics that can be applied as cranial implants [64].

8.4 The Criteria of Smart Bioceramics

The primary definition of smart bioceramic scaffolds is to design a remarkable artificial structure that will potentially promote osteoconduction as well as the opportunity to induce bone formation. The specific features were verified in clinical studies from the European program within the concept of "smart osteogenic scaffolds" [65]. Macroporous Biphasic Calcium Phosphate (MBCP) Technology in combination with human hMSC suspension is another approach to osteoinductive smart bioceramics [36].

The definition of osteoconduction is that the bioceramic is able to promote bone formation and conform to its surface. Bone tissue provides an optimum environment for the scaffold where it can regenerate bone. On the other hand, according to the definition of osteoinduction, it stimulates the cells to phenotypically transform into osteoprogenitor cell types that can form bone in any tissue. BMP injection and/or implantation are excellent examples of inducing osteoinduction (Fig. 8.4) [66, 67]. Scaffolds can induce osteogenic cell differentiation through nano, micro and macrostructure [36]. Osteogenic is defined as the ability for living cells to form bone such as osteocytes or osteoblasts [68].

The advantages of smart bioceramic scaffolds are their enhanced support for the recovery and reconstruction of bone tissue. Such absorbable medical devices have versatile properties that can be found as solid bioceramics or moldable injectable materials. In the clinical community, this smart scaffold innovative approach is referred to as a part of the "Smart Biomaterials Diamond Concept". The three features included in this concept are volume restoration, stabilization and bone regeneration [69].

Although several studies have recently shown osteoinductive properties for some micro and macro scale CaP bioceramics with addition of growth factor and/or cells [70], the major focus is on the nano and micro scale interconnectivity of smart materials as depicted in Fig. 8.7. Therefore, the creation of a "smart scaffold" to assist in the latest tissue engineering approach is challenging. The originality of the scaffold is determined by the macro-to-nanostructure biomimetic approach associated with biological fluids infiltration, protein adhesion, cells and tissue colonization and bone reconstruction [71].

In order to meet the criteria and functions essential for the treatment of bone defects, the smart scaffold must contain the features presented in Table 8.3. There are several ways for the biomaterial to be resorbed by surrounding tissues such as the process of chemical degradation, the resorption via osteoclasts and macrophagic

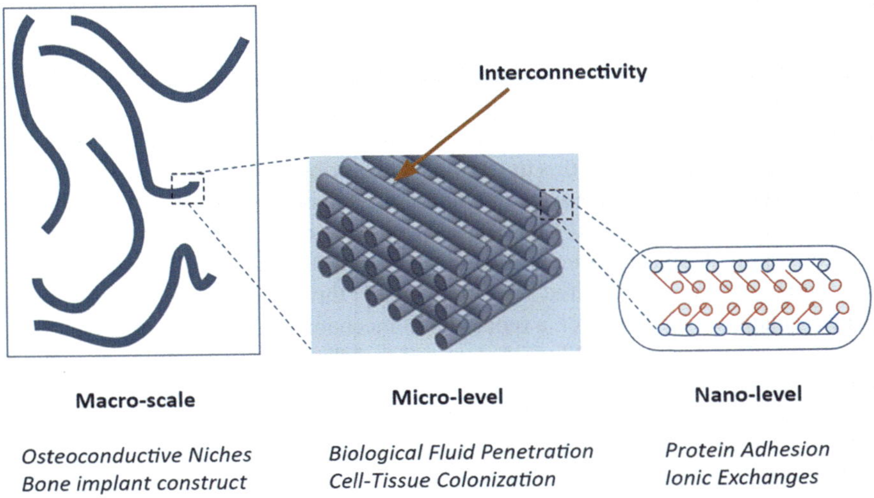

Fig. 8.7 Macro to nano-level bioceramics representations and their biological advantages

phagocytosis of the bioceramic fragments [72]. Another requirement is the multidimensional 3D porosity. The first of these is the macroporosity that act as an osteoconductive niche for cell attachment, growing and development. The second requirement is the micro-level porosity for volume interconnectivity that provide enhance surface area and great degradation rate for adsorption of biomolecules [65] (Fig. 8.3). Functional performance of growth factors and peptides, which are osteoinductive agents, can be enhanced by covalent implantation to the bone, causing their continuous interaction with the bone microenvironment [73]. Interconnected 3D structures suitable for the circulation of body fluids are essential to establish a biologically active medium to cell stimulation and accumulation, protein adsorption and ion exchange. The assembly of macro-micro-nano structures may provide extremely reduced density bioceramic structure with enhanced osteointegration capacity [72, 74].

In terms of bioactivity, the presence of Ca and P ions released from the nutritive apatite plays an important role in the continuation of apatitic nucleation and re-precipitation. Another remarkable feature is that body fluid interacts with the adsorbed biomolecules to provide cell adhesion and osseointegration. In addition, bone biomimicry should be considered in terms of microstructure and chemistry, and other key factors include osteoinductivity or conductivitiy, as well as non-toxicity and non-immunogenicity, in addition to biocompatibility [75, 76].

8.5 Smart Bioceramics

8.5.1 Porosity

A hierarchical porosity, which includes macro to nanoscale pore sizes as shown in Fig. 8.7, which is similar to the unique bone structure, and the cascaded pores are needed in tissue reconstruction [1]. The nano-structures and high percent of porosity increase the surface area, and this tremendously contributes to the ion homeostasis of the bone. The lattice structure sizes smaller than 1 μm is responsible for bioactivities including protein interactions [77]. Highly porous structures allow the movement and the growth of osteoblasts, osteoprogenitor cells and MSCs, as well as the delivery of mineral and oxygen required for angiogenesis throughout the process of bone tissue formation [78]. However, a reduction in mechanical strength is observed with larger pore size. A study by Yuan et al. stated that the Young's modulus of cortical bone is between 7.7–21.8 GPa, while for cancellous bone this value is reported to vary between 0.01 and 1.57 GPa [79]. Zaharin et al. reported that Young's modulus with lattice dimensions with pore size 300 μm could be considered as giving similar properties to that of natural bone [80]. In addition, nano bioceramics, when assembled by electrospinning or 3D printing, could possibly obtain appropriate pore size with increased mechanical strength. Various lattice structures might influence the mobility of bone cells, which can serve as a system to control cell behavior in the ECM environment [81].

Looking at the synthetic bioceramics manufactured using conventional techniques, inadvertent porosity causes low performance for bone repair and regeneration [82]. The fabrications of biomimetic hierarchical lattice structure (macro to nano) bone implants are therefore essential in triggering greater bone regeneration potential [83].

8.5.2 Trace Elements with Bioceramics

Trace elements can be integrated to HAp to improve properties such as bioactivity, thermal stability, solubility, cell response and degradation, and mechanical properties [18, 44]. HAp can incorporate with anionic and cationic substitutes for phosphates and calcium, respectively. While zinc (Zn^{+2}), magnesium (Mg^{+2}), strontium (Sr^{+2}), boron (B), sodium (Na^{+1}), silver (Ag^{+1}), barium (Ba^{+2}), yttrium (Y^{+3}) and titanium (Ti^{+4}) may interchange with calcium (Ca^{+2}), while the phosphates are substituted with carbonates, silicates, vanadates, borates and manganates [75, 84].

There are two main kinds of calcium, which affect the crystallinity and interactions with different trace elements in a unit cell of the apatite [18]. Ca^{+2} (I) includes four Ca^{+2} ions surrounded by nine oxygen atoms, Ca^{+2} (II) includes six Ca^{+2} ions, which are bigger in size and volume than Ca^{+2} (I), for each unit cell which provides the motion of anions. Phosphate atom consists of $PO_4{}^{3-}$ surrounded by four oxygen atoms [85].

The growth of the human skeleton requires silicon to increase bone mineralization and regeneration with triggering gene activity [86, 87]. SiO_4 interacts with PO_4 that prevents grain growth and enhance surface area/volume ratio that affects topography of cell attachment [88]. The bioactivity of HAp can be enhanced through the incorporation various elements such as magnesium, strontium, boron [46], and silicon [89]. As a mineral, the amount of magnesium contained in bone, dentin and enamel are 0.72%, 1.23% and 0.44%, respectively. Similar to magnesium, strontium is another element found in bone and they promote osteoblastic activity while it decreases osteoclastic resorption. Both minerals are required for bone mineralization at different stages. While magnesium is substituted with Ca^{+2} (I), strontium reacts with Ca^{+2} (II) apatite region, in this way, these substitutions restricts stabilization and ion release, while promoting β-TCP transition through thermal transformation [90]. Geng et al. suggested that eggshell are a promising source of Ca^{+2} and doping it with Sr^{+2} offers improved performance compared to strontium-free HAp in terms of cell growth and differentiation. Alkaline phosphatase (ALP) activation for protein expression is also prominent with the incorporation of Sr^{2+} [91]. Zinc, which interacts with Ca^{+2} (II) and therefore exhibits a similar initiator effect on crystallinity with magnesium and strontium. These features enhance protein adsorption and cofactor enzymes. On the other hand, a better environment for osteoblastic adhesion is offered by yttrium than compared to magnesium and zinc [18].

Protein adsorption is associated with the surface charge, which in turn impacts the osteoinductivity and bioactivity of HAp. Therefore, the negatively charged HAp surface is responsible for the collection of basic proteins. The release of protein can be controlled through the incorporation of zinc and magnesium [92]. Furthermore, silicon-HAp, which is a higher negatively charged bioceramic compared to pure HAp, provides a lot more protein adsorption on the surface because of addition of silicate [93]. Integrins that are ECM receptors promote osteoblast attachment by giving off signals [44]. After attachment, these receptors impact the cellular phenotype and cytoskeletal molecules including vinculin and actin [8].

β-TCP requires improvement for expediting bone growth, and this can be achieved through the addition of MgO, ZnO, SrO, and SiO_2 [94]. The integration of such elements may influence and change the physical, chemical and biological behavior of β-TCP. Increasing osteogenic activity was recorded in MgO/SrO doped β-TCP [95]. Chou et al. reported that β-TCP doped with zinc bioceramics enhanced growth of MSCs and ALP efficiency. Osteogenic differentiation and ECM formation occurred and the calcium level established a positive difference after 14 days [96].

Gizer et al. investigated the effect of boron and nanoscale-HAp composite on osteogenic activities of MSCs and osteoblast cells. The results showed increased MSCs proliferation and ALP efficiency with boron-containing-nano-HAp scaffolds in comparison to boric acid and nano-HAp composite [46]. In addition, the study by Tunçay et al. showed that when comparing boron-doped-nano-HAp composite with nano-HAp, the boron-doped-nano-HAp composite enhanced the adhesion and growth potential of the MC3T3-E1 osteoblastic cells [97].

8.5.3 Nanoscale Bioceramics

The dimensions of nanoceramics should be less than 100 nm. There have been changes in the manufacturing approaches from the micro-scale to nano-scale with novel nanotechnologies. Nanoceramics will highly interact with biological tissues and with nucleic acids, protein and others. Materials of this scale have been integrated and adapted to biomedical devices. The biggest reason for this is that many biological systems such as viruses, protein complexes, and membranes display natural nanostructures. In addition, improvements in mechanical properties and bioactivity are described as other advantages of nanomaterials over micro-scaled ones [44].

Cells can be triggered precisely based on the environmental topography at nanoscale, and in this way, cell response increases [98]. Surface nano-topography will positively influence the cell adhesion, cell signaling and some gene expression. The nano-scale structure is similar to the bone hierarchical architecture from micro-scale to nano-scale [99, 100]. The endocytosis of nano-HAp particles into the MSCs cytoplasm by scanning electron microscope is shown in Fig. 8.8. This finding strengthened the opinion regarding the strong biocompatibility of such bioceramics.

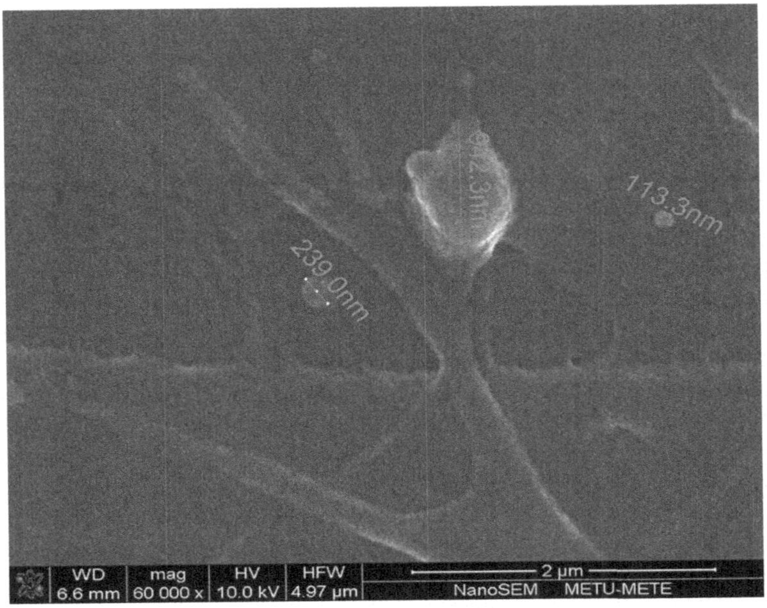

Fig. 8.8 The scanning electron micrograph of a mesenchymal stem cell presenting the nano-HAp particles within the cytoplasmic processes, 60,000x

8.6 Smart Bioceramics for Orthopaedic Applications

8.6.1 Bioceramic Surface Coatings

If the cellular recognition and excellent chemical bonds with the hosted tissue are absent from surface of the implant, attachment of the implant will be inadequate and micro-motions will occur at the bone-implant interface. This situation will lead to a failure of the implant system. Along with the wear of the implant material, corrosion provokes the release of metal ions or particles and causes infection and immune response. These limitations are addressed by the surface coating method that protects the implant material and at the same time increases their surface biocompatibility [101, 102].

The most important requirement of any implant (metallic, polymeric, ceramic) is to prevent infection and bone resorption, as well as to increase osteointegration and biological stabilization. The deposition of nanocoatings and nanocomposite coatings on these implants is to increase biological activity, corrosion resistance, and protect the body from adverse reactions due to metal ion release. These coatings also allow the surface properties of the implant to be altered to achieve improvements in clinical reliability, longevity, and performance [30].

Calcium phosphate has been used as a porous coating on metal implants in orthopedic applications. The development of good interfacial strength between the bioceramic coating and bone tissue is the effect of biochemical interference of released calcium and phosphate ions [30]. Calcium phosphates are classified according to their solubilities such as when attached to surrounding tissues together with their ability to be degraded and replaced by advancing bone growth. Surface ions of calcium phosphate are replaced with the aqueous solution when in contact with body fluid. Various ions and molecules (protein and collagen), on the other side, can be adsorbed onto the surface of calcium phosphate bioceramics (Fig. 8.4) [103].

Nanoscale coatings that improve the bonding between surfaces are gaining worldwide attention. The principle of using bioceramics as bioactive coatings is due to their fast and excellent biochemical bonding to the bone. The description of biological fixation is an action by which an implant or prosthetic are tightly attached to host bone through bone ingrowth and without the assistance of any form of physical fixation such as the application of adhesive [103]. Quality bioceramic coating promotes speedy recovery and greater adhesion to the bone. Several implants coated with a thick bioceramic layer show comprehensive bone attachment in vivo. Properties of the calcium phosphate coating such as porosity, surface roughness, crystallinity, component phase and thickness determine its long-term longevity [31, 103].

In the literature, Fujino et al. reported that silicate glass can be successfully coated to Co-Cr alloy and formed hydroxyapatite on the glass surface after immersion in SBF solution for 30 days [104]. In addition, Drnovšek et al. stated that nanoscale bioglass on titanium implant stimulates osseointegration with interactions between apatite layer and tissues [105]. Furthermore, Vitale-Brovarone et al. postulated that the use of porous bioactive glass coating on alumina substrates could enhance osteogenesis on bone defect repair [106].

8.6.2 Bone Graft

In addition to applications in orthopedic and maxillofacial surgery, bone grafts are also utilized for mandibular reconstruction, filling defects, and non-union treatments. The use of synthetic bone graft is preferred instead of the natural autograft and allograft option. Although the autografts are accepted as the first and well-accepted choice in bone recovery because of procuring osteoconductive and osteoinductive growth factors, problems such as limited tissue suitability, high cost, nerve damage, hematoma, pain to the patient, long operation time [107, 108] and infections at the autograft removal site [109] makes the use of allograft and synthetic grafts a preferred alternative option. The osteogenic activity of allograft is not as high as autografts. In addition, the immunological reactions and disease transmission risk for allograft is completely opposite. However, a reduction in osteoinductive feature and structural strength can be observed in frozen allografts [68].

The fabrication of calcium phosphates (HAp and TCP) and their combinations are produced under controlled conditions to attain pure fine-grained powder bioceramics for bone grafts. Customized bone grafts can be in various forms such as powder, granules, or bulk ceramics [110]. The synthetic ceramic bone grafts are osteoconductive similar to autografts; however, they do not show the osteoinductive feature that stimulates fresh bone formation. Owing to smart technologies, osteoconductive and osteoinductive properties are incorporated through osteogenic cells and/or inductive proteins. Besides, bone growth factors have been integrated into ceramics to trigger protein synthesis, bone formation, and defect repair [44].

BMP is isolated and obtained from the demineralized bone matrix (DBM) which contains a transforming growth factor family (TGF-β) [108]. BMP-2 and BMP-7 stimulates osteogenesis or angiogenesis with differentiation of MSCs for new bone formation. Meeting the requirements of bone remodeling can be achieved through the applications of BMP delivered using CaP ceramics and composites to the intended defect sites as well as in the therapeutic applications in osteoporosis [44]. Commercially available DBM is used in the treatment of non-union fractures. Bovine collagen is another material that is used for bone graft substitute in conjunction with HAp at the fracture site. Although the graft is osteogenic, osteoconductive, and osteoinductive by having high biological features and no transmissible diseases, it is weak however in terms its mechanical strength [111].

8.6.3 Scaffolds

Bioceramic scaffolds are generally preferred for non-load bearing and compressive load orthopedic applications due to its unfavorable mechanical properties [6]. Smart synthetic scaffolds for bone tissue engineering should compensate for essential requirements such as osteoinduction, 3D porosity (nano-micro level), and mechanical stability. It is worthy to mention that these requirements should be different to the criteria of other bioceramics [36]. The hierarchical porous structure provides the combination of nano- and macro-porosity with interconnectivity for the production of synthetic scaffold. If the porosity is categorized based on biological functions, then pore sizes between 150 and 400 μm are suitable for cell aggregation and vascularization. Pore sizes below 20 μm are ideal for capillary ingrowth factors, peptides, and biological apatite [1]. Figure 8.6 shows the nano and micrometer dimensions of biological structures. The microscale to nanoscale approaches accelerated due to the strong interactions between nanostructures and nucleid acids and proteins [44].

A study by Lu et al. examined the anti-microbial/anti-inflammation effects of nanoporous HAp/Polyamide 66 scaffold with 2.35 wt% TiO_2 and various amounts of silver ions (0.22% or 0.64 wt%). Using an animal model, they observed the scaffolds exhibited effective antibacterial activities against *E. coli* and *S. aureus* as well as supporting cell adhesion and proliferation of pre-osteoblasts, and stimulate osteogenic regulator/marker expression. More importantly, scaffold containing 0.64 wt% Ag^+ displayed more proficient antibacterial/anti-inflammation effects in vivo and promotes bone formation at the lesion site of osteomyelitis [112].

In addition, improvements in smart bioceramics should combine traditional properties of 3-D porous bioactive scaffolds with therapeutic ion release, as well as providing antibacterial properties and supporting osteoinductivity via doping with different trace elements such as strontium, copper, and zinc [5]. Surface reactions of bioglass increase with a porous texture and larger surface area. Consequently, acceleration in the amount of ions released during the degradation of the glass is observed and the rate in which apatite is formed on the glass increases. Nonetheless, the degradation rate can be controlled simply by changing the pore size, structure, and the composition of the glass [113]. A study by Naik et al. noticed that porous 45S5 bioglass/polyethylene composite orbital prosthesis enhances neovascularization compared to porous synthetic HAp, alumina, bovine and coralline scaffolds [114].

8.6.4 Injectable Bioceramics

Injectable bone substitutes are used in complicated surgeries and injected into regions where the bone structure is uneven as they can be molded into shape. In other words, they present an alternative concept to the minimum invasive surgery procedure [115]. Although calcium sulphate and calcium phosphate injectable bone cement have been clinically applied in dental, maxillofacial surgeries, and orthopedic fracture treatments at the moment, there are serious drawbacks as they degrade before healing is completed and losses its mechanical support functions [116]. As a result, cracking in the cancellous bone will occur under loading and the formation of injectable cement fragments can result in inflammation [117]. Xu et al. investigated the use of an injectable composite consisting of magnesium-doped TCP nanoparticles (43.8 $\pm$ 9 nm) and calcium sulphate hemihydrate microparticles between 5–21 μm in size for bone defect repair instead of pure calcium sulphate bone cement [118]. The results of mechanical and biological testing demonstrated that the injectable composite possess appropriate compression strength (between 2.28–6.33 MPa) and no cytotoxicity when the composite was cultured with MC3T3-E1 osteoblast-like cells in vitro. Another significant outcome is the acceptable initial setting time of between 11.7–19.2 min. In vivo bone defect repair using an animal model could be successful using the injectable composite. Moreover, nanoparticles degrade faster than microparticles and are more favorable for cellular uptake.

A biomimetic and injectable hydrogel scaffold was synthesized by Huang et al. consisted of chitosan, nano-hydroxyapatite, and collagen that is capable of forming a stable gel at body temperature [119]. The injectable scaffold demonstrated characteristics of natural bone both in terms of microstructure and composition due to the presence of collagen fibers and nano HAp. Xu et al. developed a moldable/injectable self-setting calcium phosphate cement using tetracalcium phosphate and dicalcium phosphate [120]. The scaffold was mechanically strong and can be used for bone regeneration and delivery of osteogenic cells and growth factors.

Malik et al. [121] suggested that nanobioglass/polymer composite semi-solid scaffolds could enhance bioactivity and mechanical properties with the generation of new apatite on the bovine dental tooth tissue cavity under ambient conditions. During the experiment, bioglass nanoparticles assisted in the continuous cell proliferation and growth through the highly interconnected porous structure and the enhancing surface area.

8.7 Conclusion

Autografts are accepted as the best bone replacement and regeneration alternative. Complications at the donor site and their limited availability increased the search for smart synthetic materials. Osteogenesis can be supported with synthetically obtainable intelligent bioceramic materials and design approaches. Research into smart bioceramic to replace and regenerate bone is an active and ongoing field of study. Surgical needs can be best overcome by injectable materials that attain biomimetic porosity and biomechanical properties after application. Nano-ceramics possess advantages such as increased surface areas for cell attachment, proliferation, differentiation, and ECM formation. Such nano-bioceramics can deliver signaling molecules and cells/cell components to the regeneration site. However, smart bioceramics that induce bone formation must meet physicochemical and bioactivity criteria for osteoconduction and osteoinduction. The osteoinductive feature is a key function for potential new bone formation. Various trace elements can be integrated into nano-bioceramics to enhance osteointegration. Biomimetic porosity and hierarchical bone structure can be regained when such biomaterial is combined with functional polymers. Active signaling molecules, stem, progenitor or mature cells combined with smart bioceramics are open to research. Such molecules and growth factors function in a time and dose-dependent manner. Smart bioceramics and especially their nano-forms can also be used in 3D printing. This technology will soon be available to produce implants in extreme conditions to replace and regenerate bone tissue.

Acknowledgements This study was supported by the cooperation of Scientific and Technological Research Council of Turkey (Grant agreement number 120N943), and National Research Foundation of Korea (Grant agreement number 2020K2A9A1A06108513).

References

1. Salinas AJ, Esbrit P, Vallet-Regí M (2013) A tissue engineering approach based on the use of bioceramics for bone repair. Biomater Sci 1:40–51
2. Vallet-Regí M, Ruiz-Hernández E (2011) Bioceramics: from bone regeneration to cancer nanomedicine. Adv Mater 23:5177–5218
3. Cowin SC, Cardoso L (2015) Blood and interstitial flow in the hierarchical pore space architecture of bone tissue. J Biomech 48:842–854
4. Renders GA, Mulder L, van Ruijven LJ et al (2007) Porosity of human mandibular condylar bone. J Anat 210:239–248
5. Baino F, Novajra G, Vitale-Brovarone C (2015) Bioceramics and scaffolds: a winning combination for tissue engineering. Front Bioeng Biotechnol 3:202. https://doi.org/10.3389/fbioe.2015.00202
6. Hench LL (1998) Biomaterials: a forecast for the future. Biomaterials 19:1419–1423
7. Baino F, Vitale-Brovarone C (2011) Three-dimensional glass-derived scaffolds for bone tissue engineering: current trends and forecasts for the future. J Biomed Mater Res A 97:514–535
8. Korkusuz P, Korkusuz F (2003) Hard tissue-biomaterial interactions. In: Yaszemski MJ (ed) Biomaterials in orthopedics, 2nd edn. CRC Press, Boca Raton, pp 1–40
9. Bommala VK, Krishna MG, Rao CT (2019) Magnesium matrix composites for biomedical applications: A review. J Magnes Alloy 7:72–79
10. Alizadeh-Osgouei M, Li Y, Wen C (2018) A comprehensive review of biodegradable synthetic polymer-ceramic composites and their manufacture for biomedical applications. Bioact Mater 4:22–36
11. Dziadek M, Stodolak-Zych E, Cholewa-Kowalska K (2017) Biodegradable ceramic-polymer composites for biomedical applications: A review. Mater Sci Eng C Mater Biol Appl 71:1175–1191
12. Albrektsson T, Johansson C (2001) Osteoinduction, osteoconduction and osseointegration. Eur Spine J 10 Suppl 2:S96-S101
13. Sushmita VP, Chethan Kumar JS, Hegde C et al (2019) Interaction of dental pulp stem cells in bone regeneration on titanium implant. An in vitro study. J Osseointegration 11:553–560
14. Korkusuz P, Kose S, Kopru CZ (2016) Biomaterial and stem cell interactions: histological biocompatibility. Curr Stem Cell Res Ther 11:475–486
15. Kim TI, Jang JH, Kim HW et al (2008) Biomimetic approach to dental implants. Curr Pharm Des 14:2201–2211
16. Kim SH, Yeon YK, Lee JM et al (2018) Precisely printable and biocompatible silk fibroin bioink for digital light processing 3D printing. Nat Commun 9:1620. https://doi.org/10.1038/s41467-018-03759-y
17. Lee JM, Chae T, Sheikh FA et al (2016) Three dimensional poly(ε-caprolactone) and silk fibroin nanocomposite fibrous matrix for artificial dermis. Mater Sci Eng C Mater Biol Appl 68:758–767
18. Köse S, Kankilic B, Gizer M et al (2018) Stem cell and advanced nano bioceramic interactions. Adv Exp Med Biol 1077:317–342
19. Shekaran A, Garcia AJ (2011) Nanoscale engineering of extracellular matrix-mimetic bioadhesive surfaces and implants for tissue engineering. Biochim Biophys Acta 1810:350–360
20. Kon T, Cho TJ, Aizawa T (2001) Expression of osteoprotegerin, receptor activator of NF-kappaB ligand (osteoprotegerin ligand) and related proinflammatory cytokines during fracture healing. J Bone Miner Res 16:1004–1014
21. Debnath S, Yallowitz AR, McCormick J et al (2018) Discovery of a periosteal stem cell mediating intramembranous bone formation. Nature 562:133–139
22. Marsell R, Einhorn TA (2011) The biology of fracture healing. Injury 42:551–555
23. Gebraad A, Kornilov R, Kaur S et al (2018) Monocyte-derived extracellular vesicles stimulate cytokine secretion and gene expression of matrix metalloproteinases by mesenchymal stem/stromal cells. FEBS J 285:2337–2359

24. Lehmann W, Edgar CM, Wang K et al (2005) Tumor necrosis factor alpha (TNF-alpha) coordinately regulates the expression of specific matrix metalloproteinases (MMPS) and angiogenic factors during fracture healing. Bone 36:300–310
25. Shimizu E, Tamasi J, Partridge NC (2012) Alendronate affects osteoblast functions by crosstalk through EphrinB1-EphB. J Dent Res 91:268–274
26. Amini AR, Laurencin CT, Nukavarapu SP (2012) Bone tissue engineering: recent advances and challenges. Crit Rev Biomed Eng 40:363–408
27. Fan H, Guo Z (2020) Bioinspired surfaces with wettability: biomolecule adhesion behaviors. Biomater Sci 8:1502–1535
28. Ferrari M, Cirisano F, Morán MC (2019) Mammalian cell behavior on hydrophobic substrates: influence of surface properties. Colloids Interfaces 3:48. https://doi.org/10.3390/colloids3020048
29. Menzies KL, Jones L (2010) The impact of contact angle on the biocompatibility of biomaterials. Optom Vis Sci 87:387–399
30. Choi AH, Cazalbou S, Ben-Nissan B (2015) Nanobiomaterial coatings in dentistry. In: Deb S (ed) Biomaterials for oral and craniomaxillofacial applications. Frontiers of oral biology series, vol 17. Karger Publisher, Basel, pp 49–61
31. Kämmerer TA, Palarie V, Schiegnitz E et al (2017) A biphasic calcium phosphate coating for potential drug delivery affects early osseointegration of titanium implants. J Oral Pathol Med 46:61–66
32. Hench LL (1991) Bioceramics: from concept to clinic. J Am Ceram Soc 74:1487–1510
33. Asri RI, Harun WS, Hassan MA et al (2016) A review of hydroxyapatite-based coating techniques: Sol-gel and electrochemical depositions on biocompatible metals. J Mech Behav Biomed Mater 57:95–108
34. Chang HH, Yeh CL, Wang YL et al (2020) Neutralized dicalcium phosphate and hydroxyapatite biphasic bioceramics promote bone regeneration in critical peri-implant bone defects. Materials 13:823. https://doi.org/10.3390/ma13040823
35. Zafar MJ, Zhu D, Zhang Z (2019) 3D Printing of bioceramics for bone tissue engineering. Materials 12:3361. https://doi.org/10.3390/ma12203361
36. Daculsi G (2015) Smart scaffolds: the future of bioceramic. J Mater Sci Mater Med 26:154. https://doi.org/10.1007/s10856-015-5482-7
37. Tian Y, Lu T, He F et al (2018) β-tricalcium phosphate composite ceramics with high compressive strength, enhanced osteogenesis and inhibited osteoclastic activities. Colloids Surf B Biointerfaces 167:318–327
38. Bouler JM, Pilet P, Gauthier O et al (2017) Biphasic calcium phosphate ceramics for bone reconstruction: A review of biological response. Acta Biomater 53:1–12
39. Parent M, Baradari H, Champion E et al (2017) Design of calcium phosphate ceramics for drug delivery applications in bone diseases: A review of the parameters affecting the loading and release of the therapeutic substance. J Control Release 252:1–17
40. Shuai C, Li P, Liu J et al (2013) Optimization of TCP/HAP ratio for better properties of calcium phosphate scaffold via selective laser sintering. Mater Charact 77:23–31
41. Schmidleithner C, Malferarri S, Palgrave R et al (2019) Application of high resolution DLP stereolithography for fabrication of tricalcium phosphate scaffolds for bone regeneration. Biomed Mater 14:045018. https://doi.org/10.1088/1748-605x/ab279d
42. Łączka M, Cholewa-Kowalska K, Osyczka AM (2016) Bioactivity and osteoinductivity of glasses and glassceramics and their material determinants. Ceram Int 42:14313–14325
43. Hench LL (2006) The story of Bioglass. J Mater Sci Mater Med 17:967–978
44. Huang J (2017) Design and Development of Ceramics and Glasses. In: Vishwakarma A, Karp JM (eds) Biology and Engineering of Stem Cell Niches. Academic Press, Massachusetts, p 315–329
45. Day RM (2005) Bioactive glass stimulates the secretion of angiogenic growth factors and angiogenesis in vitro. Tissue Eng 11:768–777
46. Gizer M, Köse S, Karaosmanoglu B et al (2020) The effect of boron-containing nano-hydroxyapatite on bone cells. Biol Trace Elem Res 193:364–376

47. Abou Neel EA, Pickup DM, Valappil SP et al (2009) Bioactive functional materials: A perspective on phosphate-based glasses. J Mater Chem 19:690–701
48. Gerhardt LC, Boccaccini AR (2010) Bioactive glass and glass-ceramic scaffolds for bone tissue engineering. Materials 3:3867–3910
49. Jones JR (2013) Review of bioactive glass: from Hench to hybrids. Acta Biomater 9:4457–4486
50. Li R, Clark AE, Hench LL (1991) An investigation of bioactive glass powders by sol-gel processing. J Appl Biomater 2:231–239
51. Mala R, Ruby Celsia AS (2018) Bioceramics in orthopaedics: A review. In: Thomas S, Balakrishnan P, Sreekala MS (eds) Fundamental biomaterials: ceramics. Woodhead Publishing Series in Biomaterials. Woodhead Publishing, Cambridge, pp 195–221
52. Arcos D, Izquierdo-Barba I, Vallet-Regí M (2009) Promising trends of bioceramics in the biomaterials field. J Mater Sci Mater Med 20:447–455
53. Ojansivu M, Vanhatupa S, Björkvik L et al (2015) Bioactive glass ions as strong enhancers of osteogenic differentiation in human adipose stem cells. Acta Biomater 21:190–203
54. Saravanapavan P, Hench LL (2001) Low-temperature synthesis, structure, and bioactivity of gel-derived glasses in the binary CaO-SiO_2 system. J Biomed Mater Res 54:608–618
55. Kokubo T (1991) Bioactive glass ceramics: properties and applications. Biomaterials 12:155–163
56. Shi M, Chang J, Wu C (2016) Bioactive glasses: advancing from micro to nano and its potential application. In: Marchi J (ed) Biocompatible glasses. Advanced structured materials, vol 53. Springer, Cham, pp 147–181
57. Yuan H, de Bruijn JD, Zhang X et al (2001) Bone induction by porous glass ceramic made from Bioglass (45S5). J Biomed Mater Res 58:270–276
58. Lee JH, Rim NG, Jung HS et al (2010) Control of osteogenic differentiation and mineralization of human mesenchymal stem cells on composite nanofibers containing poly[lactic-co-(glycolic acid)] and hydroxyapatite. Macromol Biosci 10:173–182
59. Zhou H, Lawrence JG, Bhaduri SB (2012) Fabrication aspects of PLA-CaP/PLGA-CaP composites for orthopedic applications: a review. Acta Biomater 8:1999–2016
60. Lock J, Nguyen TY, Liu H (2012) Nanophase hydroxyapatite and poly(lactide-co-glycolide) composites promote human mesenchymal stem cell adhesion and osteogenic differentiation in vitro. J Mater Sci Mater Med 23:2543–2552
61. Li Y, Zhou J, Pavanram P et al (2018) Additively manufactured biodegradable porous magnesium. Acta Biomater 67:378–392
62. He SY, Sun Y, Chen MF et al (2011) Microstructure and properties of biodegradable β-TCP reinforced Mg-Zn-Zr composites. T Nonferr Metal Soc 21:814–819
63. Qi T, Weng J, Yu F et al (2021) Insights into the role of magnesium ions in affecting osteogenic differentiation of mesenchymal stem cells. Biol Trace Elem Res 199:559–567
64. Engstrand T, Kihlström L, Neovius E et al (2014) Development of a bioactive implant for repair and potential healing of cranial defects. J Neurosurg 120:273–277
65. Regenerative bone defects using new biomedical engineering approaches, REBORNE 2010–2014, 7th PCRD EC program. www.reborne.org
66. Bal Z, Kaito T, Korkusuz F et al (2020) Bone regeneration with hydroxyapatite-based biomaterials. Emergent Mater 3:521–544
67. Bal Z, Kushioka J, Kodama J et al (2020) BMP and TGFβ use and release in bone regeneration. Turk J Med Sci 50:1707–1722
68. Heness G, Ben-Nissan B (2004) Innovative Bioceramics. Mater Forum 27:104–114
69. Giannoudis PV, Einhorn TA, Marsh D (2007) Fracture healing: the diamond concept. Injury 38:S3-S6
70. Ginebra MP, Espanol M, Maazouz Y et al (2018) Bioceramics and bone healing. EFORT Open Rev 3:173–183
71. Chen FM, Liu X (2016) Advancing biomaterials of human origin for tissue engineering. Prog Polym Sci 53:86–168

72. Daculsi G, Miramond T, Borget P (2012) Smart Calcium Phosphate Bioceramic Scaffold for Bone Tissue Engineering. Key Eng Mater 529–530:19-23
73. Manzano M, Lozano D, Arcos D et al (2011) Comparison of the osteoblastic activity conferred on Si-doped hydroxyapatite scaffolds by different osteostatin coatings. Acta Biomater 7:3555–3562
74. Coathup MJ, Hing KA, Samizadeh S et al (2012) Effect of increased strut porosity of calcium phosphate bone graft substitute biomaterials on osteoinduction. J Biomed Mater Res A 100:1550–1555
75. Albulescu R, Popa AC, Enciu AM et al (2019) Comprehensive *in vitro* testing of calcium phosphate-based bioceramics with orthopedic and dentistry applications. Materials 12:3704
76. Eliaz N, Metoki N (2017) Calcium phosphate bioceramics: a review of their history, structure, properties, coating technologies and biomedical applications. Materials 10:334
77. Frieß W, Warner J. (2002) Biomedical applications. In: Schüth F, Sing KSW, Weitkamp J (eds) Handbook of porous solids. Wiley-VCH Verlag GmbH, Weinheim, pp 2923–2970
78. Chen Y, Frith JE, Dehghan-Manshadi A et al (2017) Mechanical properties and biocompatibility of porous titanium scaffolds for bone tissue engineering. J Mech Behav Biomed Mater 75:169–174
79. Yuan L, Ding S, Wen C (2018) Additive manufacturing technology for porous metal implant applications and triple minimal surface structures: A review. Bioact Mater 4:56–70
80. Zaharin HA, Abdul Rani AM, Azam FI et al (2018) Effect of Unit Cell Type and Pore Size on Porosity and Mechanical Behavior of Additively Manufactured Ti6Al4V Scaffolds. Materials 11:2402
81. Zhu L, Luo D, Liu Y (2020) Effect of the nano/microscale structure of biomaterial scaffolds on bone regeneration. Int J Oral Sci 12:6. https://doi.org/10.1038/s41368-020-0073-y
82. Xia L, Zhang N, Wang X et al (2016) The synergetic effect of nano-structures and silicon-substitution on the properties of hydroxyapatite scaffolds for bone regeneration. J Mater Chem B 4:3313–3323
83. Wu L, Zhou C, Zhang B, et al (2020) Construction of biomimetic natural wood hierarchical porous-structure bioceramic with micro/nanowhisker coating to modulate cellular behavior and osteoinductive activity. ACS Appl Mater Interfaces 12:48395–48407
84. Shepherd JH, Shepherd DV, Best SM (2012) Substituted hydroxyapatites for bone repair. J Mater Sci Mater Med 23:2335–2347
85. Šupová M (2015) Substituted hydroxyapatites for biomedical applications: A review. Ceram Int 41:9203–9231
86. Jugdaohsingh R (2007) Silicon and bone health. J Nutr Health Aging 11:99–110
87. Patel N, Best SM, Bonfield W et al (2002) A comparative study on the *in vivo* behavior of hydroxyapatite and silicon substituted hydroxyapatite granules. J Mater Sci Mater Med 13:1199–1206
88. Munir G, Koller G, Di Silvio L et al (2011) The pathway to intelligent implants: osteoblast response to nano silicon-doped hydroxyapatite patterning. J R Soc Interface 8:678–688
89. Wang W, Yeung KWK (2017) Bone grafts and biomaterials substitutes for bone defect repair: A review. Bioact Mater 2:224–247
90. Ratnayake JTB, Mucalo M, Dias GJ (2017) Substituted hydroxyapatites for bone regeneration: A review of current trends. J Biomed Mater Res B Appl Biomater 105:1285–1299
91. Geng Z, Cheng Y, Ma L et al (2018) Nanosized strontium substituted hydroxyapatite prepared from egg shell for enhanced biological properties. J Biomater Appl 32:896–905
92. Zhu XD, Zhang HJ, Fan HS et al (2010) Effect of phase composition and microstructure of calcium phosphate ceramic particles on protein adsorption. Acta Biomater 6:1536–1541
93. Botelho CM, Lopes MA, Gibson IR et al (2002) Structural analysis of Si-substituted hydroxyapatite: zeta potential and X-ray photoelectron spectroscopy. J Mater Sci Mater Med 13:1123–1127
94. Ke D, Tarafder S, Vahabzadeh S et al (2019) Effects of MgO, ZnO, SrO, and SiO_2 in tricalcium phosphate scaffolds on in vitro gene expression and in vivo osteogenesis. Mater Sci Eng C Mater Biol Appl 96:10–19

95. Banerjee SS, Tarafder S, Davies NM et al (2010) Understanding the influence of MgO and SrO binary doping on the mechanical and biological properties of beta-TCP ceramics. Acta Biomater 6:4167–4174
96. Chou J, Hao J, Hatoyama H et al (2015) Effect of biomimetic zinc-containing tricalcium phosphate (Zn-TCP) on the growth and osteogenic differentiation of mesenchymal stem cells. J Tissue Eng Regen Med 9:852–858
97. Tunçay EÖ, Demirtaş TT, Gümüşderelioğlu M (2017) Microwave-induced production of boron-doped HAp (B-HAp) and B-HAp coated composite scaffolds. J Trace Elem Med Biol 40:72–81
98. Alidadi S (2020) Nanoscale bioceramics in bone tissue engineering- an overview. Indian J Ver Sci Biotechnol 16:7–11
99. Zheng H, Tian Y, Gao Q et al (2020) Hierarchical micro-nano topography promotes cell adhesion and osteogenic differentiation via integrin α2-PI3K-AKT signaling axis. Front Bioeng Biotechnol 8:463
100. Yin C, Zhang Y, Cai Q et al (2017) Effects of the micro-nano surface topography of titanium alloy on the biological responses of osteoblast. J Biomed Mater Res A 105:757–769
101. Bartkowiak A, Suchanek K, Menaszek E et al (2018) Biological effect of hydrothermally synthesized silica nanoparticles within crystalline hydroxyapatite coatings for titanium implants. Mater Sci Eng C Mater Biol Appl 92:88–95
102. Sirin HT, Vargel I, Kutsal T et al (2016) Ti implants with nanostructured and HA-coated surfaces for improved osseointegration. Artif Cells Nanomed Biotechnol 44:1023–1030
103. Choi AH, Ben-Nissan B (2017) Calcium phosphate nanocomposites for biomedical and dental applications: recent developments. In: Thakur VK, Thakur MK, Kessler MR (eds) Handbook of composites from renewable materials. John Wiley and Sons, Inc, New Jersey, p 423–450
104. Fujino S, Tokunaga H, Saiz E et al (2004) Fabrication and characterization of bioactive glass coatings on Co-Cr implant alloys. Mater Trans 45:1147–1151
105. Drnovšek N, Novak S, Dragin U et al (2012) Bioactive glass enhances bone ingrowth into the porous titanium coating on orthopaedic implants. Int Orthop 36:1739–1745
106. Vitale-Brovarone C, Baino F, Tallia F et al (2012) Bioactive glass-derived trabecular coating: a smart solution for enhancing osteointegration of prosthetic elements. J Mater Sci Mater Med 23:2369–2380
107. Ekin O, Calis M, Aliyev A et al (2916) Poly(L-Lactide)/Poly(ε-Caprolactone) and collagen/β-tricalcium phosphate scaffolds for the treatment of critical-sized rat alveolar defects: a microtomographic, molecular-biological, and histological study. Cleft Palate Craniofac J 53:453–463
108. Urist MR (1965) Bone: formation by autoinduction. Science 150:893–899
109. Kankilic B, Dede EC, Korkusuz P et al (2017) Apatites for orthopedic applications. In: Kaur G (ed) Clinical applications of biomaterials. Springer, Cham, pp 65–90
110. Korkusuz F, Timuçin M, Korkusuz P (2014) Nanocrystalline apatite-based biomaterials and stem cells in orthopaedics. In: Ben-Nissan B. (ed) Advances in calcium phosphate biomaterials. Springer series in biomaterials science and engineering, vol 2. Springer, Heidelberg, pp 373–390
111. Ben-Nissan B, Pezzotti G (2002) Bioceramics: processing routes and mechanical evaluation. J Ceram Soc Jpn 110:601–608
112. Lu M, Liao J, Dong J et al (2016) An effective treatment of experimental osteomyelitis using the antimicrobial titanium/silver-containing nHP66 (nano-hydroxyapatite/polyamide-66) nanoscaffold biomaterials. Sci Rep 6:39174
113. Wu C, Chang J (2014) Multifunctional mesoporous bioactive glasses for effective delivery of therapeutic ions and drug/growth factors. J Control Release 193:282–295
114. Naik MN, Murthy RK, Honavar SG (2007) Comparison of vascularization of Medpor and Medpor-Plus orbital implants: a prospective, randomized study. Ophthalmic Plast Reconstr Surg 23:463–467
115. Sadiasa A, Sarkar SK, Franco RA et al (2014) Bioactive glass incorporation in calcium phosphate cement-based injectable bone substitute for improved in vitro biocompatibility and in vivo bone regeneration. J Biomater Appl 28:739–756

116. Liu H, Li H, Cheng W et al (2006) Novel injectable calcium phosphate/chitosan composites for bone substitute materials. Acta Biomater 2:557–565
117. No YJ, Roohani-Esfahani SI, Zreiqat H (2014) Nanomaterials: the next step in injectable bone cements. Nanomedicine 9:1745–1764
118. Xu Q, Liang J, Xue H et al (2020) Novel injectable and self-setting composite materials for bone defect repair. Sci China Mater 63:876–887
119. Huang Z, Feng Q, Yu B et al (2011) Biomimetic properties of an injectable chitosan/nano-hydroxyapatite/collagen composite. Mater Sci Eng C 31:683–687
120. Xu HH, Weir MD, Simon CG (2008) Injectable and strong nano-apatite scaffolds for cell/growth factor delivery and bone regeneration. Dent Mater 24:1212–1222
121. Malik QUA, Iftikhar S, Zahid S et al (2020) Smart injectable self-setting bioceramics for dental applications. Mater Sci Eng C Mater Biol Appl 113:110956

Fatma Nur Depboylu Ms. Depboylu is Ph.D. scholar in the Department of Bioengineering at Hacettepe University Institute of Science and Technology, Turkey. She has completed her B.Sc. in biomedical engineering with a full scholarship at FMV Işık University, Turkey, and later at Newcastle University in UK where she graduated from a M.Sc. biomedical engineering program. Her master thesis was based on 3D printing of porous polymer structures for tissue engineering. She continues her Ph.D. research on additive manufacturing of metal implants and their coating with bioceramics composites for orthopedic applications. Specifically, she has been working on the porous implant design, use of additive manufacturing and coating techniques for osteosynthesis mini plates.

Petek Korkusuz Prof. P. Korkusuz received her medical degree in 1991 from Gazi University Faculty of Medicine in Turkey. In 1997, she completed her residency in histology and embryology at Hacettepe University Faculty of Medicine. She received her Ph.D. in 2000 from the Gazi University Faculty of Medicine Department of Medical Pharmacology. Presently, she is the head of Histology and Embryology Department at Hacettepe University Faculty of Medicine. Her secondary appointments are in the Department of Nanomedicine and Nanotechnology, Department of Bioengineering, Department of Stem Cell Sciences and Department of Oral Biology. She has 142 publications and 2217 citations in Science Citation Index (WOS) with an h index of 25 (January 2021). She owns 3 patents, several international and national book chapters and awards. Her research is focused on biomaterials and stem cells in soft/hard tissue engineering and regenerative medicine and recently somatic and germ stem cell microenvironments.

Evren Yasa Dr. Yasa graduated in mechanical engineering from Istanbul Technical University, Turkey and completed her master's degree at the University of British Columbia on volumetric error modeling and compensation of machine tool errors. She received her Ph.D. degree on "Combined Process of Selective Laser Melting and Selective Laser Erosion/Laser Re-melting" at the Catholic University of Leuven in Belgium under the supervision of Prof.Dr. Jean-Pierre Kruth and won the "Emerald Outstanding Doctoral Study-Highly commended" award with her doctoral dissertation. After her Ph.D., she worked as a senior engineer at TEI, a GE-joint venture company specializing in manufacturing aero-engine parts, where she led Additive Manufacturing (AM) projects and activities. In 2017, she joined Eskisehir Osmangazi University in Turkey as an assistant professor. She has also been working as an independent expert in laser-based manufacturing on behalf of the European Commission in FP7 and Horizon 2020 projects and as a consultant to aerospace and defense companies for adopting and developing AM processes since 2014.

Feza Korkusuz Prof. F. Korkusuz completed his medical specification in Orthopedic Surgery and Traumatology at the Gazi University Medical Faculty in Turkey between 1987 and 1992 after graduating from Ankara University Medical Faculty in 1986. Between 1989 and 1990, he studied at the Department of Orthopaedic Surgery, Osaka University Faculty of Medicine in Japan where he graduated with a postgraduate diploma. His clinical and research studies focused on spine surgery and sports medicine, and in particular, he focused on biomechanics of biomaterials, bioceramics and their interaction with cells and tissues. In 1999, he received the METU Parlar Foundation young investigator and the TUBITAK promotion award in medical sciences in 2000. Currently, he is the Head of the Department of Hacettepe University Medical Faculty, Department of Sports Medicine and works as the Advisor to the President of TUBITAK.

Chapter 9
Bone Morphogenic Proteins and Bioceramic Scaffolds in Orthopedics

Howa Begam, Subhasis Roy, Prasenjit Mukherjee, Abhijit Chanda, Biswanath Kundu, and Samit Kumar Nandi

Abstract Bone morphogenetic proteins (BMPs) as osteoinductive components originated from bone, play vital roles in a wide variety of processes during the formation of bone, cartilage, blood vessels, etc. BMPs fall within the superfamily of the transforming growth factor-β (TGF-β) and signal transduction occurs through type I and type II serine-threonine kinase receptors. BMPs are indispensable mediators of osseous tissue creation concerned with the regulation of segregation of osteoprogenitor cells into osteoblasts. However, their sustained and targeted delivery to a particular site is of concern to researchers, scientists, and clinicians. Bioceramic is considered an outstanding candidate for novel scaffold material because of its excellent biocompatibility, bioactivity, and osteoconductivity but it lacks osteoinductivity. On the contrary, BMPs are osteoinductive and their combination can result in a scaffold with enhanced quality in orthopedics.

Keywords Bone morphogenetic proteins (BMPs) · Signaling cascades of BMPs · Carriers of BMPs · Bone regeneration · Cartilage regeneration

---

H. Begam
Centre for Healthcare Science and Technology, Indian Institute of Engineering Science and Technology, Shibpur, India

S. Roy · P. Mukherjee
Department of Veterinary Clinical Complex, West Bengal University of Animal and Fishery Sciences, Mohanpur, Nadia, India

A. Chanda
Department of Mechanical Engineering, Jadavpur University, Kolkata, India

B. Kundu
Bioceramic and Coating Division, CSIR-Central Glass & Ceramic Research Institute, Jadavpur, Kolkata, India

S. K. Nandi (✉)
Department of Veterinary Surgery and Radiology, West Bengal University of Animal and Fishery Sciences, Kolkata, India

© The Author(s), under exclusive license to Springer Nature Singapore Pte Ltd. 2022

A. H. Choi and B. Ben-Nissan (eds.), *Innovative Bioceramics in Translational Medicine II*, Springer Series in Biomaterials Science and Engineering 18,
https://doi.org/10.1007/978-981-16-7439-6_9

9.1 Introduction

Bone has the inherent capability of rejuvenation as part of the repair process against the threat of injury, as well as during skeletal expansion or constant remodeling during adult life. Bone regeneration follows well-orchestrated sequences of biological events involving bone induction and conduction, connecting numerous cell types and intracellular and extracellular molecular signaling pathways.

From a different perspective, there are various biological factors responsible for bone formations and these include: (i) definite bone cell categories viz. mesenchymal stem cells (MSCs) and osteoclasts; (ii) appearance of soluble molecules (cytokines, growth factors, hormones, ions, vitamins); (iii) the scaffold (calcium hydroxyapatite, extracellular matrix molecules, etc.); and (iv) several mechanical stimuli [1].

There are five types of bone cells present in human bone tissues such as pre-osteoblast, osteoblast, osteoclast, osteocyte, and bone lining cell. In response to new stress, a constant bone remodeling process is facilitated by those cells. Above all, osteocyte cells play a vital task during the remodeling process [2].

Several key molecules have been identified which regulates the fracture healing and bone regeneration process (Fig. 9.1). There are three types of signaling molecules that governs the healing process: (i) the pro-inflammatory cytokines, (ii) the transforming growth factor-beta (TGF-β) superfamily and other growth factors, and (iii) the angiogenic factors [1].

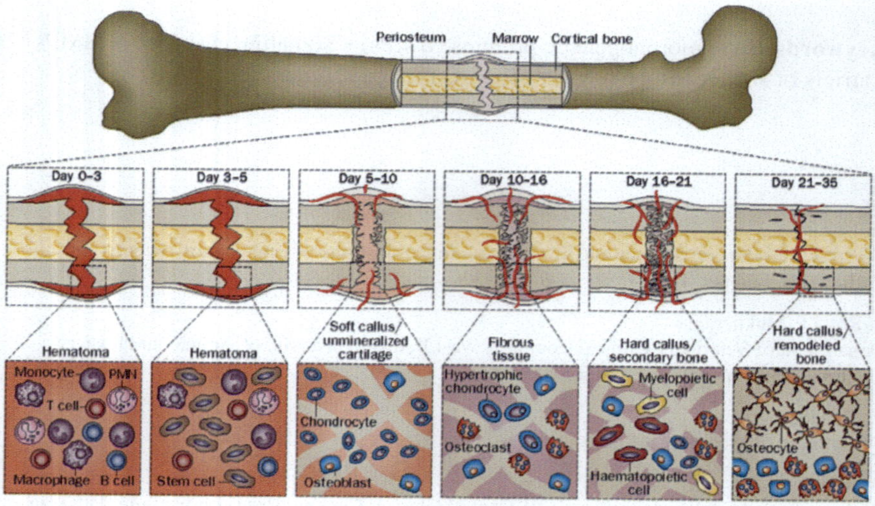

Fig. 9.1 Graphical representation shows three principal stages of normal bone healing, inflammatory stage, endochondral bone formation, and coupled remodeling (Reprint with permission from [3])

- **Pro-inflammatory cytokines**: Interleukin- 1 (IL-1), interleukin-6 (IL-6), and tumor necrosis factor-alpha (TNF-a) play a significant role in the commencement of the repair cascade in addition to bone remodeling.
- **Growth and differentiation factor**: It includes the role of the transforming growth factor-beta (TGF-β) Superfamily, the platelet-derived growth factor (PDGF), fibroblast growth factors (FGFs), and the insulin-like growth factors (IGFs).
- **Metalloproteinases and angiogenic factors**: Matrix metalloproteinases cause degradation of cartilage and bone and permit angiogenesis at the last stages of endochondral ossification and during the remodeling phase. Angiogenesis is regulated by two different dependent pathways namely the vascular endothelial growth factor (VEGF) and the angiopoietin.

A well-designed three-dimensional scaffold is one of the vital tools to monitor in vitro and in vivo tissue formation during the healing process. It is essential that appropriate scaffolds capable of providing adequate support to the cells be selected so that cells can perform in a preferred manner during the formation of tissues and organs with desired shape and size. Furthermore, the addition of growth factors like bone morphogenetic proteins (BMPs) or other bioactive elements into scaffolds will effectively increase bone tissue response and improve bone fracture healing [4]. BMPs inspire the differentiation of mesenchymal stem cells and other osteogenic cells and boost the differentiation function of osteoblasts. This chapter plans to review the present understanding of BMP signaling within the osteoclast lineage, its function in bone resorption, and osteoblast-osteoclast pairing. In addition, subsequent clinical implications in cartilage and bone repair using bioceramic scaffolds and BMPs are discussed.

9.2 Bone Morphogenetic Proteins and Their Classification

The transforming growth factor-beta (TGF-β) superfamily contains a number of various growth and differentiation factors including BMPs, TGF-β, growth differentiation factors (GDFs), activins, inhibins, and Müllerian inhibiting substance [5].

BMPs are created during fracture by mesenchymal cells, osteoblast cells, and chondrocyte cells. After the fracture, these cells in combination with individual BMP or with other members of the TGF-β superfamily stimulate the bone and cartilage formation process. Different cellular processes were promoted like angiogenesis, chemotaxis, mesenchymal cell propagation, and the creation of an extracellular matrix [6].

BMPs are produced and accumulated as large dimeric proteins in the cytoplasm and slashed by proteases throughout the secretion. BMP is a dimeric molecule with two polypeptide chains containing 431 amino acids held together by a single disulfide bond. The crystal structure of BMP 2 is a “hand-shaped structure” containing two

Table 9.1 Classification of bone morphogenetic proteins [8]

BMP classification	Function
BMP-2	Supports osteoinduction, osteoblast differentiation, and apoptosis
BMP-3 (osteogenin)	Most abundant BMP in bone in addition to inhibiting osteogenesis
BMP-4	Assists in osteoinduction in addition to lung and eye development
BMP-5	Supports chondrogenesis
BMP-6	Supports osteoblast differentiation and chondrogenesis
BMP-7 (osteogenic protein-1)	Assists in osteoinduction as well as the development of kidney and eye
BMP-8 (osteogenic protein-1)	Supports osteoinduction
BMP-9	Assists in the development of the nervous system and hepatic reticuloendothelial system
BMP-10	Supports cardiac development
BMP-11 (growth/differentiation factor-8)	Assists in the development of neuronal tissues
BMP-12 (growth/differentiation factor-7)	Supports tendon iliac tissue formation
BMP-13 (growth/differentiation factor-6)	Supports tendon and ligament-like tissue formation
BMP-14 (growth/differentiation factor-5)	Activates tendon healing and bone formation
BMP-15	Helps in follicle stimulating hormone activity

fingers of antiparallel β-strand and an α-helical region at the heel of the palm [7]. BMPs are classified based on the amino acid residues sequence and are shown in Table 9.1 [8].

9.3 Receptors of BMPs

BMPs express their actions through two types of serine/threonine kinase transmembrane receptors: type I and type II receptors [9]. In mammals, three types of type II receptors such as the BMP type II receptor (BMPRII), the activin type II receptor (ActRII), and the activin type IIB receptor (ActRIIB) are present. These receptors attach mostly to BMP ligands and they regulate the binding inclinations of BMP to type I receptors. In addition, these type II receptors phosphorylate the glycine-serine domain of the type I receptor. Type I receptors generally control the specificity of intracellular signals. For most BMPs, type I receptors are- activin receptor-like kinase (ALK)-3 (BMPRIA), ALK-6 (BMPRIB), ALK-2, and ALK-1.

There are eight Smad proteins present in mammals (from Smad 1 to Smad 8). The activated receptor kinases phosphorylate the transcription factors Smad1, 5, and 8. The phosphorylated and activated R-Smad proteins form complexes with Smad4 and

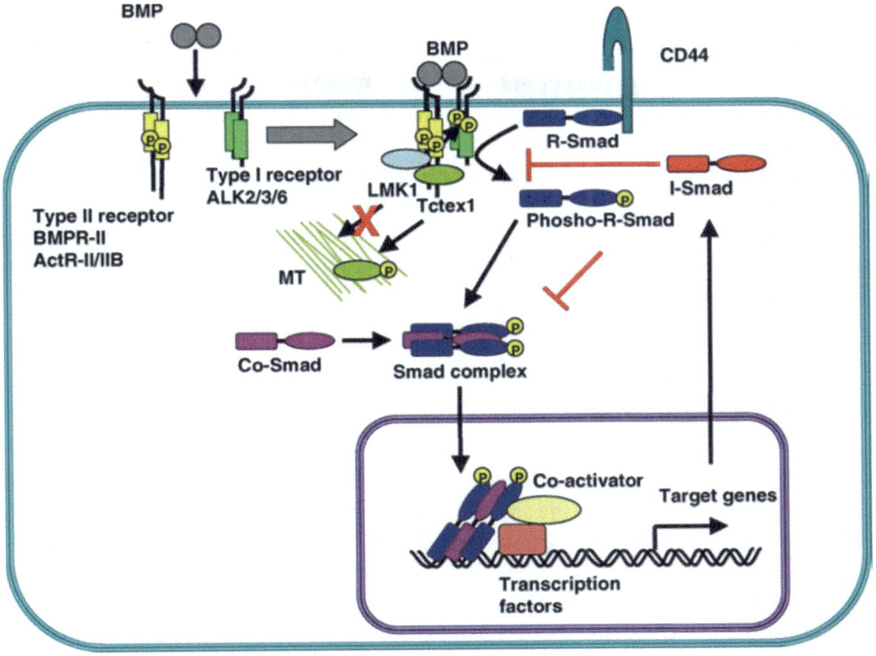

Fig. 9.2 Schematic representation of BMP signaling pathway (Reprint with permission from [10])

move into the nucleus and control transcription of target genes attaching to specific DNA sequences, networking with other DNA-binding proteins, and employing transcriptional co-activators (Fig. 9.2) [10].

9.4 Signaling Cascades of BMPs

BMPs perform their function as growth and differentiation factors and chemotactic agents. At the time of embryonic development, BMPs regulate neurogenesis and erythropoiesis and induce mesmerization. They encourage angiogenesis and relocation, propagation, and differentiation of mesenchymal stem cells into osteoblasts and cartilage. Clinical studies revealed that BMPs are also capable of motivating bone fracture healing, controlling the formation of different morphological characteristics of mammal bone systems [8, 11].

Due to small quantities of bone morphogenetic proteins identified within the extracellular matrix of bone, recombinant technologies play a crucial role and they have been utilized to generate BMPs for therapeutic purposes. Since the recognition of the chemical structures of several human BMPs, it is likely to employ DNA probes to obtain human complementary DNA sequences. The human cDNA is cloned and merged into a viral expression vector. Consequently, recombinant-BMP produced offers the optimal capacity for clinical purposes. In 2002, The US Food and Drug Administration (FDA) permitted the utilization of BMP-2 and BMP-7 for application in bone regeneration [8]. Recombinant human BMPs (rhBMPs) are primarily intended for applications in orthopedics and in oral and maxillofacial surgery. They are also equally important as an adjunct treatment for the management of few musculoskeletal disorders. The FDA-approved rhBMP-2 and rhBMP-7 are being used for anterior lumbar interbody fusion, non-union, and open tibial fractures. Spinal fusion using iliac crest bone graft may become a successful procedure nowadays as a result of the advancement of bone morphogenetic proteins [12].

9.5 Carriers for BMPs

A suitable carrier is required to express the biological potency of BMP. Carriers help to confine BMPs within the injured treatment area for a longer period of time so that it is possible to recruit regenerative tissue forming cells and for these tissue forming cells to propagate and differentiate. Moreover, this matrix also assists in cell infiltration during the tissue repairing process.

The indispensable necessities for ideal carriers of BMPs are listed below [13, 14]:

- The capacity of preventing potential inflammatory responses;
- The formation of a boundary with the neighboring biological tissue;
- Perfect porosity for cell infiltration and vascular in-growth;
- Should have satisfactory mechanical strength;
- Should have adequate biodegradability property and in the same time it must also protect BMPs from degradation for a sufficient time to stimulate bone mass at the treatment site;
- Carriers should be produced on a large scale effortlessly and economically;
- Sustained and controlled release of the incorporated proteins.

The incorporation of BMPs in a delivery system can be achieved through adsorption, entrapment or immobilization, in addition to covalent binding. Amid the various ways, adsorption of rhBMPs to the surface of the implant is the simplest mode of delivery. However, it leads to a less sustained release of the protein due to the initial burst release. Although, rhBMPs are attached to the implant surfaces for a longer time, the covalent bonding hinders free diffusion of the protein within the environment. Entrapment and encapsulation of rhBMPs are the most popular way of delivery of rhBMP, which avoids the concerns of a fast release. However, complication can crop up during the loading of BMPs within certain materials as pH or temperature conditions often lead to denaturation of the protein. Therefore, researchers are searching for a more effective and optimal way to develop a specific carrier system for BMPs without the loss of functional activity [14].

Table 9.2 Categories of carrier materials and some of their examples

Categories	Examples
Natural polymers	Collagen, Chitosan, Silk fibroin, Alginate, Gelatin
Synthetic Polymers	Polylactic acid (PLA), Poly (glycolic) acid (PGA), Polyethylene glycol (PEG), Poly lactic-co-glycolic acid (PLGA)
Bioceramics	Hydroxyapatite, β-Tricalcium phosphate (β-TCP), Biphasic calcium phosphate (BCP)
Composites	Hydroxyapatite-Collagen, Hydroxyapatite-Chitosan, Tricalcium phosphate-Collagen, PLGA-Collagen

There are four major categories of carrier materials for the delivery of BMPs such as natural and synthetic polymers, bioceramics, and their combinations in the form of composites. Examples of the types of carrier materials currently being utilized for BMP delivery are shown in Table 9.2.

9.6 Calcium Phosphate as a Carrier of BMP

Calcium phosphate bioceramics are considered to be an excellent candidate for novel scaffold material due to its excellent biocompatibility, bioactivity, and osteoconductivity properties [15, 16]. The biocompatibility of calcium phosphate ceramics is similar to the structural similarity of bone (Fig. 9.3) [17] and its degradation

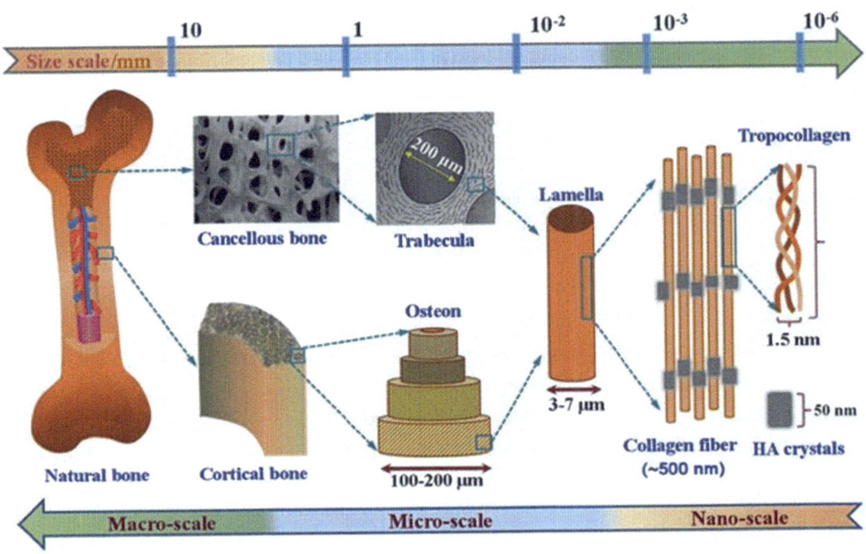

Fig. 9.3 Structure of human bone (Reprint with permission from [17])

products can contribute to bio-mineralization [18]. The chemical analysis of human bone revealed it is made up of both organic and inorganic components. The organic constituent of bone is the main collagen (33%), while the inorganic constituent (67%) is consisted of calcium, phosphate, magnesium, sodium, carbonate, etc.

Collagen, the matrix of bone, is found in form of fibers and they produce a net-like structure. The diameter of fibers ranges from 100 to 200 nm. Calcium phosphates are the main inorganic part of bone that provides the stiffness to the bone. Calcium phosphates are present in the form of crystallized hydroxyapatite having needle of plate shape with a length of 40 to 60 nm, 20 nm in width, and a thickness ranging between 1.5–5 nm [19]. The hydroxyapatite crystals are oriented in the direction of collagen fibers within the bone [20]. Calcium phosphates have a high linking affinity for growth factors, which makes them appropriate candidate for the delivery of growth factors. There are three types of calcium phosphates such as hydroxyapatite, tricalcium phosphate and biphasic calcium phosphate, and they have been used in orthopedic implants. Tricalcium phosphate/hydroxyapatite composite material displayed a very high resemblance to rhBMP-2. The growth factor-ceramic relations is dependent on the hydroxyl, amine, and carboxyl groups present in the BMP and on the number of divalent Ca^{2+} ions present in the bioceramic materials [1].

Autogenous bone graft is 'gold standard' material as bone graft substitutes. There are numbers of studies examining the the use of calcium phosphate-based materials for delivery of BMP as shown in Table 9.3 [21–28]. Majid et al. investigated the use of calcium phosphate coating as a BMP delivery system in spinal arthrodesis [29]. Calcium phosphate showed sustained and localized delivery of BMP at fusion site when coated with rhBMP [29]. Hirakawa et al. demonstrated the tendon-to-bone repair process, which was performed by use of BMP-2 delivered by β-TCP [22].

Table 9.3 The delivery of BMP using calcium phosphate-based carriers [21–28]

Investigator	Year	Materials	Growth Factor	Subject
Han et al. [21]	2020	HAp/Collagen	BMP-2	Rat
Hirakawa et al. [22]	2018	β-TCP	BMP-2	Rabbit
Huang et al. [23]	2018	Chondroitin sulfate-functionalized calcium phosphate cement	BMP-2	Mice
Ji et al. [24]	2018	Calcium phosphate	BMP6	Mice
Tenkumo et al. [25]	2018	Nano HAp/Collagen	BMP-2	Rat
Ding et al. [26]	2016	Mg modified calcium phosphate cement	rhBMP-2	Rat
Tsuzuki et al. [27]	2012	Gelatin-β-TCP sponge	BMP-2	Horse
Louis-Ugbo et al. [28]	2002	BCP and BCP/Collagen	rhBMP-2	Rabbit

9.7 Role of BMP in Bone Regeneration and Repair

The enhancement of fracture repair and regeneration of critical-sized bone defect remains an important challenge in the arena of orthopedic and maxillofacial surgery. Although autologous bone grafts are considered the gold standard as previously mentioned, the issues of limited availability, donor site morbidity, in addition to enhanced cost have compelled researchers to search for alternatives to augment bone restoration and repair [30, 31]. Hence, three new strategies have been evolved [32]:

(1) **Gene therapy**: gene transduction at repair site by encoding cytokines that have osteogenic properties.
(2) **Stem cell therapy**: host bone marrow-derived osteogenic cells transplantation at the defect site.
(3) **Protein therapy**: osteoinductive growth factor application at the defect site.

Amongst these, gene therapy and stem cell therapy are most promising, but still there are some drawbacks with respect to their efficacy and safety for human application [33–37].

The idea of using BMPs for bone regeneration started after the studies that depicted demineralized bone matrix is capable of inducing ectopic bone formation in subcutaneous and intramuscular spaces as well in rodents [38, 39]. Subsequently, many investigators reported that BMPs have the potentiality to produce osteogenic cells by inducing differentiation of mesenchymal stem cells, leading to bone formation [40]. As discussed above, BMPs belong to the superfamily transforming growth factor-β (TGF-β). The actual mechanism by which BMPs augment bone formation remains unclear until recent studies that were carried out [41]. It was postulated that the mechanism starts with the activation of a BMP signaling cascade at the time BMPs adhere to the surface receptors of mesenchymal cells; subsequently a specific protein sends the signal to the nucleus that eventually elicits gene expression leading to the production of specific macromolecules responsible for chondrogenesis and osteogenesis. As a result, mesenchymal cells differentiate into chondrocyte or an osteoblast.

Demineralized bone matrix acts as a repository of BMPs. The process of bone morphogenesis is considered as a sequential phenomenon of three phases: (1) firstly, chemotaxis and mitotic division of mesenchymal cells, (2) secondly, mesenchymal cell differentiation into cartilage, (3) thirdly, and finally, cartilage replacement into bone [42]. To be more precise, the phenomenon is initiated with the attachment of plasma fibronectin with the demineralized bone matrix [43] and attachment of mesenchymal cell as well as enhanced proliferation after three days. Mesenchymal cell differentiation into chondroblasts is mostly evident after five days. Mostly, chondrogenesis is evident after seven to eight days. Hypertrophic cartilage with resultant mineralization is seen on day nine. Angiogenesis and osteoblastic differentiation are mostly seen after ten to eleven days. Thus, remodeling of new endochondral bone occurs and hematopoiesis [44, 45].

In recent years, BMPs, more precisely BMP-2 and BMP-7, have shown promise as a substitute for various pathological conditions of bone-like compound tibial fracture, non-union and spinal fusion, and they are also capable of expediting osseous tissue healing [46, 47]. Acceleration of fracture healing and new osseous tissue formation is the primary goal in a majority of clinical studies using various pharmaceutical agents [48, 49]. However, BMPs combined with osteoconductive agents [50] and ultrasonic device Exogen® are the only FDA approved drug and/ or delivery systems that can be used to treat patients in a clinical environment. There are several in vivo studies where it is evident that BMPs along with various growth factors like transforming growth factor-beta (TGF-β), insulin-like growth factor (IGF), fibroblast growth factor (FGF), platelet-derived growth factor (PDGF), and vascular endothelial growth factor (VEGF) are responsible for normal bone healing [51–58]. On the contrary, the clinical utilization of rhBMP-2 and rhBMP-7 is governed by specific regulatory bodies of various countries in which their application is restricted to only treating non-union fracture of the tibia (rhBMP-7) in the treatment of tibial compound fracture and spinal fusion (rhBMP-2) [59–61]. Nevertheless, there is still a considerable quantity of BMPs in demineralized bone matrix despite the fact that platelet rich plasma and demineralized bone matrix can be used in a clinical setting [62, 63] (Table 9.4).

9.8 Role of BMP in Cartilage Repair

Typically, osteogenic BMPs induce stimulates undifferentiated mesenchymal cells to differentiate into chondrocytes that produce type II collagen and various proteoglycans within seven days following implantation [97].

Articular cartilage possesses the minimum capacity to repair itself after an injury [98]. Even so, it cannot hold the same strength to withstand the joint mobility after it has been repaired, as the repaired tissue is biochemically and structurally different to that of normal cartilage. Most of the time, the repaired cartilage will degenerate as a result of the mechanical forces applied at the joints. Degeneration of articular cartilage will eventually lead to the development of osteoarthritis [99]. Following an injury to cartilage, chondrocytes are activated by chemokines [100, 101] and lead to the degradation of the damaged tissue. Yet, the exact healing mechanism of cartilage remains unknown [102].

Since the mid-1980s, a number of clinical approaches and trials have been conducted for the repair of articular cartilage that includes chondrocyte transplantation, perichondrium, meniscal allograft, periosteum, osteochondral grafts, microfracture, and abrasions arthroplasty [103–113]. Healing of articular cartilage defect utilizing stem cells to well-differentiated chondrocytes [114] as well as allogeneic muscle-derived cells [115] have been demonstrated experimentally with promising results. Efficacy of many growth factors like insulin-like growth factor-1, basic fibroblast growth factor, transforming growth factor-β, BMPs, have been extensively investigated for its chondrogenic properties both in vivo and in vitro [116–126].

The fascinating work carried out by Seller and co-workers [98] confirmed that the incorporation rhBMP-2 into collagen sponge is capable of accelerating the repair of full-thickness cartilage defect. It has been postulated that the accelerated healing

Table 9.4 Carriers used to deliver BMPs in various experimental and clinical studies

Growth factors	Delivery carriers
BMP-2	Absorbable gelatin sponge [64, 65]
	Autogenic graft [66]
	Chemically crosslinked absorbable collagen, de-hydrothermally crosslinked collagen sponge [64]
	Chitosan [67]
	Collagen [68, 69]
	Fibrin matrix [70–72]
	Gelatin capsules loaded with PDLLGA microparticles and demineralized freeze-dried bone allografts [73]
	Glutaraldehyde cross-linked gelatin [74]
	Human demineralized bone matrix, thermoashed bone mineral, non-demineralized bone particles, and irradiated cancellous chips [64]
	Hydroxyapatite/tricalcium phosphate bone graft covered with heparinized collagen membrane [75]
	Hydroxyapatite porous particles and coral-replicated porous tablets [76]
	PLA coating [77, 78]
	Poly (α-hydroxy acids) [64, 79, 80]
	Poly (DL-lactic-co-glycolic acid) (PDLLGA)/gelatin microcapsules [81, 82]
	Poly (ε-caprolactone) scaffolds [83]
	Polyester nanocapsules [84]
	Rat demineralized bone matrix, and de-lipidated bovine bone matrix [64]
	Si–Ca-P porous glass (xerogels) [85]
	Synthetic and bovine-derived hydroxyapatite particles and coral-derived hydroxyapatite [64]
	Tricalcium phosphate [64]
BMP-7	Calcium phosphate-coated nanofiber mesh/PRP matrix [86]
	Chitosan [67]
	Chitosan microparticles/vancomycin and cefazolin [87]
	Collagen [88]
	Hydroxyapatite/tricalcium phosphate bone graft covered with heparinized collagen membrane [75]
	Mesoporous bioglass/silk fibrin scaffolds [89]
	PGLA microspheres/3D collagen gel [90]
	Poly (ε-caprolactone) scaffolds [83]
	Polyester nanocapsules [84]
	Polylactide (PLGA) microspheres [91]
	Polyphosphate [92]

(continued)

Table 9.4 (continued)

Growth factors	Delivery carriers
BMP-9	Adenoviral vector (ADGBMP9) [93, 94]
	Citrate-based thermoresponsive hydrogel (PPCNg) [95]
	Collagen/Chitosan [96]

might be either due to recruitment of more numbers of required cells in the defect sites or due to initiation of repair process or the combination of both processes. Other studies have demonstrated that the repeated administration may be required as the proteins have a very short half-life which again predispose to osteophyte formation and hypertrophy despite the positive effects of direct injection of such proteins [127]. A study by Kuroda et al. demonstrated that MDSCs can be used as an adequate carrier of therapeutic gene as well as possessing the capability of delivering a proper amount of BMP-4 protein to the damaged cartilage site [114]. Furthermore, another study by Mattioli-Belmonte et al. examined the efficacy of chitosan as a delivery agent of BMP for cartilage healing and the results are promising [128].

A cell-free technology of subchondral implantation of a triple composite of interconnected porous hydroxyapatite, synthetic polymer (PLA-PEG), and bone morphogenetic protein-2 (rhBMP-2) for the regeneration of induced rabbit articular cartilage defect model was studied by Tamai et al., with very promising results with respect to early healing [129].

In another study, overexpression of BMP-7 along with accelerated healing and greater quantity of hyaline cartilage deposition were observed in an experimental cartilage defect in the patelleo-femoral joint of horses [130]. Similarly, greater hyaline cartilage regeneration was also observed in a study by Yang et al. in which they examined the theory that the long-term delivery of BMP-2 to cartilage defects subjected to microfracture results in regeneration of high-quality hyaline-like cartilage [131]. The role of mesenchymal stem cells and BMP in cartilage repair has been extensively studied by various workers and have logically been presented in a review by Scarfi [132] and are shown in Table 9.5.

Table 9.5 Utilization of mesenchymal stem cells and bone morphogenetic proteins in cartilage defects in vivo

Conditions	Scaffold
MSCs transfected with BMP7	Bioresorbable polymer scaffold [133]
MSCs transfected with BMP7 and TGF-β1	Bilayered osteochondral scaffold [134]
MSCs + TGF-β1, PDGF and BMP2	Bilayer scaffold with platelet rich plasma [135]
MSCs + TGF-β1, TGF-β3, BMP2, 4 and 7	Osteochondral allograft with extracellular matrix proteins [136]

9.9 Conclusion

Bone tissue engineering involves the combined use various materials, cells, growth factors and it holds great potential in the repair and management of bone disorders. However, composite scaffolds have not been extensively utilized in the clinical setting on a wide scale. Autologous bone graft is still the 'gold standard' in orthopedic surgery but its ample availability within a required timeframe is a great concern for orthopedic surgeons. As a member of the multi-functional cytokines, BMP are involved in a multitude of molecular cascades and signaling pathways. Apart from their significant role during bone remodeling, BMPs also persuade osteoclast homeostasis. Despite a considerable amount of time has passed, we still have not fully understood how the signals from the family of BMPs regulate the formation and maintenance of various organs in vivo. The local and sustained delivery of BMPs is also a major concern and to address this issue, the application of bioceramic scaffolds is worth considering not only to deliver BMPs in a sustained manner but also to serve as bone graft substitutes.

Acknowledgements The authors gratefully acknowledge the supports rendered by the Honorable Vice Chancellor, West Bengal University of Animal and Fishery Sciences, Kolkata, India.

References

1. Begam H, Nandi SK, Kundu B et al (2017) Strategies for delivering bone morphogenetic protein for bone healing. Mater Sci Eng C Mater Biol Appl 70:856–869
2. Bharadwaz A, Jayasuriya AC (2021) Osteogenic differentiation cues of the bone morphogenetic protein-9 (BMP-9) and its recent advances in bone tissue regeneration. Mater Sci Eng C Mater Biol Appl 120:111748
3. Paxton NC, Wong CS, Desselle MR et al (2020) Bone morphogenetic protein-assisted bone regeneration and applications in biofabrication. In: Vrana NE, Knopf-Marques H, Barthes J (eds) Biomaterials for Organ and Tissue Regeneration. Woodhead Publishing, England, pp 363–391
4. Preethi Soundarya S, Haritha Menon A, Viji Chandran S et al (2018) Bone tissue engineering: Scaffold preparation using chitosan and other biomaterials with different design and fabrication techniques. Int J Biol Macromol 119:1228–1239
5. Ai-Aql ZS, Alagl AS, Graves DT et al (2008) Molecular mechanisms controlling bone formation during fracture healing and distraction osteogenesis. J Dent Res 87:107–118
6. Han SH, Lee J, Lee KM et al (2020) Enhanced healing of rat calvarial defects with 3D printed calcium-deficient hydroxyapatite/collagen/bone morphogenetic protein 2 scaffolds. J Mech Behav Biomed Mater 108:103782
7. Vaibhav B, Nilesh P, Vikram S et al (2007) Bone morphogenic protein and its application in trauma cases: a current concept update. Injury 38:1227–1235
8. Rao SM, Ugale GM, Warad SB (2013) Bone morphogenetic proteins: periodontal regeneration. N Am J Med Sci 5:161–168
9. Katagiri T, Watabe T (2016) Bone Morphogenetic Proteins. Cold Spring Harb Perspect Biol 8:a021899
10. Miyazono K, Maeda S, Imamura T (2005) BMP receptor signaling: transcriptional targets, regulation of signals, and signaling cross-talk. Cytokine Growth Factor Rev 16:251–263

11. Yang J, Shi P, Tu M et al (2014) Bone morphogenetic proteins: relationship between molecular structure and their osteogenic activity. Food Sci Hum Well 3:127–135
12. Calori GM, Mazza E, Colombo M et al (2011) The use of bone-graft substitutes in large bone defects: any specific needs? Injury 42:S56–S63
13. Lo KW, Ulery BD, Ashe KM et al (2012) Studies of bone morphogenetic protein-based surgical repair. Adv Drug Deliv Rev 64:1277–1291
14. Bessa PC, Casal M, Reis RL (2008) Bone morphogenetic proteins in tissue engineering: the road from laboratory to clinic, part II (BMP delivery). J Tissue Eng Regen Med 2:81–96
15. Anushika, Sharma P, Trivedi A et al (2019) Synthesis and characterization of pure and titania doped hydroxyapatite. Mater Today Proc 16:302–307
16. Begam H, Mandal S, Chanda A et al (2014) Effect of zinc doping on biological properties of biphasic calcium phosphate ceramics in orthopaedic animal model. Trans Indian Ceram Soc 73:284–292
17. Gao C, Peng S, Feng P et al (2017) Bone biomaterials and interactions with stem cells. Bone Res 5:17059
18. Sun H, Yang HL (2015) Calcium phosphate scaffolds combined with bone morphogenetic proteins or mesenchymal stem cells in bone tissue engineering. Chin Med J (Engl) 128:1121–1127
19. Kalita SJ, Bhardwaj A, Bhatt HA (2007) Nanocrystalline calcium phosphate ceramics in biomedical engineering. Mater Sci Eng C 27:441–449
20. Bhattacharjee P, Begam H, Chanda A et al (2014) Animal trial on zinc doped hydroxyapatite: a case study. J Asian Ceram Soc 2:44–51
21. Han SH, Cha M, Jin YZ et al (2020) BMP-2 and hMSC dual delivery onto 3D printed PLA-Biogel scaffold for critical-size bone defect regeneration in rabbit tibia. Biomed Mater 16:015019
22. Hirakawa Y, Manaka T, Orita K et al (2018) The accelerated effect of recombinant human bone morphogenetic protein 2 delivered by β-tricalcium phosphate on tendon-to-bone repair process in rabbit models. J Shoulder Elbow Surg 27:894–902
23. Huang B, Wu Z, Ding S et al (2018) Localization and promotion of recombinant human bone morphogenetic protein-2 bioactivity on extracellular matrix mimetic chondroitin sulfate-functionalized calcium phosphate cement scaffolds. Acta Biomater 71:184–199
24. Ji W, Kerckhofs G, Geeroms C et al (2018) Deciphering the combined effect of bone morphogenetic protein 6 and calcium phosphate on bone formation capacity of periosteum derived cells-based tissue engineering constructs. Acta Biomater 80:97–107
25. Tenkumo T, Vanegas Sáenz JR, Nakamura K et al (2018) Prolonged release of bone morphogenetic protein-2 in vivo by gene transfection with DNA-functionalized calcium phosphate nanoparticle-loaded collagen scaffolds. Mater Sci Eng C Mater Biol Appl 92:172–183
26. Ding S, Zhang J, Tian Y et al (2016) Magnesium modification up-regulates the bioactivity of bone morphogenetic protein-2 upon calcium phosphate cement via enhanced BMP receptor recognition and Smad signaling pathway. Colloids Surf B Biointerfaces 145:140–151
27. Tsuzuki N, Otsuka K, Seo J et al (2012) In vivo osteoinductivity of gelatin β-tri-calcium phosphate sponge and bone morphogenetic protein-2 on an equine third metacarpal bone defect. Res Vet Sci 93:1021–1025
28. Louis-Ugbo J, Kim HS, Boden SD et al (2002) Retention of 125I-labeled recombinant human bone morphogenetic protein-2 by biphasic calcium phosphate or a composite sponge in a rabbit posterolateral spine arthrodesis model. J Orthop Res 20:1050–1059
29. Majid K, Tseng MD, Baker KC et al (2011) Biomimetic calcium phosphate coatings as bone morphogenetic protein delivery systems in spinal fusion. Spine J 11:560–567
30. Bishop GB, Einhorn TA (2007) Current and future clinical applications of bone morphogenetic proteins in orthopaedic trauma surgery. Int Orthop 31:721–727
31. Geiger M, Li RH, Friess W (2003) Collagen sponges for bone regeneration with rhBMP-2. Adv Drug Deliv Rev 55:1613–1629
32. Rose FR, Hou Q, Oreffo RO (2004) Delivery systems for bone growth factors—the new players in skeletal regeneration. J Pharm Pharmacol 56:415–427

33. D'Mello S, Atluri K, Geary SM et al (2016) Bone regeneration using gene-activated matrices. AAPS J 19:43–53
34. Asatrian G, Pham D, Hardy WR et al (2015) Stem cell technology for bone regeneration: current status and potential applications. Stem Cells Cloning 8:39–48
35. Fischer J, Kolk A, Wolfart S et al (2011) Future of local bone regeneration—Protein versus gene therapy. J Craniomaxillofac Surg 39:54–64
36. Gómez-Barrena E, Rosset P, Müller I et al (2011) Bone regeneration: stem cell therapies and clinical studies in orthopaedics and traumatology. J Cell Mol Med 15:1266–1286
37. Arvidson K, Abdallah BM, Applegate LA et al (2011) Bone regeneration and stem cells. J Cell Mol Med 15:718–746
38. Urist MR (1965) Bone: formation by autoinduction. Science 150:893–899
39. Van de Putte KA, Urist MR (1965) Osteogenesis in the interior of intramuscular implants of decalcified bone matrix. Clin Orthop Relat Res 43:257–270
40. Granjeiro JM, Oliveira RC, Bustos-Valenzuela JC et al (2005) Bone morphogenetic proteins: from structure to clinical use. Braz J Med Biol Res 38:1463–1473
41. Reddi AH (2001) Bone morphogenetic proteins: from basic science to clinical applications. J Bone Joint Surg Am 83-A Suppl 1:S1-S6
42. Reddi AH, Anderson WA (1976) Collagenous bone matrix-induced endochondral ossification hemopoiesis. J Cell Biol 69:557–572
43. Weiss RE, Reddi AH (1980) Synthesis and localization of fibronectin during collagenous matrix-mesenchymal cell interaction and differentiation of cartilage and bone in vivo. Proc Natl Acad Sci U S A 77:2074–2078
44. Reddi AH (1998) Role of morphogenetic proteins in skeletal tissue engineering and regeneration. Nat Biotechnol 16:247–252
45. Reddi AH (1994) Symbiosis of biotechnology and biomaterials: applications in tissue engineering of bone and cartilage. J Cell Biochem 56:192–195
46. Nauth A, Ristevski B, Li R et al (2011) Growth factors and bone regeneration: how much bone can we expect? Injury 42:574–579
47. Lieberman JR, Daluiski A, Einhorn TA (2002) The role of growth factors in the repair of bone. Biology and clinical applications. J Bone Joint Surg Am 84:1032–1044
48. Einhorn TA, Gerstenfeld LC (2015) Fracture healing: mechanisms and interventions. Nat Rev Rheumatol 11:45–54
49. Dent-Acosta RE, Storm N, Steiner RS et al (2012) The tactics of modern-day regulatory trials. J Bone Joint Surg Am 94 Suppl 1:39–44
50. Kostenuik P, Mirza FM (2017) Fracture healing physiology and the quest for therapies for delayed healing and nonunion. J Orthop Res 35:213–223
51. Mayer H, Bertram H, Lindenmaier W et al (2005) Vascular endothelial growth factor (VEGF-A) expression in human mesenchymal stem cells: autocrine and paracrine role on osteoblastic and endothelial differentiation. J Cell Biochem 95:827–839
52. Zimmermann G, Henle P, Küsswetter M et al (2005) TGF-beta1 as a marker of delayed fracture healing. Bone 36:779–785
53. Kloen P, Di Paola M, Borens O et al (2003) BMP signaling components are expressed in human fracture callus. Bone 33:362–371
54. Okazaki K, Jingushi S, Ikenoue T et al (2003) Expression of parathyroid hormone-related peptide and insulin-like growth factor I during rat fracture healing. J Orthop Res 21:511–520
55. Bostrom MP (1998) Expression of bone morphogenetic proteins in fracture healing. Clin Orthop Relat Res. https://doi.org/10.1097/00003086-199810001-00013
56. Andrew JG, Hoyland JA, Freemont AJ et al (1995) Platelet-derived growth factor expression in normally healing human fractures. Bone 16:455–460
57. Bourque WT, Gross M, Hall BK (1993) Expression of four growth factors during fracture repair. Int J Dev Biol 37:573–579
58. Joyce ME, Jingushi S, Bolander ME (1990) Transforming growth factor-beta in the regulation of fracture repair. Orthop Clin North Am 21:199–209

59. Burkus JK, Heim SE, Gornet MF et al (2003) Is INFUSE bone graft superior to autograft bone? An integrated analysis of clinical trials using the LT-CAGE lumbar tapered fusion device. J Spinal Disord Tech 16:113–122
60. Govender S, Csimma C, Genant HK et al (2002) Recombinant human bone morphogenetic protein-2 for treatment of open tibial fractures: a prospective, controlled, randomized study of four hundred and fifty patients. J Bone Joint Surg Am 84:2123–2134
61. Friedlaender GE, Perry CR, Cole JD et al (2001) Osteogenic protein-1 (bone morphogenetic protein-7) in the treatment of tibial nonunions. J Bone Joint Surg Am 83-A Suppl 1:S151-S158
62. Wildemann B, Kadow-Romacker A, Haas NP et al (2007) Quantification of various growth factors in different demineralized bone matrix preparations. J Biomed Mater Res A 81:437–442
63. Bae HW, Zhao L, Kanim LE et al (2006) Intervariability and intravariability of bone morphogenetic proteins in commercially available demineralized bone matrix products. Spine 31:1299–1306; discussion 1307-1308
64. Uludag H, D'Augusta D, Palmer R et al (1999) Characterization of rhBMP-2 pharmacokinetics implanted with biomaterial carriers in the rat ectopic model. J Biomed Mater Res 46:193–202
65. Boyne PJ, Marx RE, Nevins M et al (1997) A feasibility study evaluating rhBMP-2/absorbable collagen sponge for maxillary sinus floor augmentation. Int J Periodontics Restorative Dent 17:11–25
66. Issa JP, Gonzaga M, Kotake BG et al (2016) Bone repair of critical size defects treated with autogenic, allogenic, or xenogenic bone grafts alone or in combination with rhBMP-2. Clin Oral Implants Res 27:558–566
67. Yilgor P, Tuzlakoglu K, Reis RL et al (2009) Incorporation of a sequential BMP-2/BMP-7 delivery system into chitosan-based scaffolds for bone tissue engineering. Biomaterials 30:3551–3559
68. Lee SS, Huang BJ, Kaltz SR et al (2013) Bone regeneration with low dose BMP-2 amplified by biomimetic supramolecular nanofibers within collagen scaffolds. Biomaterials 34:452–459
69. Sumanasinghe RD, Bernacki SH, Loboa EG (2006) Osteogenic differentiation of human mesenchymal stem cells in collagen matrices: effect of uniaxial cyclic tensile strain on bone morphogenetic protein (BMP-2) mRNA expression. Tissue Eng 12:3459–3465
70. Kaipel M, Schützenberger S, Schultz A et al (2012) BMP-2 but not VEGF or PDGF in fibrin matrix supports bone healing in a delayed-union rat model. J Orthop Res 30:1563–1569
71. Schützenberger S, Schultz A, Hausner T et al (2012) The optimal carrier for BMP-2: a comparison of collagen versus fibrin matrix. Arch Orthop Trauma Surg 132:1363–1370
72. Lucarelli E, Beretta R, Dozza B et al (2010) A recently developed bifacial platelet-rich fibrin matrix. Eur Cell Mater 20:13–23
73. Boyan BD, Lohmann CH, Somers A et al (1999) Potential of porous poly-D,L-lactide-co-glycolide particles as a carrier for recombinant human bone morphogenetic protein-2 during osteoinduction in vivo. J Biomed Mater Res 46:51–59
74. Yamamoto M, Tabata Y, Ikada Y (1998) Ectopic bone formation induced by biodegradable hydrogels incorporating bone morphogenetic protein. J Biomater Sci Polym Ed 9:439–458
75. Jo JY, Jeong SI, Shin YM et al (2015) Sequential delivery of BMP-2 and BMP-7 for bone regeneration using a heparinized collagen membrane. Int J Oral Maxillofac Surg 44:921–928
76. Kuboki Y, Takita H, Kobayashi D et al (1998) BMP-induced osteogenesis on the surface of hydroxyapatite with geometrically feasible and nonfeasible structures: topology of osteogenesis. J Biomed Mater Res 39:190–199
77. Chen Z, Zhang Z, Feng J et al (2018) Influence of mussel-derived bioactive BMP-2-decorated PLA on MSC behavior in vitro and verification with osteogenicity at ectopic sites in vivo. ACS Appl Mater Interfaces 10:11961–11971
78. Kandziora F, Bail H, Schmidmaier G et al (2002) Bone morphogenetic protein-2 application by a poly(D,L-lactide)-coated interbody cage: in vivo results of a new carrier for growth factors. J Neurosurg 97:40–48
79. Liu HW, Chen CH, Tsai CL et al (2007) Heterobifunctional poly(ethylene glycol)-tethered bone morphogenetic protein-2-stimulated bone marrow mesenchymal stromal cell differentiation and osteogenesis. Tissue Eng 13:1113–1124

80. Hollinger JO, Leong K (1996) Poly(α-hydroxy acids): carriers for bone morphogenetic proteins. Biomaterials 17:187–194
81. Alipour M, Firouzi N, Aghazadeh Z et al (2021) The osteogenic differentiation of human dental pulp stem cells in alginate-gelatin/Nano-hydroxyapatite microcapsules. BMC Biotechnol 21:6
82. Isobe M, Yamazaki Y, Mori M et al (1999) The role of recombinant human bone morphogenetic protein-2 in PLGA capsules at an extraskeletal site of the rat. J Biomed Mater Res 45:36–41
83. Yilgor P, Sousa RA, Reis RL et al (2010) Effect of scaffold architecture and BMP-2/BMP-7 delivery on in vitro bone regeneration. J Mater Sci Mater Med 21:2999–3008
84. Yilgor P, Hasirci N, Hasirci V (2010) Sequential BMP-2/BMP-7 delivery from polyester nanocapsules. J Biomed Mater Res A 93:528–536
85. Santos EM, Radin S, Shenker BJ et al (1998) Si-Ca-P xerogels and bone morphogenetic protein act synergistically on rat stromal marrow cell differentiation in vitro. J Biomed Mater Res 41:87–94
86. Berner A, Boerckel JD, Saifzadeh S et al (2012) Biomimetic tubular nanofiber mesh and platelet rich plasma-mediated delivery of BMP-7 for large bone defect regeneration. Cell Tissue Res 347:603–612
87. Mantripragada VP, Jayasuriya AC (2016) Effect of dual delivery of antibiotics (vancomycin and cefazolin) and BMP-7 from chitosan microparticles on Staphylococcus epidermidis and pre-osteoblasts in vitro. Mater Sci Eng C Mater Biol Appl 67:409–417
88. Ripamonti U, Van Den Heever B, Sampath TK et al (1996) Complete regeneration of bone in the baboon by recombinant human osteogenic protein-1 (hOP-1, bone morphogenetic protein-7). Growth Factors 13:273–289
89. Zhang Y, Cheng N, Miron R et al (2012) Delivery of PDGF-B and BMP-7 by mesoporous bioglass/silk fibrin scaffolds for the repair of osteoporotic defects. Biomaterials 33:6698–6708
90. Gavénis K, Klee D, Pereira-Paz RM et al (2007) BMP-7 loaded microspheres as a new delivery system for the cultivation of human chondrocytes in a collagen type-I gel. J Biomed Mater Res B Appl Biomater 82:275–283
91. Gavenis K, Schneider U, Groll J et al (2010) BMP-7-loaded PGLA microspheres as a new delivery system for the cultivation of human chondrocytes in a collagen type I gel: the common nude mouse model. Int J Artif Organs 33:45–53
92. Renier ML, Kohn DH (1997) Development and characterization of a biodegradable polyphosphate. J Biomed Mater Res 34:95–104
93. Li JZ, Hankins GR, Kao C et al (2003) Osteogenesis in rats induced by a novel recombinant helper-dependent bone morphogenetic protein-9 (BMP-9) adenovirus. J Gene Med 5:748–756
94. Varady P, Li JZ, Cunningham M et al (2001) Morphologic analysis of BMP-9 gene therapy-induced osteogenesis. Hum Gene Ther 12:697–710
95. Zhao C, Zeng Z, Qazvini NT et al (2018) Thermoresponsive citrate-based graphene oxide scaffold enhances bone regeneration from BMP9-stimulated adipose-derived mesenchymal stem cells. ACS Biomater Sci Eng 4:2943–2955
96. Bergeron E, Leblanc E, Drevelle O et al (2012) The evaluation of ectopic bone formation induced by delivery systems for bone morphogenetic protein-9 or its derived peptide. Tissue Eng Part A 18:342–352
97. Wang EA, Rosen V, D'Alessandro JS et al (1990) Recombinant human bone morphogenetic protein induces bone formation. Proc Natl Acad Sci U S A 87:2220–2224
98. Sellers RS, Zhang R, Glasson SS et al (2000) Repair of articular cartilage defects one year after treatment with recombinant human bone morphogenetic protein-2 (rhBMP-2). J Bone Joint Surg Am 82:151–160
99. Deng ZH, Li YS, Gao X et al (2018) Bone morphogenetic proteins for articular cartilage regeneration. Osteoarthritis Cartilage 26:1153–1161
100. Wong VW, Gurtner GC, Longaker MT (2013) Wound healing: a paradigm for regeneration. Mayo Clin Proc 88:1022–1031
101. Burleigh A, Chanalaris A, Gardiner MD et al (2012) Joint immobilization prevents murine osteoarthritis and reveals the highly mechanosensitive nature of protease expression in vivo. Arthritis Rheum 64:2278–2288

102. Huey DJ, Hu JC, Athanasiou KA (2012) Unlike bone, cartilage regeneration remains elusive. Science 338:917–921
103. Katsube K, Ochi M, Uchio Y et al (2000) Repair of articular cartilage defects with cultured chondrocytes in Atelocollagen gel. Comparison with cultured chondrocytes in suspension. Arch Orthop Trauma Surg 120:121–127
104. Minas T, Nehrer S (1997) Current concepts in the treatment of articular cartilage defects. Orthopedics 20:525–538
105. Shortkroff S, Barone L, Hsu HP et al (1996) Healing of chondral and osteochondral defects in a canine model: the role of cultured chondrocytes in regeneration of articular cartilage. Biomaterials 17:147–154
106. Sumen Y, Ochi M, Ikuta Y (1995) Treatment of articular defects with meniscal allografts in a rabbit knee model. Arthroscopy 11:185–193
107. Brittberg M, Lindahl A, Nilsson A et al (1994) Treatment of deep cartilage defects in the knee with autologous chondrocyte transplantation. N Engl J Med 331:889–895
108. Bert JM (1993) Role of abrasion arthroplasty and debridement in the management of osteoarthritis of the knee. Rheum Dis Clin North Am 19:725–739
109. Matsusue Y, Yamamuro T, Hama H (1993) Arthroscopic multiple osteochondral transplantation to the chondral defect in the knee associated with anterior cruciate ligament disruption. Arthroscopy 9:318–321
110. Hoikka VE, Jaroma HJ, Ritsilä VA (1990) Reconstruction of the patellar articulation with periosteal grafts. 4-year follow-up of 13 cases. Acta Orthop Scand 61:36–39
111. Grande DA, Pitman MI, Peterson L et al (1989) The repair of experimentally produced defects in rabbit articular cartilage by autologous chondrocyte transplantation. J Orthop Res 7:208–218
112. Johnson LL (1986) Arthroscopic abrasion arthroplasty historical and pathologic perspective: present status. Arthroscopy 2:54–69
113. Amiel D, Coutts RD, Abel M et al (1985) Rib perichondrial grafts for the repair of full-thickness articular-cartilage defects. A morphological and biochemical study in rabbits. J Bone Joint Surg Am 67:911–920
114. Kuroda R, Usas A, Kubo S et al (2006) Cartilage repair using bone morphogenetic protein 4 and muscle-derived stem cells. Arthritis Rheum 54:433–442
115. Adachi N, Sato K, Usas A et al (2002) Muscle derived, cell based *ex vivo* gene therapy for treatment of full thickness articular cartilage defects. J Rheumatol 29:1920–1930
116. Steinert A, Weber M, Dimmler A et al (2003) Chondrogenic differentiation of mesenchymal progenitor cells encapsulated in ultrahigh-viscosity alginate. J Orthop Re 21:1090–1097
117. Gelse K, von der Mark K, Aigner T et al (2003) Articular cartilage repair by gene therapy using growth factor-producing mesenchymal cells. Arthritis Rheum 48:430–441
118. Kramer J, Hegert C, Guan K et al (2000) Embryonic stem cell-derived chondrogenic differentiation in vitro: activation by BMP-2 and BMP-4. Mech Dev 92:193–205
119. Semba I, Nonaka K, Takahashi I et al (2000) Positionally-dependent chondrogenesis induced by BMP4 is co-regulated by Sox9 and Msx2. Dev Dyn 217:401–414
120. Glansbeek HL, van Beuningen HM, Vitters EL et al (1997) Bone morphogenetic protein 2 stimulates articular cartilage proteoglycan synthesis in vivo but does not counteract interleukin-1alpha effects on proteoglycan synthesis and content. Arthritis Rheum 40:1020–1028
121. Sellers RS, Peluso D, Morris EA (1997) The effect of recombinant human bone morphogenetic protein-2 (rhBMP-2) on the healing of full-thickness defects of articular cartilage. J Bone Joint Surg Am 79:1452–1463
122. Humbel RE (1990) Insulin-like growth factors I and II. Eur J Biochem 190:445–462
123. Tyler JA (1989) Insulin-like growth factor 1 can decrease degradation and promote synthesis of proteoglycan in cartilage exposed to cytokines. Biochem J 260:543–548
124. Cuevas P, Burgos J, Baird A (1988) Basic fibroblast growth factor (FGF) promotes cartilage repair in vivo. Biochem Biophys Res Commun 156:611–618

125. Morales TI, Roberts AB (1988) Transforming growth factor beta regulates the metabolism of proteoglycans in bovine cartilage organ cultures. J Biol Chem 263:12828–12831
126. Redini F, Galera P, Mauviel A et al (1988) Transforming growth factor beta stimulates collagen and glycosaminoglycan biosynthesis in cultured rabbit articular chondrocytes. FEBS Lett 234:172–176
127. Allen JB, Manthey CL, Hand AR et al (1990) Rapid onset synovial inflammation and hyperplasia induced by transforming growth factor beta. J Exp Med 171:231–247
128. Mattioli-Belmonte M, Gigante A, Muzzarelli RA et al (1999) N,N-dicarboxymethyl chitosan as delivery agent for bone morphogenetic protein in the repair of articular cartilage. Med Biol Eng Comput 37:130–134
129. Tamai N, Myoui A, Hirao M et al (2005) A new biotechnology for articular cartilage repair: subchondral implantation of a composite of interconnected porous hydroxyapatite, synthetic polymer (PLA-PEG), and bone morphogenetic protein-2 (rhBMP-2). Osteoarthritis Cartilage 13:405–417
130. Hidaka C, Goodrich LR, Chen CT et al (2003) Acceleration of cartilage repair by genetically modified chondrocytes over expressing bone morphogenetic protein-7. J Orthop Res 21:573–583
131. Yang HS, La WG, Bhang SH et al (2011) Hyaline cartilage regeneration by combined therapy of microfracture and long-term bone morphogenetic protein-2 delivery. Tissue Eng Part A 17:1809–1818
132. Scarfì S (2016) Use of bone morphogenetic proteins in mesenchymal stem cell stimulation of cartilage and bone repair. World J Stem Cells 8:1–12
133. Grande DA, Mason J, Light E et al (2003) Stem cells as platforms for delivery of genes to enhance cartilage repair. J Bone Joint Surg Am 85-A Suppl 2:111–116
134. Chen J, Chen H, Li P et al (2011) Simultaneous regeneration of articular cartilage and subchondral bone in vivo using MSCs induced by a spatially controlled gene delivery system in bilayered integrated scaffolds. Biomaterials 32:4793–4805
135. Seo JP, Tanabe T, Tsuzuki N et al (2013) Effects of bilayer gelatin/β-tricalcium phosphate sponges loaded with mesenchymal stem cells, chondrocytes, bone morphogenetic protein-2, and platelet rich plasma on osteochondral defects of the talus in horses. Res Vet Sci 95:1210–1216
136. Geraghty S, Kuang JQ, Yoo D et al (2015) A novel, cryopreserved, viable osteochondral allograft designed to augment marrow stimulation for articular cartilage repair. J Orthop Surg Res 10:66

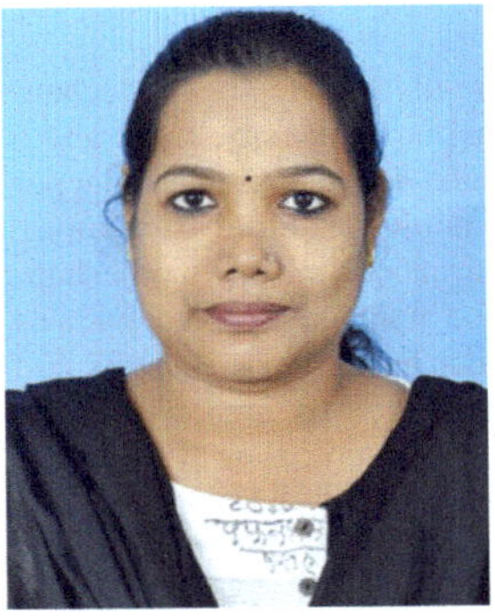

Howa Begam Dr. Begam is a Research Associate (ICMR) of Centre for Healthcare Science and Technology, Indian Institute of Engineering Science and Technology, Shibpur, Howrah, India. She completed her Ph.D. (Engineering) in 2016 and M.E. in Biomedical Engineering in 2009 from Jadavpur University. She received CSIR Senior Research Fellowship for conduction of her Ph.D. thesis work. Her research interests are calcium phosphate-based scaffold for bone tissue engineering, polymeric scaffold for drug delivery system etc. She published six number of SCI/Scopus indexed journals, two book chapter in Lecture Notes in Bioengineering. Springer, and fourteen numbers of conference proceedings.

Subhasis Roy Dr. Roy is an Assistant Professor (Veterinary Surgery & Radiology) at the West Bengal University of Animal & Fishery Sciences, where he has been a faculty since 2019. He obtained his B.V.Sc. & A.H. degree in 2002 from the same university. He achieved his M.V.Sc. and Ph.D. in Veterinary Surgery & Radiology consecutively in 2004 and 2008. He has obtained PGDAEM in the year 2012. His area of expertise is oncology and orthopaedic surgery. He has got one patent on marine biomaterials for its use in cancer treatment. He has been serving as a member of Institutional Animal Ethical committee of Chittaranjan National Cancer Institute and Calcutta Medical College for last two years. He has achieved CSIR-CGCRI Diamond Jubilee Award for Best paper Award on Material Science in 2011. He has got Appreciation Award in Orthopaedic Surgery Session for paper presentation in XXXV Annual Congress of Indian Society for Veterinary Surgery in the year 2012.

Prasenjit Mukherjee Dr. Mukherjee is an Assistant Professor (Veterinary Surgery & Radiology) at the West Bengal University of Animal & Fishery Sciences, where he has been a faculty since 2013. He obtained his B.V.Sc. & A.H. degree in 2002 from the same university. He achieved his M.V.Sc. and Ph.D. in Veterinary Surgery & Radiology consecutively in 2004 and 2008. He has worked as Senior Research Fellow in DST funded project in collaboration with Central Glass and Ceramic Research Institute, Jadavpur, Kolkata for 2 years. He has also worked as a Veterinary Officer in Govt. of West Bengal since 2009-2013. He has achieved CSIR-CGCRI Diamond Jubilee Award for Best paper Award on Material Science in 2011. His area of expertise is orthopaedic surgery. He has contributed book chapters, international and national review and research articles on tissue engineering, biomaterials, drug delivery related to orthopaedic applications in several books and journals. Presently, he is attached to DBT funded research project on peptide-based nanomedicine for blood clotting and wound healing in collaboration with IISER, Kolkata as co-principal investigator.

Abhijit Chanda Dr. Chanda is currently a Professor of Mechanical Engineering Department, Jadavpur University, India and a former Director, School of Bioscience & Engineering, Jadavpur University. He has authored two books on Solid Mechanics, sixty research articles in reputed journals, number of book chapters. He is regular reviewer of good number of peer reviewed international journals. He has guided nine Ph.D. students and more than forty ME/ M Tech students. His area of interest includes biomaterials, mechanical characterization of materials and tribology.

Biswanath Kundu Dr. Kundu is a principal scientist from CSIR-Central Glass and Ceramic Research Institute, Kolkata, India where his main field of activities include bioceramics, more specifically to synthesis, characterization of different nano calcium-phosphate based materials, bioactive glass materials, etc. which are used to develop new and tailor made materials with specific end use, for example, bone graft, bone-tissue engineering, soft-tissue engineering, orbital implants, drug delivery system for local, sustained releasing, active or passive targeting, etc. He has extensive research experience in the areas of design and fabrication of different implants both in dense and porous form by different techniques which are utilized to develop tailor-made implants for huge numbers of ailing patients of India. He has received many prestigious awards and recognitions from Government of India which include MRSI Young Scientist Awards, CSIR Best Technology Award, Best Technology Transfer Awards including many poster and oral research papers. Dr. Kundu has published more than 85 SCI journal papers and 9 book chapters. He is also holding 5 Indian patents related to biomedical products.

Samit Kumar Nandi Dr. Nandi is a Full Professor and Former Head of the Department of Veterinary Surgery and Radiology at West Bengal University of Animal & Fishery Sciences (WBUAFS), Kolkata, India. He worked as Adjunct Faculty at the School of Mechanical and Materials Engineering, Washington State University, Pullman, USA. He received several National Awards from his country. He has supervised/supervising over 45 Master's and 12 Ph.D. students in Veterinary Surgery & Radiology and allied subjects. He has contributed more than 150 scientific research papers in National and International journals of repute, 3 books, 18 International book chapters and granted patents. He is a reviewer of number of journals all over the Globe. He has given several invited and lead presentations in various National and International symposia, National Institutes and labs. Dr. Nandi is Fellow of National Academy of Agricultural Sciences, Indian Association for the Advancement of Veterinary Research, Indian Society for Veterinary Surgery, Society of Applied Biotechnology, and Member of National Academy of Sciences, India and National Academy of Veterinary Sciences, India.

Chapter 10
Spine Surgery—Part I: Biomechanics, Materials, and 3-D Printing Technology: Surgical Perspective and Clinical Impact

Samuel H. Brill, Jee Ho Chong, Dongyoung Kim, and Woojin Cho

Abstract Due to the prevalence of spine pathologies, biomechanical research has become essential in the clinical setting. Spinal stability is a requirement for the spine to achieve its biomechanical goals and therefore surgical intervention is often required to maintain a normal spine pathology. Interventions designed to promote spinal stability require various implants such as pedicle screws and interbody cages. The required hardware comes in different biomaterials which each have advantages and disadvantages based on their material and structural properties. 3D printing has been proposed as a method for accurate preoperative planning, patient and trainee education, intraoperative guidance systems, and intraoperative implants.

Keywords Spinal surgery · Biomechanics · Biomaterials · Elastic modulus · Polyetheretherketone (PEEK) · Glass ceramic · Three-dimensional printing · Tissue engineering

10.1 Biomechanics of the Spine

Clinical biomechanics of the spine refer to the study of both normal and pathologic mechanical functions of the human vertebral column caused by mechanical insult. Biomechanical research on the spine has increased at a rapid pace due to the prevalence of spinal pathologies in our society. The understanding of spine biomechanics has therefore become essential as it underlies the general methodologies of intervention in the clinical setting.

S. H. Brill · J. H. Chong
Tufts University, Medford, MA, USA

D. Kim
Rutgers New Jersey Medical School, Newark, NJ, USA

W. Cho (✉)
Department of Orthopaedic Surgery, Albert Einstein College of Medicine/Montefiore Medical Center, 3400 Bainbridge Ave, 6th Fl, Bronx, NY 10461, USA
e-mail: woojinchomd@aol.com

© The Author(s), under exclusive license to Springer Nature Singapore Pte Ltd. 2022

A. H. Choi and B. Ben-Nissan (eds.), *Innovative Bioceramics in Translational Medicine II*, Springer Series in Biomaterials Science and Engineering 18,
https://doi.org/10.1007/978-981-16-7439-6_10

The basic biomechanical goals of the spine include structural support, trunk movement, and protection of neural elements. Spine stability is a requirement for these biomechanical goals and is clinically important to prevent early mechanical deterioration of spinal components. Clinical spine stability is defined as the ability to maintain a normal pattern of displacement under physiological loads. Clinical instability is therefore the inability to limit excess or abnormal spinal displacement to prevent deformity, neurological injuries, or pain [1]. The Denis Three Column Theory suggests the stabilizing roles of the structures along the neutral axis of the spine can be divided into three areas: middle column (posterior longitudinal ligament, dorsal annulus fibrosus, dorsal wall of vertebral body), anterior column (anterior longitudinal ligament, ventral annulus fibrosus, ventral wall of vertebral body), and the posterior column (facet joints, supraspinous ligament, interspinous ligaments, ligamentum flavum) [2].

10.1.1 Biomechanics of Normal Spine

The spine consists of 33 stacked vertebrae and is categorized into five regions: 7 cervical, 12 thoracic, 5 lumbar, 5 sacral, and 4 coccygeal vertebrae. However, the sacral and coccygeal vertebrae are of lesser biomechanical importance because no motion is permitted between the vertebrae as they are fused together. The functional spinal unit (FSU) is the basic building block of the spine and consists of two adjacent vertebrae, an intervertebral disc, facet joints, and spinal ligaments [3].

The vertebra is a complex structure of bone and its relevant anatomy consists of a vertebral body, neural arch, and bony processes. As a vital structural component, one of the primary purposes of the vertebral body is to resist compressive loads. The posterior sides of the vertebral body and the neural arch form the vertebral canal as well, serving to house and protect the spinal cord along with other essential blood vessels. From the neural arch, spinous processes protrude to provide mechanical advantages to attached muscles and resist compressive loads especially during hyperextension [4]. These processes also form facet joints which are of great biomechanical importance because they are a major determinant of movement within individual spinal segments. Facet joints channel the movement of spinal segments and prevent abnormal movements of the spine such as rotational torsion and shear. Therefore, facet joints are not only vital posterior stabilizing structures but also contribute to the degree and plane of motion in the spine depending on its anatomical orientation and region [5].

Intervertebral discs are located between each vertebral body and the relevant anatomy of intervertebral discs includes an outer ring of fibrous tissue (annulus fibrosus) and the core of the disc (nucleus pulposus). The primary purpose of these discs is to transfer loads from one vertebra to the other. The core of the disc consists of highly viscoelastic mucoprotein gel and has a high resistance to compression due to its high fluid content. On the other hand, the outer ring of intervertebral discs consists of concentric and crisscrossing bands of collagenous tissue—these tissues prevent abnormal movements of the spine because they are resistant to torsion, shear, and rotational strains [3, 6].

The ligaments contribute to the stability of moving spinal segments as they respond to tensile forces and protect the spinal cord by limiting the motion of the spine. The intrinsic strength of ligaments varies based on the anatomical region and the length of the lever arm. There are several ligaments that contribute to the biomechanics of the spine: Vertebral bodies are connected through strong anterior longitudinal ligaments and weaker posterior longitudinal ligaments, and supraspinous ligaments are connected to spinous processes. Adjacent vertebrae are connected through interspinous ligaments, intertransverse ligaments, and the ligamenta flava. Most ligaments are composed of collagen fibers that provide minimal movement but the ligamentum flavum is unique due to its high proportion of elastic fibers. These fibers are constantly in tension, therefore enhancing overall stability [7].

The motion and stability at each spinal level vary due to the unique anatomy of each FSU. Overall, the spine contains four curves in the sagittal plane: 2 primary curves which concave posteriorly (thoracic and sacral), and 2 secondary curves which concave anteriorly (cervical and lumbar). The curves serve a mechanical advantage by absorbing shock and other loads as compared to a straight spine [8]. In addition to the curves of the spine, recent studies have shown that the rib-cage accounts for a great percentage of thoracic stability as well [9]. Generally, there is a progressive increase in the size of vertebrae and intervertebral discs from the cervical to the lumbar region of the spine—this serves a functional purpose because the increase in surface area of each vertebra reduces the stress subjected upon it. The vertebrae in the lower regions of the spine must bear more compressive loads than any other region and therefore are the largest [3].

Each FSU is also considered to be a motion segment and each has three joints. Although the permitted motion per motion segment is generally limited, spinal movements always involve several motion segments. As a whole, the spine allows movement in all three planes of movement and can be organized into four categories: extension, flexion, rotation, and lateral bending [10].

The range of motion (ROM) is the greatest in the cervical spine. C1 and C2 levels are specialized due to their unique anatomy and function. The atlas-occipital joint, the region between the occiput of the skull and the first cervical vertebrae, is extremely stable and provides a large degree of flexion and extension. However, motion in any other plane is not permitted in this joint. Rather the atlanto-axial joint, the joint between C1 and C2, provides a large degree of axial rotation as well as lateral bending and extension to a certain degree [6, 11].

The ROM in the thoracic and lumbar region is much more limited than the cervical region, but still provides motion in all three planes to a certain degree. The extension and flexion of the trunk can mainly be attributed to the range of motion of the lumbar region and the thoracic region to a lesser degree. On the other hand, lateral flexion and spinal rotation can mainly be attributed to the thoracic region and the lumbar region to a lesser degree. Lateral flexion is higher in lower segments of the thoracic region whereas rotation is more permitted in the upper segments [3, 5, 6].

10.1.2 Biomechanics of Abnormal Spine: Spinal Instability

Spinal instability is a highly prevalent issue in our society and common issues such as generalized lower back pain is experienced by 85% of the population [3]. Spinal stability and its natural biomechanics can be affected adversely by congenital causes, degeneration, and trauma to spinal components. However, the etiology of spinal instability is not well understood and the best diagnostic/treatment approaches are up to debate.

Spinal instability can occur due to congenital factors especially when it comes to spinal deformity. Proper spinal curvature is essential for bearing loads on the spine and to minimize deterioration of spinal components [8]. However, spinal curvature varies among individuals and can be caused congenitally although pathologic conditions and other environmental factors can be factors as well. Abnormal spine curvatures include lordosis, kyphosis, and scoliosis. Lordosis is characterized as the excessive curvature of the lumbar region, kyphosis is characterized as the excessive curvature of the thoracic region, and scoliosis is characterized by lateral deviations of the spine coupled with rotational deformities of affected vertebrae. Abnormal spinal curves are of clinical importance as improper curves can cause asymmetric and increased stress on spinal components. For example, significant lordotic curves can increase compressive stress on the posterior spine and are a risk factor for lower back pain [8, 12].

Besides congenital deformities, spinal instability can also be attributed to the degeneration of the spine and is caused by age-related factors, a multitude of environmental factors, pathological conditions, and trauma. Specifically, the lumbar region of the spine is often the subject of degeneration as it experiences the most mechanical stress and compressive loads compared to any other region of the spine. The deterioration of spinal components can occur naturally with age. An age-related decrease in water content of discs and bone density limits the range of the spine and its load-carrying capacity [10, 13]. Although there are many environmental factors contributing to the degeneration of the spine as well, they are generally due to excessive mechanical stress on the spine. Excessive stress can be caused by carrying large or repetitive loads, excessive movements, and bad postural habits, which is why the incidence of lower back pain is very high among athletes, physically demanding occupations, and obese patients [3, 14].

Common degenerative pathologies include disc degeneration, facet joint degeneration, and adjacent segment disease (ASD). Disc degeneration is characterized by the formation of tears in the annulus fibrosus, fraying, and dehydration of the nucleus pulposus. Degeneration of discs occurs naturally but also can also be expedited by alteration of mechanical forces distribution across the FSU [10, 13]. The biomechanics of instability of the spine caused by disc degeneration are complex, however, it is usually followed by instability, abnormal joint movement, and increased ROM. Disc degeneration also causes an increase in loads on facet joints and often precedes facet joint degeneration. Increased demands on facet joints can produce metaplasia, capsular hypertrophy, and bone spurs. Osteoarthritis caused by degeneration of facet joints can result in abnormal motion of motion segments and alter mechanical forces experienced by intervertebral discs. Long-term instability of the spine can result in more complex issues such as ASD, which is characterized by general degeneration of mobile spinal segments [15]. A multitude of issues can be caused by unstable motion segments such as disc herniations, listhesis, stenosis, etc. The relationship between the degeneration of various spinal components is complex. However, there is evidence that degenerative spinal instability originates from disorders of movement triggered by disc degeneration, followed by bony or articular abnormalities. In turn, these abnormalities extend to other joints at the same level and then to adjacent segments [16].

Traumatic injuries can also contribute to the deterioration of the spine. Common traumatic injuries of the spine include soft-tissue injuries, vertebral fractures, and disc herniations. Soft-tissue injuries are the most common cause of back-pain. These types of injuries include contusions, muscle strains, and ligament strains and are caused by a blow or overloading of the muscles [3]. Vertebral fractures can occur in many different ways. Compression and burst fractures occur through large compressive loads and can cause deterioration of vertebral end-plates [17]. The transverse and spinous processes are prone to acute fractures as they are the most exposed and can be caused by excessive use of attached muscles or trauma. Stress fractures often occur in the pars interarticularis, the weakest point of the neural arch. A fracture of the pars is referred to as spondylolysis. A bilateral separation of the pars results in spondylolisthesis and results in the anterior displacement of vertebrae relative to the vertebrae below it. Disc herniations can also be caused by trauma and is caused by the protrusion of the nucleus pulposus from the annulus [10]. Other factors such as cancer tissues and osteoporosis can contribute to the risk of fractures. Although there are many different biomechanical outcomes from traumatic injuries, in the clinical setting, instability is determined by trauma in at least two of three columns according to the Denis three-column theory [17].

10.1.3 Clinical Interventions

The non-surgical intervention of the treatment of back pain includes rest, physical therapy, and anti-inflammatory medications. However, when conservative treatment fails to treat pain or abnormal movement, surgical intervention is often required to regain a functional lifestyle. Surgical interventions include fusion, motion-preservation, decompression, and minimally invasive surgeries although interventions involving the use of instrumentation have the most drastic biomechanical impact on the spine. The basic goals of instrumentation include realigning the vertebrae, maintaining stability, and promoting fusion [18].

Surgical interventions including instrumentation can be categorized into fusion and motion preservation. In the last few decades, surgical stabilization of the spine through spinal fusion has become an increasingly common practice. The complexity of fusion has produced novel challenges where understanding the biomechanical nature of the spine has become essential. Indications of spinal fusion include joint pain, spine deformity, and degeneration of bony elements, soft tissues, and intervertebral discs [19]. Spinal fusion can be done in many different ways but can generally be categorized by posterior and anterior instrumentation. Instrumentation used for fusion is typically composed of longitudinal elements, vertebral attachments, and elements that join longitudinal elements together. Features such as natural mobility, facet joint orientation, size, and bone density must be considered before instrumentation [2, 20].

Common instrumentation used in fusion surgeries includes plates or rods with pedicle screws, facet screws, transverse connectors, and interbody cages. Pedicle-screw-based instrumentation is the most common form of instrumentation for fusion [21]. Posteriorly, screws are placed in the pedicles of vertebrae that can span from single to multiple segments and are connected by rods or plates. Crossbars and transverse connectors can be used for additional stability. Facet screws can also be used as an alternative for posterior instrumentation although it is not as effective [22]. In order to support the anterior column, interbody cages are commonly used and have replaced the need for anterior fusion due to their efficacy. Interbody cages are placed along the weight-bearing axis of the spine. They provide anterior column support to fusion sites and re-establish load transmissions along the spinal column [23–25].

Although fusion remains a gold standard for spine stabilization, it can result in many complications such as ASD, long postoperative rehabilitation, and other morbidities [26]. Therefore, many non-fusion techniques have been developed for motion preservation. Total disc arthroplasty is a procedure that completely replaces degenerative discs with an artificial implant. The purpose of doing so is to restore disc height, avoid ASD, faster recovery, and preservation of motion [27]. Dynamic stabilization systems have also been developed that mechanically restrict the motion of segments within a safe range [2].

10.2 Biomaterials

The term biomaterials encompasses all synthetic and natural materials which are used during orthopaedic procedures. Each biomaterial has structural and material properties. Structural characteristics differ from strength characteristics, as they not only depend on the material used, but also on the structural configuration of the object. The first term to consider is a load is a force, which acts on a body. Second is stress, which is defined as the intensity of an internal force and is calculated by force divided by area with the units Pascals or Newtons/Meters^2. Third is strain, which is defined as a relative measure of the deformation of the object and is calculated by the change in length divided by the original length of the object.

10.2.1 Biomaterials: Structural Properties

One important structural property is bending rigidity or stiffness. Stiffness is a measure of a material's ability to return to its original shape after a load is removed. This value is defined as the slope of the curve in the elastic range on a structure stress-strain curve. For a solid cylinder, bending rigidity, or stiffness is proportional to the radius of the cylinder, raised to the fourth power. For a hollow cylinder, the bending rigidity is proportional to the radius of the cylinder raised to the third power. For a rectangular object, the bending rigidity is proportional to the base multiplied by the height raised to the third power. Another structural property is the area moment of inertia (I) is closely related to bending rigidity. It is a function of structure width, thickness, and polar moment of inertia (J). The polar moment of inertia represents an object's resistance to torsion. The last structural property is deflection. Deflection is proportional to the applied force divided by the elastic modulus all multiplied by the area moment of inertia.

10.2.2 Mechanical Properties

When a load is applied to any biomaterial, two kinds of deformations can occur before fracture. The first is elastic deformation. This deformity is a reversible change in shape due to load in which the material returns to its original shape when the load is removed. The second is plastic deformation which creates an irreversible change in the shape of a material in which the material does not return to its original shape when the load is removed. When considering the type of material that is used, one must consider the elastic zone and the plastic zone which are separated on the stress versus strain curve by the yield point or proportional limit, which is defined as the transition point between elastic and plastic deformation.

Every material obeys Hooke's law when in the elastic zone as the amount of stress it can handle is proportional to the amount of strain it can handle. Each material's measure of stiffness, or ability to resist deformation is calculated by measuring the slope in the elastic zone of the material's stress versus strain graph (Fig. 10.1). This is defined by Young's modulus of elasticity which measures the stiffness, or ability to resist deformation, of a material in the elastic zone. This value can be calculated by measuring the slope of the stress versus strain curve in the elastic zone for that given material. A higher Young's modulus of elasticity directly correlates to a stiffer material. On the stress versus strain plot, the elastic zone is separated from the plastic zone by the yield point or the proportional limit.

The yield point is the point on the curve for which plastic deformation begins to occur. In the plastic zone, the material will not return to its original shape for a given amount of stress. In this region, there is a breaking point at which the object fails and breaks. Before the breaking point on the graph is the point of ultimate strength,

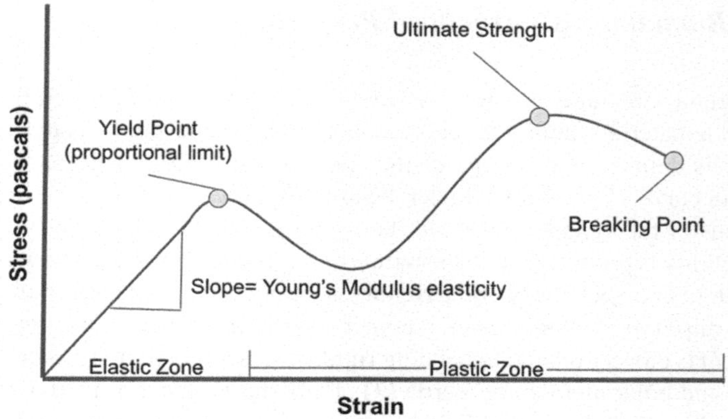

Fig. 10.1 The stress versus strain curve for a material's strength can be derived by axially loading an object into a tensile strength testing machine and plotting the curve

which is defined as the load to failure. This point represents the maximum amount of stress coupled with the strain that a material can handle before failing as a device or fracturing.

Toughness is defined as the amount of energy per volume that a material can absorb before failure. This is calculated by taking the area underneath the stress versus strain curve with the units of joules per meter cubed. Creep is defined as increased load-deformation with time under constant load. Load relaxation is a decrease in applied stress under conditions of constant strain. Hysteresis or energy dissipation is the characteristic of viscoelastic materials where the loading curve does not follow the unloading curve.

10.2.3 Material Descriptions

Based on the mechanical and structural properties of a material, one can apply various descriptions based on the behavior of the material when loads are applied. A brittle material is a material which exhibits elastic deformity and linear stress and strain curve until its yield point. For a brittle material, the yield point is analogous to the point of failure because it will break entirely. Two examples of brittle materials are polymethylmethacrylate (PMMA), which has a relatively high Young's modulus of elasticity and ceramics such as Al_2O_3 (Fig. 10.2).

Viscoelastic materials exhibit a stress-strain relationship which depends on the duration of time that the load is applied to the material and the strain rate, which is the rate that the load is applied. Viscoelastic materials include ligaments and bones. Isotropic materials possess the same mechanical properties in all directions due to their symmetry around a given point. One example of an isotropic material is a golf ball. Anisotropic materials possess different mechanical properties and are the

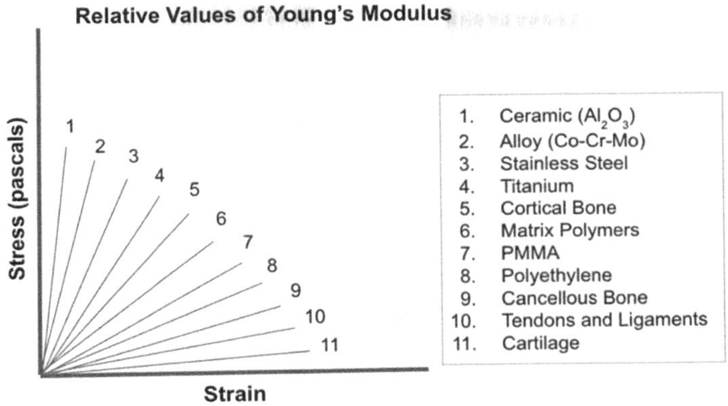

Fig. 10.2 The relative values of Young's modulus as shown on the stress versus strain graph

opposite of isotropic materials as they have different mechanical properties based on the direction of the applied load. Bones and ligaments are anisotropic for this reason. Ductile material is defined by its ability to undergo significant amounts of plastic deformation before failure. Some examples of ductile materials include metals such as cobalt alloys, stainless steel, and titanium. These are typically used for fracture plates, screws, femoral stems, and intramedullary nails.

10.2.4 Metal Characteristics

Metals also have specific terms associated with their properties. Fatigue failure is the failure of material before the ultimate tensile strength point due to repeated loads being applied to the material. This will often depend on the magnitude and frequency of the stress that has been previously applied to the material. The endurance limit is the maximal stress that a material can withstand before it is no longer immune to fatigue failure, regardless of the number of cycles.

One of the most important characteristics to consider when choosing a metal to use is corrosion or the chemical dissolving of metal. There are various types of corrosion as well. Galvanic corrosion is when dissimilar metals lead to electrochemical destruction of the material or implant. The highest risk of galvanic corrosion occurs when 316: stainless steel is mixed with cobalt-chromium. The chances of galvanic corrosion occurring can be reduced by using similar metals. Another type of corrosion is crevice corrosion which occurs in fatigue cracks due to differences in oxygen tension, or the differences in the partial pressure of oxygen, within the surrounding environment. The fatigue resistance of titanium alloys is lower than that of other available rod materials including Co-Cr [28]. 316L stainless steel is the most prone to crevice corrosion of the current metals used for biological implants. Fretting corrosion is a mode of destruction of the implant at the contact site from the relative motion of two materials or two components of the materials. This is clinically significant due to its frequency of occurrence at the head-neck junction of hip arthroplasty. Fretting corrosion is also the most common cause of mid-stem failure in modular revision type stems. There is an increased risk of fretting corrosion with the increased number of interfaces between the various metals.

Cobalt alloys present many advantages including their strength and their superior resistance to corrosion over stainless steel. Stainless steel and Co–Cr are protected by a chromium oxide layer whereas titanium is protected by titanium oxide [29]. Co–Cr alloys have also shown low ion release and better biocompatibility than other samples [30]. Co–Cr and ultrahigh strength stainless steel rods can produce the highest correction forces for rigid scoliosis, but both can plastically deform, making it important to have a know the quality of bone fixation [31]. Stainless steel alloys which are composed typically of iron-carbon alloys can also contain small amounts of chromium, molybdenum, manganese, and nickel. Stainless steel alloys are very stiff and fracture resistant which contributes to their use. However, they are susceptible to corrosion and due to their superior stiffness, they sometimes cause stress shielding of bone which can lead to a reduction in bone density. Furthermore, a remarkable chromium release without any clinical-radiologic sign was recorded in some female patients raising concern for young fertile women [32].

Titanium alloys are very biocompatible due to their corrosion resistance. This is due to the fact that they form an adherent oxide coating through self-passivation. Titanium's low modulus of elasticity also makes it more similar to biological materials such as cortical bone which increases its effectiveness. Titanium does present disadvantages such as poor resistance to wear and generates more metal debris than cobalt chrome. The fatigue resistance of titanium alloys is lower than many other commercially available rod [28]. Therefore, improvements are of paramount importance to minimize imperfections and withstand greater loads without cracking.

10.2.5 Non-metal Characteristics

Some specific non-metals have large clinical significance. One such material is PMMA which is commonly referred to as bone cement and is often used for fixation and load distribution in conjunction with orthopaedic implants. It functions by interlocking bones and can be used to fill tumor defects or minimize local recurrence. It is a two-component material with powder and liquid. It presents the advantages of reaching its ultimate strength at 24 h, high strength in compression, and a Young's modulus between cortical and cancellous bone. However, PMMA has relatively poor tensile and shear strength. In a study conducted by Cho et al. concluded that pedicle screws could be easily and safely backed out after augmentation with PMMA or calcium phosphate cement before any bony ingrowth [33].

An alternative would be silicones, which can be used for replacement in non-weight bearing joints. Silicones have less clinical significance as they have poor strength and wear capability that can cause high occurrences of synovitis. Ceramics are another material which present advantages such as high compressive strength and superior wear characteristics. Ceramics are also limited by their high Young's modulus, low tensile strength, poor crack resistance characteristics, low fracture toughness, and brittle character.

10.2.6 Surgical Implications

Despite its relatively low incidence, recent literature has discussed the drawback of bacterial adherence of biofilm formation on implants which can cause infection on various spinal implant surfaces, due to the significance of serious infection. Ha et al. published their results which state that biofilm-forming *Staphylococcus epidermis* showed heavy adhesion across all rough and smooth surfaces for both stainless steel and titanium alloys [34]. On scanning electron microscopy, there was a significantly greater number of aggregated microcolonies with thick biofilm in biofilm-forming *S. epidermidis*, and much fewer colonies in nonbiofilm-forming *S. epidermidis*. *M. tuberculosis* rarely adhered to the metal surfaces and did not show biofilm formation. Therefore, it is less likely that *M. tubuerculosis* will form infection than *S. epidermidis* on implant surfaces. Further studies have studied the effects of the antimicrobial effects of modified titanium pedicle screws with methicillin-resistant *Staphylococcus aureaus* (MRSA). The novel experimental study completed by Hazer et al. concluded that modified titanium pedicle screws were shown to have antimicrobial effects especially after additional inhibition of biofilm formation [35].

Other research has centered on implants, which can support growth around the implant margins. Polyetheretherketone (PEEK) devices have been widely used for cervical discectomy and fusion, but they are limited in the surface area for bone attachment. Furthermore, although rare, PEEK implants may cause allergic reactions [36]. PEEK implants have also reduced osteoblastic differentiation of progenitor cells and produced an inflammatory environment that favors cell death via apoptosis and necrosis [37]. In a study conducted by Sinclair *et al.*, they found that the porous tantalum implants supported bone growth into and around the implant margins more effectively than the PEEK implants [38]. Biomaterials such as PEEK may actually hinder bone healing [39]. Porous tantalum implants may be advantageous because of their open cell porous structure, which facilitates host bone ingrowth and bone bridging through these devices.

Traditionally, dexamethasone or biological signals (BMPs) have been required for osteogenic differentiation of stem cells [39]. Recent evidence has suggested that nano topography alone can alter cell morphology, adhesion, motility, and proliferation. From this, we can understand that by integrating trabecular metals, applying nanoparticulate sprays, etching metal surfaces, embedding ceramics, depositing protein onto inorganic scaffolds or implants, and implementing complex matrix changes, microenvironmental cues can be changed and improved.

When considering a material for interbody cages, one must consider their clinical outcomes. The most widely used interbody fusion method uses the insertion of a titanium or PEEK cage filled with autogenous bone into the disc space after removing the disc. However, PEEK is a hydrophobic material. This issue can be solved by using a plasma-sprayed titanium coating which can improve the mechanical properties in the cortical sites of the surgery [40]. However, these titanium coated PEEK implants can easily lose coating material, and are susceptible to impaction related wear debris [41]. Another option, as indicated by Olivares-Navarrete et al. is to modify the surface structure, as this is sufficient to create an osteogenic environment without the use of exogenous factors as this may induce better and faster bone during interbody fusion [42]. Other materials include $CaO–SiO_2–P_2O_5–B_2O_3$ glass ceramics, which can naturally improve the osteoblastic differentiation of human mesenchymal stem cells [43]. In a clinical study by Lee *et al.*, they concluded that $CaO–SiO_2–P_2O_5–B_2O_3$ glass ceramics spacer showed a similar fusion rates and clinical outcomes when compared with titanium cages [44]. Despite these viable materials, carbon fiber-reinforced polymer (CFRP) cages should be avoided due to their failure in patients with rheumatoid arthritis which can lead to significant morbidity [45].

10.3 Three-Dimensional (3D)-Printing

Three-dimensional (3D) printing has been a transformative technology for the field of medicine. Its implications can reach more specifically into orthopaedic spine surgery. 3D printing has already been used for surgical planning of complex spine surgeries since the 1990s [46]. The models created by 3D printing have since been expanded to both the preoperative and intraoperative settings.

In a preoperative setting, 3D printing is currently being used to create anatomically accurate and precise models of spinal deformities to educate residents and patients. These models are also being used for preoperative surgical planning. They allow for easier visualization and simulation for complex spinal pathologies. Although computed tomography (CT) and magnetic resonance imaging (MRI) are both used clinically, they are limited by their two-dimensional images. Tangible 3D visual representations provide a better depiction than any two-dimensional image. Intra-operatively, 3D printing can be used in other ways that can aid the surgeon. These methods include the creation of surgical guidance systems, templates, and customized patient-specific implants. 3D printing provides significant utility for surgeons.

Advantages of 3D printing include improved patient outcomes, decreased radiation exposure for patients, and increased patient and resident education. However, 3D printing currently faces challenges which prevent its broader application in clinical practice. 3D printing remains an intriguing technology for orthopaedic surgery, and it has the potential to vastly improve the practice of orthopaedic spine surgery in the near future [47].

10.3.1 Printing Techniques and Tissue Engineering Applications

The first of various useful printing techniques is additive manufacturing, colloquially referred to as 3D printing or rapid prototyping. The process begins with two-dimensional images from CT or MRI image data and are converted to 3D models using software which can create a specialized multiplanar model. The computer then uses this model to print a 3D image using thin cross-sections layered on top of one another. Since the earliest version of 3D printing, technology has developed significantly. Additional processes have been developed for a wide range of applications including non-clinical settings as well as surgical simulations and personalized surgery and tissue engineering.

The technique of 3D printing involves building a model from cross sectional layers. The technique has been advantageous in fabricating scaffold with microscopic internal channels because the surrounding powders act as the supporting material [48]. It has allowed for phosphate and strontium-containing scaffolds which enable the stimulation of blood vessel formation and osteoblast proliferation [49, 50]. Various studies have demonstrated the efficacy of using scaffolds to support tissue development, enable cell growth, and use as potential bone implantation [51–54].

Another technique which has proven useful in 3D printing is fused deposition modeling (FDM), which deposits molten thermoplastic materials into cross-sectional layers [48]. Recently, a scaffold has been created with biocompatible building material. Such biocompatible materials have been able to support FDM-generated scaffolds in bone tissue through the use of hydroxyapatite which promoted osteoconductive properties [48, 55]. Other recently developed 3D printing techniques include selective laser sintering and 3D plotting; both of which have demonstrated excellent potential in reconstructing tissue, especially with regards to bone tissue [56–58].

10.3.2 Accuracy of 3D Printing

In order for 3D printing to have clinical utility in orthopaedic spine surgery, outstanding accuracy and precision are necessary in order to create representations of true anatomic structures [47]. Studies have confirmed the potential for 3D-printed media to provide such accuracy in representing the anatomy of the spine. Wu *et al.* compared the CT images of cervical, thoracic, and lumbar vertebrae against their respective 3D-printed models and found a strong correlation between the anatomy of the two [59]. McMenamin et al. found that 3D-printed models could accurately represent air and fluid filled negative spaces in addition to accurately representing radiologic data [60]. The various modern printing techniques allow for the creation of models which can represent the true density of bone through the use of high calcium content [61]. Modern printing techniques allow for the creation of models with high calcium content to accurately mimic the true density of bone [62].

10.3.3 Preoperative Planning Applications

One previously discussed advantage provided by 3D printing includes the use of spine models. By using these models, physicians can realistically simulate surgery in preoperative planning, which can be advantageous in preparing for complex spinal pathologies. Xiao *et al.* was able to perform complex en block resections of primary cervical tumors through using 3D printed spine models to prepare and refer to as a visual reference [62]. In another case, 3D models were utilized to study inclination of joints, false articulations, pedicle sizes, and vertebral artery courses preoperatively in craniovertebral surgery. Data from the models allowed physicians to calculate screw and plate sizes as well as the angle of screw insertion preoperatively [63]. Personalized 3D printing has also been used to create surgical planning models and patient-specific spinal instrumentation in the correction of spinal deformity in children with meningomyelocele [64]. Preoperative modeling for surgical planning was also used in surgery of a thoracolumbar fracture with dislocation by providing a model of sagittal curves. It also provided navigation templates for pedicle screws. When completed with a 3D model, the procedure was performed with shorter operation time, less intraoperative blood loss, better recovery of thoracolumbar dislocation, and better Frankel classification, all of which are clear advantages [65]. Reduced operation time and perioperative blood loss have also been reported through the use of rapid prototyping of 3D models in surgical planning for lumbar discectomy [66]. Despite these advantages, no significant differences were observed in complication rate, length of hospital stay, postoperative radiological outcomes, or pedicle screw misalignments between surgeries performed with a 3D printed model and those without one [67].

Furthermore, there are concerns and limitations associated with the introduction of 3D printing to the preoperative setting. Although studies show multiple advantages of the utility of 3D-printed models, this additional step may be unnecessary. Regardless of the fact that models may provide reduced operation time, physicians must consider the time required to design and construct the preoperative model, especially when considering reconstruction after trauma.

Advancements in the speed and availability of rapid prototyping technology may allow for more widespread use in the trauma setting. Consequently, 3D printers will be most useful in complex cases.

10.3.4 Trainee and Patient Education Applications

Before the development of 3D printing, cadaveric models were traditionally used for educational purposes and surgical simulation. However, these models are limited in their usage due to constraints regarding the quantity of models which exist in addition to health and safety issues [59]. 3D-printed spine models could provide an alternative to cadaveric models. Due to their customizability and ability to resemble complex cases, 3D-printed models will inspire more confidence in residents as they will be able to perform those surgeries much more effectively. These 3D-printed models have a promising future in supplying surgical residents and practitioners with excellent training simulations of surgical experiences [68]. Hughes et al. reported the utility of 3D models in navigating complex deformities and improving anatomical understanding for training [69]. 3D-printed spine models have also been successfully used in the informed consent process through the use of patient-specific 3D-printed spine models to help guide patients in making decisions by educating them on their own spinal pathology and potential surgery [70, 71].

10.3.5 Intraoperative Applications: Guidance Systems

One of the primary clinical applications of 3D printing in spine surgery is the creation of intraoperative guidance systems for the insertion of pedicle screws. Computer software can design navigation templates for pedicle screw fixation for intraoperative use that can be 3D-printed preoperatively through the use of CT thin-slice data of the patient's spine [72, 73]. 3D-printed templates also present advantages in reducing the risk of complications during fixation of cervical pedicle screws (CPS) due to narrow clearance and increased risk for neural and vascular injury. Multiple studies comparing pedicle screw insertion via 3D template versus fluoroscopy indicate increased accuracy and precision of placement, which in turn leads to reduced complications. 3D template guided placement of pedicle screws has also been conducted on thoracic and lumbar pedicle screw fixation and has proven to be equally accurate, safe, and convenient [74–76]. As a result, authors agree that the reduced operating time, ease of use, moderate cost, and ability to insert cervical screws without radiation provide additional benefits for intra-operative use [72–74].

Despite the benefits, there are multiple limitations preventing the widespread use of 3D printing in the intraoperative setting. Using 3D technology requires a specialist who understands managing 3D software to perform segmentation of a 3D model for printing. As previously mentioned, creating these devices is also a time intensive process which has limited its use in a hospital setting. Furthermore, the modeling of these devices requires an additional cost for the patient for surgery. As technology becomes more user-friendly, the process for device creation will become more streamlined. Although imaging processing and printing can take several hours before surgery, the reduced time spent in the operating room is extremely valuable. One review of intraoperative 3D guidance systems concluded that 10 min saved in the operating room is equivalent in monetary terms to one hour spent creating the guidance devices [77].

10.3.6 Intraoperative Applications: Implants

Another application of 3D-printing in spine surgery can be through the use of implants. Because commercially available implants will not always match with the patient specifically, 3D printing encourages the advancement of personalized surgery. Multiple cases have reported successful use of patient specific 3D-printed implants which have been used in spinal fusion for complex spinal pathologies. These 3D implants include posterior fixation implants and anterior and posterior intervertebral fusion cage implants [78–81]. The implants all showed improved load bearing surface, lowered the rate of implant dislocation and subsidence, provided excellent primary stabilization, shortened time of procedure, improved correction of deformity, decreased blood loss, and reduced risk of neurovascular compromise [64, 77, 81]. Although the technology allows the creation of patient-specific spinal implants, there is very limited literature on the subject.

There are limitations to the use of 3D-printed implants which include a lack of long-term follow-up data for post-surgical clinical outcomes of surgeries. Data which supports the use of 3D printing in implants must be used in order to justify its utilization, and for the safety of patients, the Food and Drug Administration must provide approval before inserting 3D-printed implants into the patients' body.

References

1. Izzo R, Guarnieri G, Guglielmi G et al (2012) Biomechanics of the spine. Part I: spinal stability. Eur J Radiol 82:118–126
2. Nouh MR (2012) Spinal fusion-hardware construct: Basic concepts and imaging review. World J Radiol 4:193–207
3. Hall SJ (2012) The biomechanics of the human spine. In: Hall SJ (ed) Basic Biomechanics, 6th edn. McGraw-Hill, New York, pp 275–318
4. Inoue N, Orías AAE, Segami K (2019) Biomechanics of the Lumbar Facet Joint. Spine Surg Relat Res 4:1–7
5. Jaumard NV, Welch WC, Winkelstein BA (2011) Spinal facet joint biomechanics and mechanotransduction in normal, injury and degenerative conditions. J Biomech Eng 133:071010
6. Oxland TR (2016) Fundamental biomechanics of the spine—What we have learned in the past 25 years and future directions. J Biomech 49:817–832
7. Sharma M, Langrana NA, Rodriguez J (1995) Role of ligaments and facets in lumbar spinal stability. Spine 20:887–900
8. Le Huec JC, Thompson W, Mohsinaly Y et al (2019) Sagittal balance of the spine. Eur Spine J 28:1889–1905
9. Brasiliense LB, Lazaro BC, Reyes PM et al (2011) Biomechanical contribution of the rib cage to thoracic stability. Spine 36:E1686-E1693
10. Adams MA, Dolan P (2005) Spine biomechanics. J Biomech 38:1972–1983
11. Swartz EE, Floyd RT, Cendoma M (2005) Cervical spine functional anatomy and the biomechanics of injury due to compressive loading. J Athl Train 40:155–161
12. Lamartina C, Berjano P (2014) Classification of sagittal imbalance based on spinal alignment and compensatory mechanisms. Eur Spine J 23:1177–1189
13. Singh K, Phillips FM (2005) The biomechanics and biology of the spinal degenerative cascade. Semin Spine Surg 17:128–136

14. Zielinska N, Podgórski M, Haładaj R et al (2021) Risk factors of intervertebral disc pathology-a point of view formerly and today-a review. J Clin Med 10:409
15. Iorio JA, Jakoi AM, Singla A (2016) Biomechanics of degenerative spinal disorders. Asian Spine J 10:377–384
16. Izzo R, Guarnieri G, Guglielmi G et al (2013) Biomechanics of the spine. Part II: spinal instability. Eur J Radiol 82:127–138
17. Whyne CM, McLachlin S, Burke M et al (2017) Biomechanics of vertebral fracture. In: Manfrè L (ed) Vertebral Lesions. Springer, Cham, pp 31–61
18. Cho W, Wang W, Bucklen B (2020) The role of sagittal alignment in predicting major failure of lumbopelvic instrumentation: a biomechanical validation of lumbopelvic failure classification. Spine Deform 8:561–568
19. Reid PC, Morr S, Kaiser MG (2019) State of the union: a review of lumbar fusion indications and techniques for degenerative spine disease. J Neurosurg Spine 31:1–14
20. Dall BE, Eden SV, Cho W et al (2019) Biomechanical analysis of motion following sacroiliac joint fusion using lateral sacroiliac screws with or without lumbosacral instrumented fusion. Clin Biomech 68:182–189
21. Cho W, Cho SK, Wu C (2010) The biomechanics of pedicle screw-based instrumentation. J Bone Joint Surg Br 92:1061–1065
22. Rutherford EE, Tarplett LJ, Davies EM et al (2007) Lumbar spine fusion and stabilization: hardware, techniques, and imaging appearances. Radiographics 27:1737–1749
23. Cho W, Wang W, Bucklen B et al (2021) The role of biological fusion and anterior column support in a long lumbopelvic spinal fixation and its effect on the S1 screw-an in silico biomechanics analysis. Spine 46:E250–E256
24. Ferrara LA, Goel VK (2010) The biomechanics of spinal fusion. ArgoSpine News J 22:57–61
25. Cho W, Wu C, Mehbod AA et al (2008) Comparison of cage designs for transforaminal lumbar interbody fusion: a biomechanical study. Clin Biomech 23:979–985
26. Lee CS, Hwang CJ, Lee SW et al (2009) Risk factors for adjacent segment disease after lumbar fusion. Eur Spine J 18:1637–1643
27. Othman YA, Verma R, Qureshi SA (2019) Artificial disc replacement in spine surgery. Ann Transl Med 7(Suppl 5):S170
28. Yamanaka K, Mori M, Yamazaki K et al (2015) Analysis of the fracture mechanism of ti-6al-4v alloy rods that failed clinically after spinal instrumentation surgery. Spine 40:E767–E773
29. Yoshihara H (2013) Rods in spinal surgery: a review of the literature. Spine J 13:1350–1358
30. Xin XZ, Xiang N, Chen J et al (2012) In vitro biocompatibility of Co–Cr alloy fabricated by selective laser melting or traditional casting techniques. Mater Lett 88:101–103
31. Serhan H, Mhatre D, Newton P et al (2013) Would CoCr rods provide better correctional forces than stainless steel or titanium for rigid scoliosis curves? J Spinal Disord Tech 26:E70-E74
32. Savarino L, Greggi T, Martikos K et al (2015) Long-term systemic metal distribution in patients with stainless steel spinal instrumentation: a case-control study. J Spinal Disord Tech 28:114–118
33. Cho W, Wu C, Zheng X et al (2011) Is it safe to back out pedicle screws after augmentation with polymethyl methacrylate or calcium phosphate cement? A biomechanical study. J Spinal Disord Tech 24:276–279
34. Ha KY, Chung YG, Ryoo SJ (2005) Adherence and biofilm formation of Staphylococcus epidermidis and Mycobacterium tuberculosis on various spinal implants. Spine 30:38–43
35. Hazer DB, Sakar M, Dere Y et al (2016) Antimicrobial effect of polymer-based silver nanoparticle coated pedicle screws: experimental research on biofilm inhibition in rabbits. Spine 41:E323-E329
36. Maldonado-Naranjo AL, Healy AT, Kalfas IH (2015) Polyetheretherketone (PEEK) intervertebral cage as a cause of chronic systemic allergy: a case report. Spine J 15:e1-e3
37. Olivares-Navarrete R, Hyzy SL, Slosar PJ et al (2015) Implant materials generate different peri-implant inflammatory factors: poly-ether-ether-ketone promotes fibrosis and microtextured titanium promotes osteogenic factors. Spine 40:399–404

38. Sinclair SK, Konz GJ, Dawson JM et al (2012) Host bone response to polyetheretherketone versus porous tantalum implants for cervical spinal fusion in a goat model. Spine 37:E571-E580
39. Chaput CD (2016) Optimization of scaffolds and surface-based treatments for orthopedic applications. Spine 41 Suppl 7:S14–S15
40. Walsh WR, Bertollo N, Christou C et al (2015) Plasma-sprayed titanium coating to polyetheretherketone improves the bone-implant interface. Spine J 15:1041–1049
41. Kienle A, Graf N, Wilke HJ (2016) Does impaction of titanium-coated interbody fusion cages into the disc space cause wear debris or delamination? Spine J 16:235–242
42. Olivares-Navarrete R, Gittens RA, Schneider JM et al (2012) Osteoblasts exhibit a more differentiated phenotype and increased bone morphogenetic protein production on titanium alloy substrates than on poly-ether-ether-ketone. Spine J 12:265–272
43. Lee JH, Seo JH, Lee KM et al (2013) Fabrication and evaluation of osteoblastic differentiation of human mesenchymal stem cells on novel CaO-SiO_2-P_2O_5-B_2O_3 glass-ceramics. Artif Organs 37:637–647
44. Lee JH, Kong CB, Yang JJ et al (2016) Comparison of fusion rate and clinical results between CaO-SiO_2-P_2O_5-B_2O_3 bioactive glass ceramics spacer with titanium cages in posterior lumbar interbody fusion. Spine J 16:1367–1376
45. Sardar Z, Jarzem P (2013) Failure of a carbon fiber-reinforced polymer implant used for transforaminal lumbar interbody fusion. Global Spine J 3:253–256
46. Grant C, Izatt M, Labrom R et al (2016) Use of 3D printing in complex spinal surgery: historical perspectives, current usage, and future directions. Tech Orthop 31:172–180
47. Cho W, Job AV, Chen J et al (2018) A review of current clinical applications of three-dimensional printing in spine surgery. Asian Spine J 12:171–177
48. Chia HN, Wu BM (2015) Recent advances in 3D printing of biomaterials. J Biol Eng 9:4
49. Zhang J, Zhao S, Zhu Y et al (2014) Three-dimensional printing of strontium-containing mesoporous bioactive glass scaffolds for bone regeneration. Acta Biomater 10:2269–2281
50. Tarafder S, Davies NM, Bandyopadhyay A et al (2013) 3D printed tricalcium phosphate scaffolds: Effect of SrO and MgO doping on in vivo osteogenesis in a rat distal femoral defect model. Biomater Sci 1:1250–1259
51. Qian C, Zhang F, Sun J (2015) Fabrication of Ti/HA composite and functionally graded implant by three-dimensional printing. Biomed Mater Eng 25:127–136
52. Wang Y, Li X, Wei Q et al (2015) Study on the mechanical properties of three-dimensional directly binding hydroxyapatite powder. Cell Biochem Biophys 72:289–295
53. Simon JL, Rekow ED, Thompson VP et al (2008) MicroCT analysis of hydroxyapatite bone repair scaffolds created via three-dimensional printing for evaluating the effects of scaffold architecture on bone ingrowth. J Biomed Mater Res A 85:371–377
54. Seitz H, Rieder W, Irsen S et al (2005) Three-dimensional printing of porous ceramic scaffolds for bone tissue engineering. J Biomed Mater Res B Appl Biomater 74:782–788
55. Kutikov AB, Gurijala A, Song J (2015) Rapid prototyping amphiphilic polymer/hydroxyapatite composite scaffolds with hydration-induced self-fixation behavior. Tissue Eng Part C Methods 21:229–241
56. Van Bael S, Desmet T, Chai YC et al (2013) In vitro cell-biological performance and structural characterization of selective laser sintered and plasma surface functionalized polycaprolactone scaffolds for bone regeneration. Mater Sci Eng C Mater Biol Appl 33:3404–3412
57. Kim K, Yeatts A, Dean D et al (2010) Stereolithographic bone scaffold design parameters: osteogenic differentiation and signal expression. Tissue Eng Part B Rev 16:523–539
58. Williams JM, Adewunmi A, Schek RM et al (2005) Bone tissue engineering using polycaprolactone scaffolds fabricated via selective laser sintering. Biomaterials 26:4817–4827
59. Wu AM, Shao ZX, Wang JS et al (2015) The accuracy of a method for printing three-dimensional spinal models. PLoS One 10:e0124291
60. McMenamin PG, Quayle MR, McHenry CR et al (2014) The production of anatomical teaching resources using three-dimensional (3D) printing technology. Anat Sci Educ 7:479–486
61. West SJ, Mari JM, Khan A et al (2014) Development of an ultrasound phantom for spinal injections with 3-dimensional printing. Reg Anesth Pain Med 39:429–433

62. Xiao JR, Huang WD, Yang XH et al (2016) En bloc resection of primary malignant bone tumor in the cervical spine based on 3-dimensional printing technology. Orthop Surg 8:171–178
63. Goel A, Jankharia B, Shah A et al (2016) Three-dimensional models: an emerging investigational revolution for craniovertebral junction surgery. J Neurosurg Spine 25:740–744
64. Karlin L, Weinstock P, Hedequist D et al (2017) The surgical treatment of spinal deformity in children with myelomeningocele: the role of personalized three-dimensional printed models. J Pediatr Orthop B 26:375–382
65. Wu C, Tan L, Lin X et al (2015) Clinical application of individualized reference model of sagittal curves and navigation templates of pedicle screw by three dimensional printing technique for thoracolumbar fracture with dislocation. Zhongguo Xiu Fu Chong Jian Wai Ke Za Zhi 29:1381–1388
66. Li C, Yang M, Xie Y et al (2015) Application of the polystyrene model made by 3-D printing rapid prototyping technology for operation planning in revision lumbar discectomy. J Orthop Sci 20:475–480
67. Yang M, Li C, Li Y et al (2015) Application of 3D rapid prototyping technology in posterior corrective surgery for Lenke 1 adolescent idiopathic scoliosis patients. Medicine 94:e582
68. Rehder R, Abd-El-Barr M, Hooten K et al (2016) The role of simulation in neurosurgery. Childs Nerv Syst 32:43–54
69. Hughes A, Soden P, Abdulkarim A et al (2014) The use of rapid prototyping and 3D printing in revision hip arthroplasty. Bone Joint J 96B(SUPP 10):2
70. Liew Y, Beveridge E, Demetriades AK et al (2015) 3D printing of patient-specific anatomy: A tool to improve patient consent and enhance imaging interpretation by trainees. Br J Neurosurg 29(5):712–714
71. Madrazo I, Zamorano C, Magallón E et al (2009) Stereolithography in spine pathology: a 2-case report. Surg Neurol 72:272–275; discussion 275
72. Guo F, Dai J, Zhang J et al (2017) Individualized 3D printing navigation template for pedicle screw fixation in upper cervical spine. PLoS One 12:e0171509
73. Deng T, Jiang M, Lei Q et al (2016) The accuracy and the safety of individualized 3D printing screws insertion templates for cervical screw insertion. Comput Assist Surg 21:143–149
74. Takemoto M, Fujibayashi S, Ota E et al (2016) Additive-manufactured patient-specific titanium templates for thoracic pedicle screw placement: novel design with reduced contact area. Eur Spine J 25:1698–1705
75. Chen H, Guo K, Yang H et al (2016) Thoracic pedicle screw placement guide plate produced by three-dimensional (3-D) laser printing. Med Sci Monit 22:1682–1686
76. Chen H, Wu D, Yang H et al (2015) Clinical use of 3D printing guide plate in posterior lumbar pedicle screw fixation. Med Sci Monit 21:3948–3954
77. Martelli N, Serrano C, van den Brink H et al (2016) Advantages and disadvantages of 3-dimensional printing in surgery: A systematic review. Surgery 159:1485–1500
78. Mobbs R, Coughlan M, Thompson R et al (2017) The utility of 3D printing for surgical planning and patient-specific implant design for complex spinal pathologies: case report. J Neurosurg Spine 26:513–518
79. Phan K, Sgro A, Maharaj MM et al (2016) Application of a 3D custom printed patient specific spinal implant for C1/2 arthrodesis. J Spine Surg 2:314–318
80. Serra T, Capelli C, Toumpaniari R et al (2016) Design and fabrication of 3D-printed anatomically shaped lumbar cage for intervertebral disc (IVD) degeneration treatment. Biofabrication 8:035001
81. Spetzger U, Frasca M, König SA (2016) Surgical planning, manufacturing and implantation of an individualized cervical fusion titanium cage using patient-specific data. Eur Spine J 25:2239–2246

Samuel H. Brill is a current undergraduate student at Tufts University with a plan to attend medical school upon graduation in 2023. He has worked on Dr. Woojin Cho's research team since the summer of his senior year of high school in 2018. Since that summer, Brill has been published in print in SPINE alongside a team of researchers for Infra-Adjacent Segment Disease after Lumbar Fusion: An Analysis of Pelvic Parameters. He was born and raised in New York City, but while he is at school volunteers as a patient escort at Massachusetts General Hospital.

Jee Ho Chong was born in Yeosu, South Korea and immigrated to Queens, New York when he was 1 year old. He is currently conducting spine research under Dr. Woojin Cho at Montefiore Medical Center, specifically investigating spinal deformity and degenerative conditions. As an aspiring physician, Jee Ho is passionate about health equity and diversity/inclusion in the healthcare field. While he is at school, Jee Ho also works as an EMT in the Boston area. He currently attends Tufts University as an undergraduate student studying Biopsychology and plans on attending medical school in 2023 upon graduation.

Dongyoung Kim is a pediatric resident at Washington University/St. Louis Children's Hospital. He completed his medical training at Rutgers New Jersey Medical School. During his medical school, he enrolled in a Scholar's research year and spent one year working with the Spine Deformity research team at Montefiore Medical Center. Following his residency training, he plans to pursue a fellowship in pediatric emergency medicine.

Woojin Cho Professor Cho joined the faculty of Montefiore Medical Center in 2013, specializing in adult thoracolumbar spine deformity and degenerative conditions with expertise in the treatment with all types of surgical technique. He is the Chief of Orthopedic Spine Surgery at Montefiore and an Associate Professor at Albert Einstein College of Medicine. He earned his medical degree at Hanyang University College of Medicine in South Korea, completed his orthopedic internship and residency at Hanyang University Hospital, and became a Korean orthopedic board-certified surgeon. Following his residency, he completed three years of spine clinical fellowship under Dr. Cho and Dr. Chang and earned a Ph.D. degree at Hanyang University College of Medicine.

Dr. Cho has extensive clinical training experience in the United States, which includes a research fellowship at Twin Cities Spine Center in Minneapolis, a clinical research fellowship at Washington University School of Medicine in St. Louis, and spine fellowships at the University of Virginia Health System and the Hospital for Special Surgery, respectively. Dr. Cho is a board-eligible (ABOS) orthopedic surgeon, a faculty member of the Scoliosis Research Society, the North American Spine Society, AOSpine North America, American Academy of Orthopedic Surgeons, American Medical Association, the Korean American Spine Society, the Korean Society of Spine Surgery, the Korean Spine Arthroplasty Society, the Pacific and Asian Society of Minimally Invasive Spine Surgery, the Korean Orthopedic Association, and the Korean Medical Association. In addition to having written more than 100 articles related to spine research in peer-reviewed publications, Dr. Cho has been a guest speaker at prestigious international meetings and, to date, has made more than 250 presentations.

Chapter 11
Spine Surgery—Part II: Ceramic and Non-ceramic Bone Substitutes: A Surgical Perspective

Sanghyo Lee, Matthew T. Morris, David A. Essig, and Woojin Cho

Abstract Bone grafts have been used for decades to achieve successful bone fusion in spinal surgeries. Autograft is the most effective bone graft due to the properties of osteogenesis, osteoconduction, and osteoinduction. However, autograft may not always be available in sufficient quantities, and harvesting may cause patient morbidity. Various ceramic and non-ceramic bone graft extenders have been introduced to limit the need for autograft harvest. These bone substitutes have rapidly evolved in recent years with technological and industrial advancements. Spinal surgeons should closely follow new trends in this industry to achieve the best outcomes for their patients.

Keywords Bone grafts · Hydroxyapatite · Tricalcium phosphate (TCP) · Calcium sulfate · Bioactive glass · Spinal fusion · Autograft · Allograft

11.1 Introduction

Spinal fusion surgeries have become treatments of choice for a variety of conditions, especially degenerative spinal diseases and spinal deformity. Fusion surgeries may consist of intervertebral body fusion, posterior fusion, or posterolateral fusion, and generally consist of instrumentation and bone graft transplantation on the decorticated bone in order to encourage a solid bony fusion. There are many available options currently available for bone graft, both commercially available and harvested from the patient.

S. Lee · W. Cho (✉)
Department of Orthopaedic Surgery, Albert Einstein College of Medicine/Montefiore Medical Center, 3400 Bainbridge Ave, 6th Fl., Bronx, NY 10461, USA
e-mail: woojinchomd@aol.com

M. T. Morris · D. A. Essig
Northwell Health at Long Island Jewish Medical Center, New Hyde Park, NY, USA

D. A. Essig
Donald and Barbara Zucker School of Medicine at Hofstra/Northwell, Hempstead, NY, USA

© The Author(s), under exclusive license to Springer Nature Singapore Pte Ltd. 2022
A. H. Choi and B. Ben-Nissan (eds.), *Innovative Bioceramics in Translational Medicine II*, Springer Series in Biomaterials Science and Engineering 18,
https://doi.org/10.1007/978-981-16-7439-6_11

The ideal bone graft contains properties of osteogenesis, osteoinduction, and osteoconduction. Osteogenesis refers to the presence of osteoblast precursor cells that directly contribute to the growth of new bones in grafts. Osteoinduction refers to the presence of molecular growth factors and signaling molecules that stimulate the migration of osteoblast precursors to graft sites to mature into osteoid-producing cells and increase bone matrix production. Lastly, osteoconduction refers to structural scaffolding in which the cells that produce the matrix can deposit new bones. Autograft is often regarded as the "gold standard" bone graft, as it contains all of the three of these properties to encourage bone formation (osteogenesis, osteoinduction, and osteoconduction).

Spinal surgeons have harvested autograft from iliac crest bone, ribs, spinous processes, or laminae that are approachable during the surgery. However, the amount of autograft attainable during surgery can be insufficient. Moreover, as a result of graft harvest, patients may be placed at risk of postoperative pain, nerve injury, vessel injury, hematoma, fracture, infection, and gait disturbance [1, 2].

For these reasons, a variety of bone graft substitutes, such as ceramic, demineralized bone matrix (DBM), bone morphogenetic protein (BMP), and collagen scaffolds have been introduced, studied, and applied in practice recently [2–4]. They have been used in combination with one or more other type of bone graft, including autologous bone, in order to decrease postoperative morbidity while adding bulk to the implanted bone graft mass [5–9]. Using these bone graft substitutes in combination with others can help overcome their individual shortcomings in order to create a graft mixture that contains maximum osteoconduction, osteoinduction, and osteogenesis.

11.2 Ceramic-Based Bone Graft Substitutes

A commonly used class of material for bone graft substitutes is ceramic, one of the most commonly used synthetic products worldwide. Ceramic-based bone graft substitutes contain multi-porous structures similar to human cancellous bone, affording this substance a robust osteoconductive matrix and providing structural support to the graft. While this material lacks the osteoinduction and osteogenesis of biologic materials, native mesenchymal cells are often able to adhere the structural matrix of the ceramic. Subsequently, the mesenchymal cells may proliferate and differentiate to become mature osteoblasts, yielding bone formation [10]. Nonetheless, ceramic products combined with local autograft or bone marrow aspirate have shown higher fusion rates than those of ceramic graft only [11].

The mechanical and structural properties of ceramics are important to achieve successful bone fusion. Ceramic-based bone graft must contain an adequate number and adequate size of pores. The hardness of ceramic is inversely proportional to the number of pores. When the number of pores is increased, the ceramic becomes vulnerable to fracture due to the decreased mechanical strength and resorption rates. The adequate size of pores has been reported that is from 150 to 500 micrometers to maximize ingrowth of bone to pores, biodegradation [10].

Some have conducted animal studies for assessment of the efficacy of ceramic graft for bone fusion. They largely showed inferior fusion rates when using ceramic graft alone versus when using autograft alone in a rabbit pauci model [12].

There are several advantages of using ceramics as a bone graft substitute. First, they do not induce a host inflammatory response. Second, as they are non-allogenic materials, they carry no risk for blood-borne disease transmission. Third, they can be molded into various shapes, such as brick, granule, or powder. Lastly, bone graft substitutes made of ceramic are cheaper than other commercially available bone substitutes [10, 13].

Over fifteen years, several ceramic-based bone graft products have been introduced, including hydroxyapatite, tricalcium phosphate, and calcium sulfate. Many authors have studied the efficacy of using ceramic materials as bone graft substitutes, and have tried to find the optimal ratio of multi substitute combined ceramics [1, 14]. Many of these products and product-combinations have been introduced and commercialized [14].

11.2.1 Hydroxyapatite

Hydroxyapatite is a crystalline structure with calcium-containing pores, and accounts for a majority of bone's natural mineral component. Hydroxyapatite can be found in various forms, such as powder, granule, or brick, and can be coated on metal surfaces as well. Hydroxyapatite is similar in structure to both cancellous bone and coral found in the ocean [15]. Its porosity and similar structure to bone renders it a useful clinical substance for encouraging bone formation during fusion procedures. Furthermore, hydroxyapatite is absorbed very slowly without penetrating radiation. Therefore, the main process of fusion is to connect bone tissue formed inside of hydroxyapatite to surrounding bones.

The role of hydroxyapatite is not clearly defined in the literature regarding spinal surgeries. A prospective large randomized controlled trial assessing outcomes of PLLF (posterolateral lumbar fusion) using hydroxyapatite compared to the autologous iliac crest bone graft was conducted by Korovessis et al. in 2005 [16]. In this study, the authors compared postoperative clinical and radiographic outcomes among 60 patients divided into three groups. Each group consisted of 20 patients (iliac crest bone graft (ICBG) only *vs.* coralline hydroxyapatite granule + local autograft (LAG) in a 3:1 ratio with ICBG *vs.* coralline hydroxyapatite granule + LAG bilaterally). All three groups showed radiographic evidence of solid fusion over a year, but hydroxyapatite was found to provide an inadequate fusion of intertransverse processes during this period. Both hydroxyapatite/LAG and ICBG groups showed improvements in clinical outcomes such as Oswestry Disability Index (ODI) and Visual Analogue Scale (VAS) score. Besides, using hydroxyapatite can reduce the time of the autograft harvesting process, which reduces the surgical duration and postoperative complications.

Similar studies have been reported and support these results. In some instances, hydroxyapatite can be used alone or in combination with other substances, such as bioactive glass, bone marrow aspiration, or absorbable collagen sponge. It has been shown to improve function and reduce pain to the same extent as ICBG, while it has shown suboptimal fusion rates [17–19].

11.2.2 Tricalcium Phosphate (TCP)

Tricalcium phosphate is a synthetic ceramic material composed of calcium and phosphate. β-tricalcium phosphate (β-TCP) is another calcium-based synthetic that provides osteoconductive scaffoldings for bone proliferation and fusion. Unlike hydroxyapatite, TCP is absorbed quickly, with its absorption takes place over months. Therefore, it is easier to evaluate bone fusion with TCP based on radiography than with hydroxyapatite, which has a comparatively longer life span. The disadvantage of using TCP is that the strength is weaker compared with hydroxyapatite, and it breaks more easily. Many studies have evaluated the utility of β-TCP for aiding spinal fusion surgery with mixed results.

Dai et al. conducted an RCT comparing bone fusion rates and postoperative clinical outcomes in patients who underwent single-level PLLF with ICBG/LAG *vs.* β-TCP/LAG. Both groups showed one hundred percent bone fusion. The β-TCP/LAG group showed complete bone fusion at a year postoperatively. The group using the combination of ICBG and LAG showed complete bone fusion at two years postoperatively [6].

Significant improvements in functional clinical outcomes such as ODI and Odom scale have also been reported. The Japanese Orthopaedic Association (JOA) scores tended to improve in the β-TCP/LAG group but were not clinically significant [20, 21].

11.2.3 Calcium Sulfate

Calcium sulfate is a synthetic ceramic material that combines calcium and calcium sulfate in a 1:1 ratio. Calcium sulfate has long been used as an osteoconductive void filler for lumbar fusion. Specifically, calcium sulfate has shown that there is a possibility of selling in the market in human models and animal models [22–26].

However, shortly after these initial findings, several reports demonstrated that calcium sulfate restoration occurs as fast as 6–8 weeks after surgery, causing often pathological local inflammation without bone formation. Although using calcium sulfate has become unpopular over time since the surfacing of these reports, reports that are more recent suggest a more positive outlook of using calcium sulfate for bone fusion.

In a number of studies, using calcium sulfate with LAG showed good fusion rates (up to 88%) and non-inferiority compared with ICBG [27–29]. Besides, functional clinical outcomes like ODI seemed to be comparable between calcium sulfate + LAG versus ICBG alone [28, 29].

11.2.4 Bioactive Glass

Bioactive glass is biologically compatible synthetic material containing a variety of crystalline components. There has not been enough consensus regarding the use of bioactive glass extenders.

To evaluate the stand-alone efficacy of one particular glass-ceramic, Apatite and Wollastonite-containing Glass Ceramic(AWGC), Kasai et al. performed 35 PLLFs using a total of 20 grams of AWG and local autograft in 2:1, 1:1, and 1:2 ratios. There were no significant differences in the fusion rate, functional improvements, and complication rates for all three groups [30]. While AWGC has been shown to have moderate utility in PLLF, other varieties of glass extenders show less promise.

Frantzen et al. were able to demonstrate that a bioavailable glass extender named BAG-S53P4 composed of 53% SiO_2, 23% Na_2O, 20% CaO, 4% P_2O_5 had a lower fusion rate than that of ICBG [31].

Furthermore, Acharya et al. examined the topic of different bioactive glass-based extender, Chitra-HABg (Sree Chitra Tirunal, Trivandrum, India). It also showed a lower fusion rate than that of ICBG. Their study was terminated early due to poor patient outcomes [18].

11.3 Non-ceramic-Based Bone Graft Substitutes

11.3.1 Autograft (Iliac Crest Bone Graft, Local Autograft)

Autograft is the most commonly used bone graft in spinal fusion surgery, and consists of the patient's own bone either decompressed at the surgical site or harvested from secondary site, such as the iliac crest. Autograft contains abundant osteoprogenitor cells, osteoinductive factors, and an osteoconductive scaffold of native cortical and cancellous bone. Therefore, autograft is largely considered the "gold standard" bone graft for obtaining a robust spinal fusion, resulting in very high fusion rates in many studies [6, 16]. In addition, since the graft is harvested from the patient's own tissues, there is no risk for triggering an immune response or transmitting blood-borne disease. Autograft can be harvested from a variety of secondary sites, such as iliac crest, rib, and tibia. Harvesting sites may be decided depending on the surgical site, accessibility to the site given patient positioning, the requirement of cortical bone and cancellous bone, as well as the shape, and the length, and the degree of the requirement for structural strength. The primary problem with autograft harvesting is the incidence of complications such as pain, hematoma, and infection at harvesting sites [16, 32–35]. In addition, the amount of autograft harvesting or the nature of the bone can be insufficient or improper for bone grafting. For example, if there is fatty bone marrow present or severe osteoporosis, the graft may not be ideal for bone grafting.

11.3.2 Allograft (Demineralized Bone Matrix, Corticocancellous Allograft)

Allograft, or bone donated from another individual, may refer to corticocancellous bone chips, demineralized bone matrix (DBM), or bone morphogenic proteins (BMPs). The primary function of allograft is an osteoconductive bony matrix. Given the fact that allograft originates from another individual, the graft must be frozen and sterilized to minimize potential transmission of blood-borne disease and limit recipient inflammatory reaction. Most commercially available allografts are frozen and dried after the complete removal of bone marrow. This necessary process inherently limits the grafts osteogenic properties by removing all osteogenic cells, and may actually weaken the mechanical strength of the graft as well. However, a small amount of growth factors may remain after sterilization, so the graft retains some osteoinductive qualities. Recent advancements in tests, processes, and storing techniques for allograft harvesting have also attenuated some of these drawbacks. Above all, the biggest advantage of using allograft is that it does not require harvesting from patient, thereby reducing local harvest site complications. In addition, given that allograft is commercially available in large quantities, it can be molded or cut into nearly any shape required by the surgeon. However, many commercial allografts are quite expensive, so their use should be somewhat judicious. In short, allografts spare the patient morbidity associated with graft harvest, but this must be weighed against the less ideal graft qualities of allograft compared to autograft.

11.3.2.1 Demineralized Bone Matrix (DBM)

Demineralized bone matrix is allograft that has been treated with an acid extraction process to remove its mineralization. This demineralized bone that remains is a matrix of osteoconductive type 1 collagen, non-collagenous proteins, and osteoinductive growth factors. One of the most important of these growth factors is BMP, a protein found in bone with osteoinductive properties. Other substances such as transforming growth factor-β (TGFβ), platelet-derived growth factors (PDGF), insulin-like growth factors (IGF), and fibroblast growth factors (FGF), are found in DBM and help afford the graft osteoinductive properties. Urist et al. reported that demineralized bone grafts were found to be bone-forming when transplanted into rat musculature

[36]. Demineralized bone matrix exists in various forms such as powder, granule, etc. It can be used as solid putty and shaped during the surgery to fill bone defects or fusion sites. Demineralized bone matrix is considered an autograft extender in that it is commonly combined with autograft during surgery. While this practice is common, studies have yielded shown mixed results in terms of its clinical utility. For example, Kiely et al. recently reported no significant difference in fusion rate in a rabbit model of posterolateral spinal fusion when using DBM as a graft extender compared with autograft alone [37]. Vaccaro et al. conducted a prospective study comparing mixed DBM with ICBG (1:1) and with bone marrow aspirate (BMA) and LAG (3:1:1 ratio) for use in lumbar PLLF in 2007 [34]. The authors found that both DBM composites did not show significant differences from ICBG alone in both clinical and radiographic results.

Epstein et al. reported similar results in their study [2, 21]. The study enrolled 140 patients undergoing instrumented PLLF using LAG and DBM (1:1), evaluated by radiography and functional outcome at a year postoperatively. The authors reported similar fusion rate and functional improvement in both groups. However, despite some disagreement in the efficacy of using DBM, a large number of studies do support the use of DBM as an extender/carrier for BMA, LAG, and ICBG for augmenting spinal fusion.

Furthermore, DBM has several shortcomings that are worth noting. The structure of DBM is often considered inadequate to promote good osteoconduction on its own. In addition, it is extracted from cadaver bone, so the graft quality is dependent upon the quality of the donor bone. For this reason, it is difficult to maintain precise quality of among all individual products. In addition, although there are some growth factors in DBM naturally, the quantity of growth factors is relatively less compared to the amount of artificially concentrated recombinant human bone morphogenetic protein (rhBMPs) which is used for augmenting and accelerating solid spinal fusion.

11.3.2.2 Corticocancellous Allograft

The mineral component of allograft has excellent osteoconductive properties, and for this reason, it has been used as a bone graft extender/carrier for LAG and BMA. In 1994, Jorgenson et al. conducted a prospective study of 144 patients undergoing instrumented PLLF evaluating fusion status achieved using various bone substitutes. They found that the group using ICBG and allograft combined showed a significantly lower bone fusion rate than the group using ICBG alone [4]. Demineralized allograft, however, achieved the lowest fusion rate among the three groups. Therefore, the authors did not recommend using allograft as an extender with ICBG in PLLF [4].

In 2014, Hart et al. reported that bone fusion rate improved when using bone marrow aspiration with allograft compared with using allograft alone [38]. In 2010, Lee et al. conducted a retrospective study comparing fusion rates and clinical outcomes in patients over 65 years old undergoing PLLF with either autograft or a mixture of allograft, autograft, and growth factors [32]. Fusion status was evaluated with lumbar spine x-ray and CT scans. Clinical outcomes were assessed with a VAS score and a graded functional scale. They observed similar radiographic and clinical outcomes postoperatively. These results suggest that allograft may be cautiously used with either corticocancellous autograft or bone marrow aspiration to achieve good fusion rates and clinical outcomes in the patient undergoing PLLF, although a clear consensus has not been reached regarding its efficacy [39].

11.3.3 Bone Marrow (Bone Marrow Aspirates, Bone Marrow Concentrate)

Bone marrow is rich in osteogenic cells and osteoinductive factors that can be used to augment fusion in PLLF. The marrow is aspirated in the operating room using a large bore needle and syringe, most commonly from the patient's iliac crest. The aspirate can be introduced to the fusion site as is, or can be concentrated via centrifuge prior to implantation [40]. Bone marrow aspirate (BMA) and bone marrow aspirate concentrate (BMAC) has the advantage of providing the osteogenic and osteoinductive properties of autograft while reducing the complications associated with ICBG autograft harvest. Despite these advantages, one shortcoming is that bone marrow aspirate/concentrate lacks the structural support of corticocancellous autograft. To overcome this shortcoming, BMA can be used with other bone substitutes, such as allograft and ceramics.

Odri et al. conducted a prospective study that examined the effect of the concentration process on bone growth in PLLF [40]. Patients were fused using local autograft plus calcium phosphate with either concentrated or whole bone marrow aspirate. The aspirate was centrifuged intraoperatively using a portable centrifuging machine to reach an average concentration of osteoprogenitor cells 2.2 × that of whole BMA [40]. Graft site cortical bone volume was assessed via CT scan at one week and 3 months postoperatively. There was no significant difference in bone growth between the two groups, suggesting a lack of observable benefit obtained from concentration. However, they failed to achieve an adequate concentration of osteogenic cells in the concentrated group. Therefore, the authors suggest that the raw number of these cells present in the graft is an important factor predicting initial bone growth; if that can be optimized using concentration, then it may benefit the fusion mass. Particularly when using ceramics and allografts, which do not have any osteogenic cells, adding bone marrow aspirate has been shown to significantly benefit the fusion rate [38].

Hart et al. conducted a blinded RCT regarding comparing the degree of bone fusion in patients undergoing instrumented PLLF with either spongious allograft chips or allograft mixed with concentrated bone marrow. The degree of fusion was assessed using plain radiographs and CT scans by evidence of at least unilateral bridging bone between transverse processes. The authors observed remarkably greater successful fusion and mineralization status with the combination of bone marrow aspirate and allograft than with allograft alone at two years postoperatively. The results demonstrate the role of bone marrow aspirate as a beneficial adjunct to allograft in lumbar PLLF [38].

So far, the combination of local autograft and bone marrow aspirates is known to have at least comparable fusion rate and similar functional outcomes as ICBG, with potentially lower rate of donor-site related complications [41].

11.3.4 Growth Factors (Bone Morphogenetic Proteins, Autologous Platelet Concentrate)

11.3.4.1 Bone Morphogenetic Proteins

Bone morphogenetic protein (BMP) is known to be highly osteoinductive. They are involved in the differentiation from mesenchymal cells to osteoblasts and increasing the number of osteoblasts directly. Bone morphogenetic proteins also produce extracellular substances like type I collagen, type II collagen, fibrin, proteoglycan, which are required to be essential in bone formation. It is not clear exactly how BMPs regulate gene expression within cells. So far, it is known that BMPs attach to cell wall receptors and activate an intracellular transmitter (SMAD) to induce osteogenic gene expression within the nucleus.

Bone morphogenetic proteins belong to the TGF-β superfamily of extracellular proteins. They have been known as involving in organ development, cellular differentiation, chondrogenesis, and osteogenesis. There are about 1420 known BMPs in the human body, including BMP-2, BMP-3 (osteogenin), BMP-4, BMP-5, BMP-6, BMP-7 (OP-1), BMP-8 (OP-2), BMP-9 (GDF-2), BMP-10, BMP-11 (GDF-8), BMP-12 (GDF-7, CDMP-3), BMP-13 (GDF-6, CDMP-2), BMP-14 (GDF-5,CDMP-1), and BMP-15 (OP: Osteogenic protein, GDF: Growth/differentiation factor, CDMP: Cartilage-derived morphogenetic protein). Among these 20 BMPs, BMP-2,4,6,7,9 have been known to have osteoinductive activity. Recently a high number of studies have been introduced regarding the efficacy of BMP-2 and -7 in spinal surgery. Recombinant forms of these proteins (rhBMP-2 and rhBMP-7, or OP-1, respectively). rhBMP-2 was the first BMP to be extensively studied for use in spine surgery, and was originally FDA-approved in 2002 for anterior lumbar interbody fusions. However, it has been commonly used off-label for the last decade for PLLF surgery [42].

Dimar et al. conducted the RCT comparing ICBG *versus* rhBMP-2 with bovine collagen and β-TCP/hydroxyapatite carrier to describe the advantages of this off-label use in 2009 [43]. 239 people were enrolled and the authors assessed the person's single-level fusion rate based on radiograph, CT scan, and functional outcome based on ODI and SF-36 score. The fusion rate of the group using rhBMP-2 was higher than that of the ICBG group. In addition, subjects in the group using ICBG experienced more blood loss and longer operative times than those of the group using rhBMP-2. Nearly two-thirds of patients in the ICBG group reported donor site pain or tenderness at 2-years follow-up [43].

As reported by Dimar et al., Many researchers reported that the group using rhBMP-2/β-TCP/hydroxyapatite shows a higher fusion rate than the group using ICBG for single-level PLLF, but lower fusion rate for multiple-level (>=2) PLLF [2, 44]. Glassman et al. tried to prove a similar efficacy of rhBMP-2 compared to ICBG in smoking and non-smoking population. The fusion rates when using rhBMP-2 are higher than that of using ICBG in both groups [45, 46]. Based on these reports, a number of ceramic-based extenders have been used in conjunction with rhBMP-2.

Bae et al. performed RCT regarding the comparison of ICBG and rhBMP-2/collagen sponge/ceramic matrix. In addition, rhBMP-2/collagen sponge/ceramic matrix group showed a significantly higher solid fusion rate than that of ICBG group (95% vs. 67%). The rhBMP-2/collagen sponge/ceramic matrix group also showed improvement in ODI, SF-36, VAS score at 2-year follow-up [47]. Dawson et al. also corroborate these findings for two years. Indeed, several articles demonstrated the effective use of rhBMP-2 ceramic composites as an alternative to ICBG autograft [48]. After 2006, some individual studies have been conducted regarding the complication of using rhBMP-2.

RhBMP-7, an osteogenic growth factor related to BMP-2, was first approved by the FDA in 2001 for use as an alternative to autograft for long bone fracture repair. The approval of rhBMP-7 by the FDA was expanded to cover PLLF in 2004. Since that time, several studies have been conducted regarding the safety and efficacy of rhBMP-7.

Vaccaro et al. performed a study regarding the use of rhBMP-7 (OP-1) in conjunction with bovine collagen and carboxymethylcellulose. The OP-1 putty which is a product, a proprietary blend of rhBMP-7, bovine collagen, and carboxymethylcellulose. Using OP-1 putty with ICBG obtained a 50% solid fusion rate and 20% improvement in ODI at 1-year follow-up [49]. Afterward, in a RCT comparison of OP-1 versus ICBG, The authors were able to show significantly better fusion rate in the OP-1 using group than in the ICBG group (55% vs. 40%). They also report improvement in functional outcomes based on ODI, SF-36 in the group using OP-1 [50]. While both of these studies demonstrate the efficacy using OP-1 putty, the sample sizes of both studies were small (12 and 24, respectively), limiting their broad applicability. Still, OP-1 used with graft extenders appears to be a reasonable alternative to ICBG in the carefully selected patient.

In a 2016 Buser et al. conducted a systematic review comparing synthetic bone graft versus autograft or allograft in terms of fusion rates, patient-reported outcomes, and functional outcomes in both cervical and lumbar fusion. Although the authors admitted some selection bias, they cautiously concluded that synthetic grafts performed similarly to autologous grafts with regards to these outcomes [51].

Similarly, in their review of bone graft materials used in lumbar fusion surgery, Tuchman et al. reported similar fusion rates, pain scores, and functional outcome in patients undergoing fusion with ICBG autograft or allograft groups. However, there was not sufficient evidence to make definitive recommendations. They conclude that a choice must be made carefully regarding the type using graft within the framework of the current literature [52].

However, it is also important to note that there have been several limitations documented to using BMPs. Despite their potent osteoinductive properties, to achieve adequate osteoinductivity, a large amount of BMP is required. For this reason, BMP can be associated with high cost, host immune reaction, ectopic bone formation, excessive bone formation, and soft tissue edema. Although there are some limitations of using BMP, it has shown to achieve better fusion rate than using autograft alone. Therefore, the potential benefit of using BMP is to alleviate the complications associated with autograft harvest while maximizing fusion rate. In addition, some study groups have tried certain genetic treatments based on the gene regarding BMPs or intermediate genes.

11.3.4.2 Autologous Platelet Concentrate

Another effort to stimulate bone growth and improve fusion rates in PLLF has focused on the use of autologous serum concentrated via centrifugation. This autologous concentrate contains endogenous cytokines, platelets, and growth factors that were theorized to promote fusion via osteoinductive factors. Some surgeons have advocated for its use as a supplement to corticocancellous autograft to augment PLLF.

Carreon et al. reviewed 76 patients who underwent instrumented PLLF using ICBG and APC and compared fusion rates to a randomly selected control group who received ICBG alone. The authors found no statistically significant difference in fusion rates between groups and recommended against the use of APC to supplement ICBG in PLLF [33].

Acebal-Cortina et al. confirmed this recommendation in a prospective study of 107 patients who underwent lumbar fusion using a mixture of autologous local bone graft and TCP/hydroxyapatite with and without the addition of APC. They observed a significantly lower fusion rate with the addition of APC to the graft mix, as measured by plain radiographs at 12 and 24 months [5]. Preparation of APC marginally increases patient morbidity due to the need for blood draws. Based on the current literature, the use of APC has been shown to either fail to increase or actually to decrease fusion rates. Therefore, its use as a graft adjunct in PLLF is not recommended.

11.3.5 Collagen (Absorbable Collagen Sponge)

Absorbable collage sponge (ACS) has been used generally as a carrier for BMP, but does not have a role in contributing to fusion on its own. Absorbable collage sponge is a sponge manufactured from xenogeneic type I collagen. It serves as an osteoconductive scaffold and contributes to hemostasis at the surgical site. A benefit of the ACS is that it can be applied directly to the surgical field or placed in a cage for interbody fusion, and is absorbed easily. It is more commonly used in posterior fusion, and less commonly in the case of PLLF.

Arnold et al. conducted a non-inferiority RCT comparing fusion rate and complication rate in single level anterior cervical discectomy and fusion using a synthetic collagen fragment suspended in a hydrogel (i-Factor) with cortical ring allograft versus autograft [48]. The synthetic collagen group showed non-inferiority to autograft in terms of radiographic and clinical outcomes, as well as a comparable safety profile.

Hostin et al. conducted a retrospective study evaluating the effectiveness of using collagen and bone marrow aspirates in multi-level anterior spinal fusion surgery. Using mineralized collaged and BMA, they observed a fusion rate of 88%, relatively lower than other studies that observed fusion rates greater than 95% for similar surgeries using BMPs [53].

11.4 Future of the Bone Substitutes

The ideal graft must have properties of osteoconductivity, osteoinductivity, and osteogenicity [54]. Bone graft technology has improved rapidly in recent years, proving the robust and dynamic nature of this industry. Therefore, it is vital that practicing spine surgeons pay close attention to the latest trends in bone substitutes to achieve optimal fusion rates while minimizing morbidity and complications for their patients.

References

1. Dimar JR, Glassman SD, Burkus KJ et al (2006) Clinical outcomes and fusion success at 2 years of single-level instrumented posterolateral fusions with recombinant human bone morphogenetic protein-2/compression resistant matrix versus iliac crest bone graft. Spine 31:2534–2539
2. Epstein NE (2006) A preliminary study of the efficacy of Beta Tricalcium Phosphate as a bone expander for instrumented posterolateral lumbar fusions. J Spinal Disord Tech 19:424–429
3. Vaccaro AR, Whang PG, Patel T et al (2008) The safety and efficacy of OP-1 (rhBMP-7) as a replacement for iliac crest autograft for posterolateral lumbar arthrodesis: minimum 4-year follow-up of a pilot study. Spine J 8:457–465
4. Jorgenson SS, Lowe TG, France J et al (1994) A prospective analysis of autograft versus allograft in posterolateral lumbar fusion in the same patient. A minimum of 1-year follow-up in 144 patients. Spine 19:2048–2053
5. Acebal-Cortina G, Suárez-Suárez MA, García-Menéndez C et al (2011) Evaluation of autologous platelet concentrate for intertransverse lumbar fusion. Eur Spine J 20 Suppl 3:361–366

6. Dai LY, Jiang LS (2008) Single-level instrumented posterolateral fusion of lumbar spine with beta-tricalcium phosphate versus autograft: a prospective, randomized study with 3-year follow-up. Spine 33:1299–1304
7. Kanayama M, Hashimoto T, Shigenobu K et al (2006) A prospective randomized study of posterolateral lumbar fusion using osteogenic protein-1 (OP-1) versus local autograft with ceramic bone substitute: emphasis of surgical exploration and histologic assessment. Spine 31:1067–1074
8. Chen WJ, Tsai TT, Chen LH et al (2005) The fusion rate of calcium sulfate with local autograft bone compared with autologous iliac bone graft for instrumented short-segment spinal fusion. Spine 30:2293–2297
9. Alexander DI, Manson NA, Mitchell MJ (2001) Efficacy of calcium sulfate plus decompression bone in lumbar and lumbosacral spinal fusion: preliminary results in 40 patients. Can J Surg 44:262–266
10. Khan SN, Fraser JF, Sandhu HS et al (2005) Use of osteopromotive growth factors, demineralized bone matrix, and ceramics to enhance spinal fusion. J Am Acad Orthop Surg 13:129–137
11. Cinotti G, Patti AM, Vulcano A et al (2004) Experimental posterolateral spinal fusion with porous ceramics and mesenchymal stem cells. J Bone Joint Surg Br 86:135–142
12. Miller CP, Jegede K, Essig D et al (2012) The efficacies of 2 ceramic bone graft extenders for promoting spinal fusion in a rabbit bone paucity model. Spine 37:642–647
13. Berven S, Tay BK, Kleinstueck FS et al (2001) Clinical applications of bone graft substitutes in spine surgery: consideration of mineralized and demineralized preparations and growth factor supplementation. Eur Spine J 10 Suppl 2:S169-S177
14. Nickoli MS, Hsu WK (2014) Ceramic-based bone grafts as a bone grafts extender for lumbar spine arthrodesis: a systematic review. Global Spine J 4:211–216
15. Hsu CJ, Chou WY, Teng HP et al (2005) Coralline hydroxyapatite and laminectomy-derived bone as adjuvant graft material for lumbar posterolateral fusion. J Neurosurg Spine 3:271–275
16. Korovessis P, Koureas G, Zacharatos S et al (2005) Correlative radiological, self-assessment and clinical analysis of evolution in instrumented dorsal and lateral fusion for degenerative lumbar spine disease. Autograft versus coralline hydroxyapatite. Eur Spine J 14:630–638
17. Ploumis A, Albert TJ, Brown Z et al (2010) Healos graft carrier with bone marrow aspirate instead of allograft as adjunct to local autograft for posterolateral fusion in degenerative lumbar scoliosis: a minimum 2-year follow-up study. J Neurosurg Spine 13:211–215
18. Acharya NK, Kumar RJ, Varma HK et al (2008) Hydroxyapatite-bioactive glass ceramic composite as stand-alone graft substitute for posterolateral fusion of lumbar spine: a prospective, matched, and controlled study. J Spinal Disord Tech 21:106–111
19. Singh K, Smucker JD, Gill S et al (2006) Use of recombinant human bone morphogenetic protein-2 as an adjunct in posterolateral lumbar spine fusion: a prospective CT-scan analysis at one and two years. J Spinal Disord Tech 19:416–423
20. Park DK, Kim SS, Thakur N et al (2013) Use of recombinant human bone morphogenetic protein-2 with local bone graft instead of iliac crest bone graft in posterolateral lumbar spine arthrodesis. Spine 38:E738-E747
21. Epstein NE (2008) An analysis of noninstrumented posterolateral lumbar fusions performed in predominantly geriatric patients using lamina autograft and beta tricalcium phosphate. Spine J 8:882–887
22. Blom AW, Cunningham JL, Hughes G et al (2005) The compatibility of ceramic bone graft substitutes as allograft extenders for use in impaction grafting of the femur. J Bone Joint Surg Br 87:421–425
23. Turner TM, Urban RM, Gitelis S et al (2003) Resorption evaluation of a large bolus of calcium sulfate in a canine medullary defect. Orthopedics 26(5 Suppl):s577–s579
24. Hadjipavlou AG, Simmons JW, Yang J et al (2000) Plaster of Paris as an osteoconductive material for interbody vertebral fusion in mature sheep. Spine 25:10–15
25. Sidqui M, Collin P, Vitte C (1995) Osteoblast adherence and resorption activity of isolated osteoclasts on calcium sulphate hemihydrate. Biomaterials 16:1327–1332

26. Coetzee AS (1980) Regeneration of bone in the presence of calcium sulfate. Arch Otolaryngol 106:405–409
27. Hurlbert RJ, Alexander D, Bailey S et al (2013) rhBMP-2 for posterolateral instrumented lumbar fusion: a multicenter prospective randomized controlled trial. Spine 38:2139–2148
28. Jenis LG, Banco RJ (2010) Efficacy of silicate-substituted calcium phosphate ceramic in posterolateral instrumented lumbar fusion. Spine 35:E1058–E1063
29. Johnsson R, Strömqvist B, Aspenberg P (2002) Randomized radiostereometric study comparing osteogenic protein-1 (BMP-7) and autograft bone in human noninstrumented posterolateral lumbar fusion: 2002 Volvo Award in clinical studies. Spine 27:2654–2661
30. Kasai Y, Takegami K, Uchida A (2003) Mixture ratios of local bone to artificial bone in lumbar posterolateral fusion. J Spinal Disord Tech 16:31–37
31. Frantzén J, Rantakokko J, Aro HT et al (2011) Instrumented spondylodesis in degenerative spondylolisthesis with bioactive glass and autologous bone: a prospective 11-year follow-up. J Spinal Disord Tech 24:455–461
32. Lee KB, Taghavi CE, Hsu MS et al (2010) The efficacy of rhBMP-2 versus autograft for posterolateral lumbar spine fusion in elderly patients. Eur Spine J 19:924–930
33. Carreon LY, Glassman SD, Djurasovic M et al (2009) RhBMP-2 versus iliac crest bone graft for lumbar spine fusion in patients over 60 years of age: a cost-utility study. Spine 34:238–243
34. Vaccaro AR, Stubbs HA, Block JE (2007) Demineralized bone matrix composite grafting for posterolateral spinal fusion. Orthopedics 30:567–570
35. Sassard WR, Eidman DK, Gray PM et al (2000) Augmenting local bone with Grafton demineralized bone matrix for posterolateral lumbar spine fusion: avoiding second site autologous bone harvest. Orthopedics 23:1059–1064
36. Urist MR, Strates BS (2009) The classic: bone morphogenetic protein. Clin Orthop Relat Res 467:3051–3062
37. Kiely PD, Brecevich AT, Taher F et al (2014) Evaluation of a new formulation of demineralized bone matrix putty in a rabbit posterolateral spinal fusion model. Spine J 14:2155–2163
38. Hart R, Komzák M, Okál F et al (2014) Allograft alone versus allograft with bone marrow concentrate for the healing of the instrumented posterolateral lumbar fusion. Spine J 14:1318–1324
39. Chen CL, Liu CL, Sun SS et al (2006) Posterolateral lumbar spinal fusion with autogenous bone chips from laminectomy extended with OsteoSet. J Chin Med Assoc 69:581–584
40. Odri GA, Hami A, Pomero V et al (2012) Development of a per-operative procedure for concentrated bone marrow adjunction in postero-lateral lumbar fusion: radiological, biological and clinical assessment. Eur Spine J 21:2665–2672
41. Alsaleh KA, Tougas CA, Roffey DM et al (2012) Osteoconductive bone graft extenders in posterolateral thoracolumbar spinal fusion: a systematic review. Spine 37:E993-E1000
42. Gupta A, Kukkar N, Sharif K et al (2015) Bone graft substitutes for spine fusion: a brief review. World J Orthop 6:449–456
43. Dimar JR 2nd, Glassman SD, Burkus JK et al (2009) Clinical and radiographic analysis of an optimized rhBMP-2 formulation as an autograft replacement in posterolateral lumbar spine arthrodesis. J Bone Joint Surg Am 91:1377–1386
44. Epstein NE (2009) Beta tricalcium phosphate: observation of use in 100 posterolateral lumbar instrumented fusions. Spine J 9:630–638
45. Glassman SD, Dimar JR 3rd, Burkus K et al (2007) The efficacy of rhBMP-2 for posterolateral lumbar fusion in smokers. Spine 32:1693–1698
46. Glassman SD, Dimar JR, Carreon LY (2005) Initial fusion rates with recombinant human bone morphogenetic protein-2/compression resistant matrix and a hydroxyapatite and tricalcium phosphate/collagen carrier in posterolateral spinal fusion. Spine 30:1694–1698
47. Dawson E, Bae HW, Burkus JK et al (2009) Recombinant human bone morphogenetic protein-2 on an absorbable collagen sponge with an osteoconductive bulking agent in posterolateral arthrodesis with instrumentation. A prospective randomized trial. J Bone Joint Surg Am 91:1604–1613

48. Arnold PM, Sasso RC, Janssen ME et al (2016) Efficacy of i-Factor bone graft versus autograft in anterior cervical discectomy and fusion: results of the prospective, randomized, single-blinded food and drug administration investigational device exemption study. Spine 41:1075–1083
49. Vaccaro AR, Anderson DG, Patel T et al (2005) Comparison of OP-1 Putty (rhBMP-7) to iliac crest autograft for posterolateral lumbar arthrodesis: a minimum 2-year follow-up pilot study. Spine 30:2709–2716
50. Vaccaro AR, Patel T, Fischgrund J et al (2004) A pilot study evaluating the safety and efficacy of OP-1 Putty (rhBMP-7) as a replacement for iliac crest autograft in posterolateral lumbar arthrodesis for degenerative spondylolisthesis. Spine 29:1885–1892
51. Buser Z, Brodke DS, Youssef JA et al (2016) Synthetic bone graft versus autograft or allograft for spinal fusion: a systematic review. J Neurosurg Spine 25:509–516
52. Tuchman A, Brodke DS, Youssef JA et al (2016) Iliac crest bone graft versus local autograft or allograft for lumbar spinal fusion: a systematic review. Global Spine J 6:592–606
53. Hostin R, O'Brien M, McCarthy I et al (2016) Retrospective study of anterior interbody fusion rates and patient outcomes of using mineralized collagen and bone marrow aspirate in multilevel adult spinal deformity surgery. Clin Spine Surg 29:E384-E388
54. Morris MT, Tarpada SP, Cho W (2018) Bone graft materials for posterolateral fusion made simple: a systematic review. Eur Spine J 27:1856–1867

Sanghyo Lee Dr. Lee is a neurosurgeon from South Korea. He has been volunteering currently at Bagae Hospital for COVID management which is located in Pyeongtaek, South Korea. He formerly worked in Seoul, South Korea as a clinical fellow at Eunpyeong St. Mary's Hospital which is affiliated with the Catholic University. After graduating from Kosin University College of Medicine in 2012, He specialized in Neurosurgery at the Catholic University, and subspecialized in Spine surgery at the same university. He has served in the Korean Neurosurgery Society, and Korean Spinal Neurosurgery Society. His main research area is spine deformity, relationship between clinical prognosis and radiographic parameters.

Matthew T. Morris Dr. Morris completed his medical training at Albert Einstein College of Medicine in Bronx, NY, where he graduated with distinction in medical research in 2018. As a medical student, he published several original articles and presented at national conferences for his research in spine surgery. He is now receiving his residency training in orthopedic surgery at Northwell Health in New Hyde Park, NY, where he has continued his research. He plans to complete a fellowship in spine surgery and begin a practice focusing on treating degenerative spine conditions, deformity, and trauma.

David A. Essig Dr. Essig joined the department of Orthopaedic Surgery at Northwell Health in 2012. He received a BA in Molecular Biochemistry and Biophysics and Economics from Yale University. He earned his medical degree at the University of Toledo School of Medicine in Ohio and subsequently returned to Yale University to train in Orthopaedic Surgery. He received fellowship training in Spine and Scoliosis Surgery at the Hospital for Special Surgery in New York City. He is currently an Assistant Professor of Orthopaedic Surgery at the Donald and Barbara Zucker School of Medicine at Hofstra/Northwell along with the Director of Orthopaedic Spine Surgery at Forest Hills Hospital. Dr. Essig is a Board Certified Orthopaedic Surgeon. He has clinical interests in spinal deformity surgery, spinal infections, metastatic spinal disease, and degenerative conditions of the adult spine. He is a member for the American Academy of Orthopaedic Surgeons and the North American Spine Society. He has multiple publications in peer-reviewed journals along with multiple presentations at prestigious spine society meetings.

Woojin Cho Professor Cho joined the faculty of Montefiore Medical Center in 2013, specializing in adult thoracolumbar spine deformity and degenerative conditions with expertise in the treatment with all types of surgical technique. He is the Chief of Orthopedic Spine Surgery at Montefiore and an Associate Professor at Albert Einstein College of Medicine. He earned his medical degree at Hanyang University College of Medicine in South Korea, completed his orthopedic internship and residency at Hanyang University Hospital, and became a Korean orthopedic board-certified surgeon. Following his residency, he completed three years of spine clinical fellowship under Dr. Cho and Dr. Chang and earned a Ph.D. degree at Hanyang University College of Medicine.

Dr. Cho has extensive clinical training experience in the United States, which includes a research fellowship at Twin Cities Spine Center in Minneapolis, a clinical research fellowship at Washington University School of Medicine in St. Louis, and spine fellowships at the University of Virginia Health System and the Hospital for Special Surgery, respectively. Dr. Cho is a board-eligible (ABOS) orthopedic surgeon, a faculty member of the Scoliosis Research Society, the North American Spine Society, AOSpine North America, American Academy of Orthopedic Surgeons, American Medical Association, the Korean American Spine Society, the Korean Society of Spine Surgery, the Korean Spine Arthroplasty Society, the Pacific and Asian Society of Minimally Invasive Spine Surgery, the Korean Orthopedic Association, and the Korean Medical Association. In addition to having written more than 100 articles related to spine research in peer-reviewed publications, Dr. Cho has been a guest speaker at prestigious international meetings and, to date, has made more than 250 presentations.

Chapter 12
Orthopedic Application of Collagen-Hydroxyapatite Bone Substitutes: A Clinical Perspective

Pietro Domenico Giorgi, Giuseppe Rosario Schirò, Simona Legrenzi, and Francesco Puglia

Abstract In recent years, large skeletal defects have attached wide attention from researchers and the use of new biomaterials represents a main topic in orthopedic surgery. Bone grafting is a complex process that require an accurate choice of specific material for every host. The ideal graft is able to stimulate the bone healing and at the same time is biocompatible, bioresorbable, and generating minimal fibrous reaction like autograft. The magnesium-doped hydroxyapatite/type I collagen 3D scaffolds are an effective and safe bone substitute and seems to be useful in spinal fusion and orthopedic infection's field without adverse and inflammatory reactions. They can be securely used with autologous bone stimulating formation of new bone. Further investigations are needed to support long-term efficacy and additional indications for its use.

Keywords Spine surgery · Bone grafts · Magnesium-hydroxyapatite-collagen scaffold · Skeletal defects · Bone healing

12.1 Introduction

Bone is the most common transplantation tissue, after blood [1]. The main solid constituents of human bone are collagen and a mineral phase that is composed of hydroxyapatite. Collagen is a natural polymer, and it is the most present protein in the body, which provides flexibility. On the other hand, hydroxyapatite is a natural ceramic, represents the main component of natural bone and it is also found in teeth. These constituents are typically utilized as a biomimetic composite material in tissue engineering due to their great biocompatibility and biodegradability.

P. D. Giorgi (✉) · G. R. Schirò · S. Legrenzi · F. Puglia
Orthopedics and Traumatology Unit, Emergency and Urgency Department, A.S.S.T. Grande Ospedale Metropolitano Niguarda, Milan, Italy
e-mail: pietrodomenico.giorgi@ospedaleniguarda.it

F. Puglia
University of Milan, Milan, Italy

© The Author(s), under exclusive license to Springer Nature Singapore Pte Ltd. 2022

A. H. Choi and B. Ben-Nissan (eds.), *Innovative Bioceramics in Translational Medicine II*, Springer Series in Biomaterials Science and Engineering 18,
https://doi.org/10.1007/978-981-16-7439-6_12

Tissue engineering is an interdisciplinary field that applies the principles of engineering and the life sciences to development of biological substitutes that restore tissue function [2]. Bone-tissue engineering tries to induce bone healing using osteoconductive and biodegradable matrixes associated to osteoinductive growth factors [3]. The repair mechanisms that occur in our bodies are angiogenesis, osteogenesis, and chronic wound healing. However, there are some critical size defects beyond which the tissue will not regenerate on its own and needs surgical repair.

In the Unites States alone there are around 500,000 surgical procedures per year for bone tissue repair [4]. These bone defects can arise from trauma, tumor, infections such as osteomyelitis and orthopedic revision surgery. In these cases, we need a graft not only to "fill the gap", but it has to be able to build functional bone. Despite other human tissues which heal with a connective tissue scar, bone can heal by producing normal bone, as it does after fractures. A number of processes are involved in this mechanism: osteogenesis, osteoconduction and osteoinduction. Some materials, such as bone autograft, have osteogenic, osteoconductive and osteoinductive properties.

The ideal graft is able to stimulate the bone healing and at the same time is biocompatible, bioresorbable, and generating minimal fibrous reaction. Nowadays graft options are various, but from a biological point of view the gold standard solution remains the autologous bone graft harvested from the iliac crest (ICBG). Although very successful in many operations, autografts have the disadvantages of insufficient supply and morbidity, as well as increasing surgery times and donor site pain [5–7]. Local autograft bone (LAB) harvested near the surgical site, like the lamina and the spinous process during spinal surgery, reduced local morbidity with great rates of fusion. Nevertheless, the inadequate quantity and quality, especially in older patients, leads the focus to develop bone graft substitutes as an alternative of autologous bone. Other options for bone repair are the use allografts and xenografts. However, these approaches are associated with increased incidence of infection and ethical implications. Xenografts also carry the risk of species-to-species transmissible diseases [8].

As one of the major challenges in the field of tissue engineering, large skeletal defects have attached wide attention from researchers and the use of new biomaterials represents a main topic in orthopedic surgery. In recent years, the advances in biomaterial engineering related to various aspects of regenerative medicine, surface functionalization and composite systems have led to development of new implants. However, problems and pitfalls related to use of biotechnology products in a clinical context are still discussed. Synthetic bone graft substitutes start to spread into the market from the 1980s as an osteoconductive alternative material to fill the bone loss and contour small irregularities [5, 9–11]. They represent synthetic, inorganic, or biologically organic combinations which can be inserted for treating bone defects instead of autogenous or allogenous bone graft. These materials, derived from natural ceramics, are able to function as a cellular adhesion site permitting bone growth and mechanical stability thanks to their structural porous framework.

The use of bone graft substitutes as a valid alternative to autologous bone has been well documented and they can provide comparable fusion rate to autologous bone if used as graft extenders, avoiding donor site complication [12–16]. Biomaterials can be synthetic or natural and can be non-absorbable, partially or fully absorbable. Hydroxyapatite, tricalcium phosphate (TCP) and calcium sulphate are the most common examples of ceramic-based grafts, their composition is similar to the inorganic phase of bone. They represent partially absorbable and osteoconductive materials. Hydroxyapatite, as ceramic based alternatives, accounts nearly 70% of the mineral component of bone [17, 18] so it is the most suitable material for bone replacement. The new generation of bone graft substitutes, developed in the last years, allowed to obtain enhanced mimicry of the biophysical and biochemical characteristics of the human bone providing mechanical support until the tissue can renewed and remodeled itself naturally.

Bioresorbable grafts can avoids the problems that occur with metallic implants, such as corrosion and stress shielding. Amongst various metallic biomaterials, use of titanium and its alloys (Ti-6Al-4V) is one of the most widespread for orthopaedic applications. Good biocompatibility, favorable tissue response, adequate strength and corrosion resistance are some of the key attributes which make it an excellent choice as a biomaterial worldwide. However, incomplete osteointegration can still occur where there is insufficient contact between the metal and host bone. Methods to address these issues include modifications to increase surface roughness, oxide layer and bioactivity. Titanium particles produced from wear of hip implants were shown to suppress osteogenic differentiation of human bone marrow and stroma-derived mesenchymal cells and to inhibit the mineralization of extra cellular matrix [19].

Collagen, as a natural polymer, is increasingly being used as a device in tissue engineering. It is, for example, found in bone (Type I), cartilage (Type II) and in blood vessel walls (Type III) and has excellent biocompatible properties. Collagen is easily degraded and resorbed by the body and allows good attachment to cells. However, its mechanical properties are relatively low (Young's Modulus –100 MPa) in comparison to bone (Young's Modulus –2–50 GPa) [19] and it is therefore highly crosslinked or found in composites, such as collagen-glycoaminoglycans for skin regeneration [20], or collagen-hydroxyapatite for bone remodeling [21]. Both collagen and hydroxyapatite materials significantly inhibited the growth of bacterial pathogens, the most frequent cause of prosthesis-related infection, compared to poly-lactic-co-glycolic acid (PLGA) devices [22].

12.2 Bone Defects in Orthopedic Surgery

Trauma, infections, tumors and orthopedic revision surgery can produce bone loss. Bone tumors may require surgical excision according to oncological criteria. Depending on bone defect and instability, an instrumentation with bone grafting is usually performed. In the orthopedic oncological field, autografts are limited to small bone defects while larger bone defects need allografts. The indications for bone graft substitutes in these circumstances have to be carefully established regarding the risk of recurrence (in malignant tumors), associated oncological treatments, as well as the ability to fit to the implants [23]. Nowadays, due to the increase in the number of arthroplasty procedures in the last few decades, the treatment of bone defects in revision surgery has become more frequent. Poor bone stock in the osteoporosis and the osteolytic areas induced by mechanical stress require bone graft able to sustain the functional loading [24].

In the field of bone infections, the indications of bone graft substitutes must be carefully assessed, especially because all healing processes are impaired by infection and even osteogenetic compounds have limited or no activity in a septic environment. Therefore, in the osteomyelitis grafting must be performed only when the infection has been fully resolved and using materials with maximum biological integration, since local viability is affected by chronic inflammation [25].

A very common cause of bone defects is represented by deformities or malunions of long bones following high-energy trauma. In these cases, since the main problem is the maintaining of the correction, optimal loading resistance has to be considered when choosing the most appropriate bone substitute.

The most severe cases of bone loss are those which combine two or more of the causes previously mentioned. One example is that of septic prosthetic loosening, when the bacterial action initiates the osteolysis, thus enhancing the mechanical and biological triggers of the implant-induced chronic inflammation. In these cases, the bone defect affects not only the integrity of the joint, but even the function of the limb and the daily activities of the patient.

In the orthopedic surgery, the key aspects to choose a bone substitutes are the biocompatibility, the osteointegration and the resorption. More biocompatibility has been described for calcium phosphate, collagen composites and hydroxyapatite as osteoconductive materials [26]. The osteointegration of bone substitute depends on its main features. A number of processes are involved in its ability to regenerate to normal bone:

- Osteogenesis: formation of the new bone from progenitor bone substitute;
- Osteoconduction: the property of substitute that provides the microstructure to facilitate the in-growth of cells the produce bone;
- Osteoinduction: the capability of substitute on stimulating new bone formation.

12.3 Scaffolds in Orthopedic Surgery

The fabrication of biomimetic scaffolding is a challenging issue in tissue engineering. It is well known that the role of the scaffold is to imitate as much as possible the structure and the composition of natural tissues. The scaffolds consist of 3D structures needed to promote incorporation of cells or biomolecules with the objective to promote the generation of a new tissue and support the proliferation of specific cell type. They must be designed with micrometer precision to enable cell proliferation requiring customization based on the type of tissue. Fiber orientation, porosity and pore size can modify their integration in the host. The degradation of the scaffold and the mechanical properties are important characteristics for successful performance in vivo. Their resorption varies according to material and host environment; it's a cell-mediated mechanism influenced by proteolytic degradation and hydrolysis. A material is defined biodegradable if it loses its tensile strength within 60 days in vivo [27]. The metal ions and local pH play a relevant role in the hydrolysis process. Biomaterials used as scaffolds can be synthetic or natural. Several synthetic polymers are used for implant use, such as PGA or PLA. Pure materials lack property versatility and exhibit deficiencies in clinical practice. Combination of several biomaterials to produce new composites should provide new properties. In recent years, adequate combination of natural polymers and synthetic material are of great interest, for example the association of collagen with hydroxyapatite.

12.4 Hydroxyapatite

Hydroxyapatite is a biologically active calcium phosphate ceramic with a structural similarity to the mineral parts of natural bone. Its main characteristic is the ability to stimulate the bone growth, for this reason has been widely used in orthopedic surgery. However, it has some mechanical limitations such as a poor tensile strength and weak wear resistance, thus it is an insufficient load-carrying material [28]. The incorporation of Mg^{2+} ions induces a reduction in crystal size and an increase in solubility [29]. The effect of low crystallinity is high reactivity of bone apatite, which is reflected in bone-resorption processes.

12.5 Collagen

Collagen is a key component of the extracellular matrix that provides the tensile strength needed to tissue biomechanical features. Physically, collagen forms a rod-like triple helix. The fibers are cross-linked, and this characteristic provides mechanical extracellular matrix strength and integrity. The number of cross-links influences the tensile strength and elasticity of the tissue. The collagen biocompatibility can be correlated with its physical and structural properties along its immunogenicity and natural turnover. In recent years has been describes as the collagen triple helix can interact with many molecules and enzymes involved in its synthesis and degradation.

Nowadays, 26 different types of collagen have been described. Their chemical structures condition mechanical properties. For example, collagen Type II and XI have a strong torsional stability and extensibility and they are the main type of collagen presents in articular cartilage. In contrast, the highly cross-linked structure is typical of collagen type I, the key component of bone tissue.

12.6 Collagen-Hydroxyapatite Bone Substitutes

Skeletal bones consist primarily of collagen (mainly type I) and carbonate substituted hydroxyapatite, both osteoconductive components. Therefore, an implant fabricated from these components shall have similar characteristics. In fact, the scaffolds that structurally and compositionally resemble the natural bone would be an ideal candidate for bone regeneration [30]. Actually, both collagen type I and hydroxyapatite individually were found to improve osteoblast differentiation [31], while mixed together, the result was an acceleration of osteogenesis. A composite matrix when combined with human-like osteoblast cells, displayed better osteoconductive properties than monolithic hydroxyapatite producing calcification of equal bone matrix [32, 33]. Collagen-hydroxyapatite composites resulted to be biocompatible and performed mechanically better than individual components [33, 34]. The ductile properties of collagen contribute to increase the poor fracture toughness of hydroxyapatites. Polymeric scaffolds can take up to two years to degrade while collagen-hydroxyapatite has a more reasonable degradation rate [35]. Moreover, in relation to adherence in vitro to collagen surface, osteogenic cells performed better than PLLA (poly L-lactic acid) and PGA (polyglycolic acid) implants [36]. The results of the comparison between ceramic scaffolds and ceramic composite scaffolds, was that collagen-hydroxyapatite composites performed better than single hydroxyapatite or TCP scaffolds [19]. Indeed, the collagen added to a ceramic structure can bring a number of further advantages to surgical applications: control of shape, adaptation to space, increased particle and defect wall adhesion, and the ability to favor clot formation and stabilization [34]. Therefore, the combination of collagen and hydroxyapatite should offer advantages compared to other materials for the utilization in bone tissue repair.

In addition, as several studies have showed, a multitude of doping ions (i.e., carbonate, magnesium), substituting either calcium or phosphate ions in the crystal lattice, describe the formation of new bone [37–39]. The achievement of a desirable osteogenic microenvironment is dependent on physical and chemical properties of the implanted biomaterial, including the rate of release of metal ions. A favorable bone regeneration could potentially be achieved by facilitating sustained release of ions from biodegradable materials [40]. Moreover, incorporation of inorganic or metal ions offers advantages over the use of expensive and fragile polypeptide-based growth factors [41]. This phenomenon is deemed to be the main cause of structural issue in the mineral bone component, and it increases the chemical reactivity and dissolution ability. Moreover, it keeps a good affinity with osteoblast cells. A recent study [42] demonstrated that zinc silicate/nanohydroxyapatite/collagen (ZS/HA/Col) scaffolds considerably improve bone regeneration and angiogenesis *in vivo* in comparison with hydroxyapatite/Col scaffolds. Zinc is an important factor for bone formation and mineralization and its deficiency impairs skeletal development and can lead to osteoporosis [43].

In another study [44], a novel intrafibrillar mineralized collagen-hydroxyapatite-based scaffolds, constructed in either cellular and lamellar microstructures were established through a biomimetic method to enhance the new bone-regenerating capability of tissue engineering scaffolds. Further, two of the crucial trace elements in the body, iron and manganese have been successfully integrated into the lamellar scaffold to further increase the osteoinductivity of such biomaterials. The authors found that lamellar structure have superior osteogenic skills compared to both in-house and commercial collagen-hydroxyapatite-based cellular scaffolds in vitro and *in vivo*. Iron/manganese incorporation enlarged further the osteogenetic promotion of lamellar scaffolds. The device used in a recent study by Giorgi et al. [45] is a commercially available, porous, 3-dimensional composite bone substitute composed by type I collagen fibers (from equine source) in which nanosized (10–20 nm) crystals of biomimetic magnesium-doped hydroxyapatite (Mg-HA) are nucleated at a 40–60% ratio. The combined device is construed in order to reproduce the anatomical structure of the bone compartment as it occurs in the biological process of neo-ossification [46].

12.7 Biotechnology

These materials are made by an assembling and mineralization procedure simulating the reaction of events leading to the creation of new bone on a living model [47], using equine-derived type 1 collagen, undergone to self-assembly and simultaneous mineralization with an apatite nanophase. In this intricate procedure of biomineralization an extracellular matrix (ECM) -mimicking matrix acts as an active template for the deposition of the mineral phase and is also capable to direct mineral deposition and limit crystal growth. The start of these control mechanisms happens through the connecting of collagen functional groups (e.g., carbonyl groups) with calcium ions, which are then the nucleation sites for the mineral part. The highly controlled chemical-physical interaction between the inorganic (hydroxyapatite) and organic (type I collagen) phase is essential to permitting the creation of a compound material (bone) with unique characteristics of stiffness (minerals) and elasticity (collagen) [48, 49]. The information collected in the organic phase (type I collagen) drives the mineralization process toward the development of nanostructured apatite platelets orientated along the long axis of the collagen, which is a feature considered to promote osteoblasts' adhesion [50] (Fig. 12.1). Between the doping ions taking part to this process, Mg^{2+} ions, partially substituting calcium ions, are associated with the first fast stages of new bone formation. The presence of Mg^{2+} ions intensifies the nucleation kinetic of the novel mineral bone constituent while postponing the crystallization process, thus making it very active during remodeling [48]. For the purpose of manufacturing a 3D fibrous mineralized construct which shows a very high degree of mimicry of the natural bone tissue, the assembling and simultaneous mineralization of type I collagen fibrils with biomimetic magnesium-hydroxyapatite in aqueous media is provided, thus obtaining a newly developed 3D scaffold for bone renewal made of magnesium-doped hydroxyapatite/type I collagen (MHA/Coll) (Fig. 12.2).

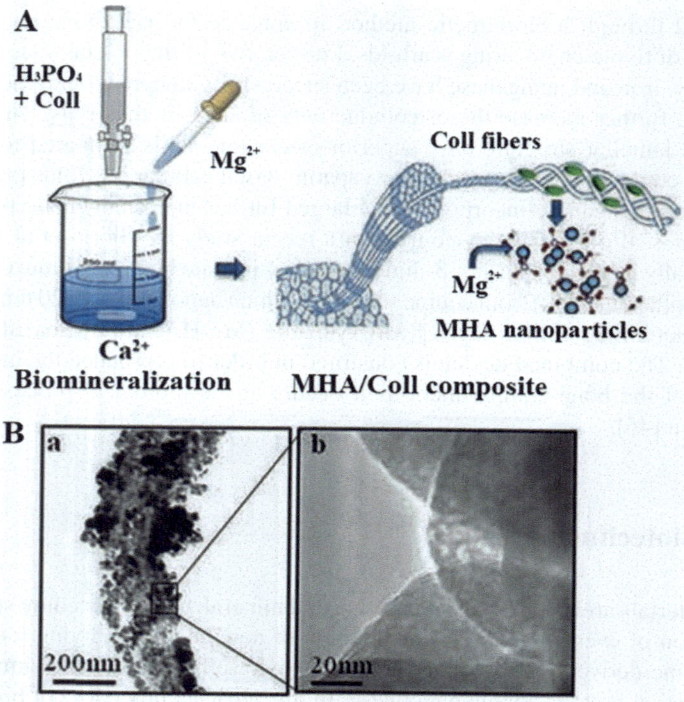

Fig. 12.1 **A** representation of biomineralization process used in the synthesis of magnesium-doped hydroxyapatite/type I collagen scaffold. **B** Transmission electron microscopy (TEM) analysis: Collagen fiber completely covered with magnesium-doped hydroxyapatite (MHA) nanoparticles **a** and detail of the disordered crystalline structure of magnesium-doped hydroxyapatite nanoparticles **b**

Fig. 12.2 The magnesium-doped hydroxyapatite/type I collagen 3D scaffold

The ability of this type of biomimetic bone graft substitutes to stimulate bone regeneration and to support new bone formation has been confirmed in preliminary animal studies [51, 52], showing good results and representing an effective alternative to ICBG in the orthopedic and traumatological field, especially for those surgical procedures that requires large amounts of bone substitute.

12.8 Pearls and Pitalls

12.8.1 Spine Surgery

More than 200.000 spine fusion procedure are performed each year in the United States [53]. Modern approaches in spinal surgery have taken advantage from the integration of classical biomechanical and pathophysiological concepts with effective and readily available prosthetic systems. The biomechanical performance of the different constructs has been considerably enhanced thanks to the usage of new materials like titanium and tantalum. Biomimetic scaffolds have attracted interest for their potential in spinal fusion applications. Biomimetic 3D collagen/hydroxyapatite scaffolds offer a robust and minimally invasive option to obtain vertebral fusion providing a structured environment to promote osteogenesis. In traumatological and degenerative spine posterior surgical procedures, these kind of bone substitutes can be associated to the rod/screw systems and intersomatic cages in order to obtain the postero-lateral or anterior arthrodesis, especially in extended approaches. Today vertebral trauma is still a significant causa of morbidity. In many cases the possibility of having bone grafts substitutes can be a game changer. No bone graft substitute is better than autologous bone graft. Nowadays, the increase of the complex spine procedures needed a most demand for high quality of bone substitutes. A basic understanding of the biology of healing in different types of spinal fusion and the differences between various bone graft substituted can help surgeons to choose the better graft substitute. As described above, autologous bone material, harvested through stripping of the laminae to obtain a certain number of lamellae is associated to limited availability of bone. Moreover, the grafts obtained from the iliac crest are an expensive procedure not only in term of time, but also in terms of morbidity of the donor site. Complications like post-op pain, hematoma, infection and blood loss (25–30%) limits its use [5]. An additional alternative is using allograft bone but this procedure can be associated with disease transmission, bacterial contamination or host-related reaction [54]. The use of magnesium-doped hydroxyapatite/type I collagen 3D scaffold [45] is effective to achieve extensive spinal fusion in the surgical treatment of adult scoliosis and can be safely mixed with autologous bone (Figs. 12.3 and 12.4). The scaffold shows physicochemical, structural, morphological and ultrastructural features similar to the newly formed bone tissue (Figs. 12.5 and 12.6) [55]. This biomimetic scaffold demonstrated great potential for osteogenic activities both in vitro and in vivo mouse model, showing high rates of blood vessels

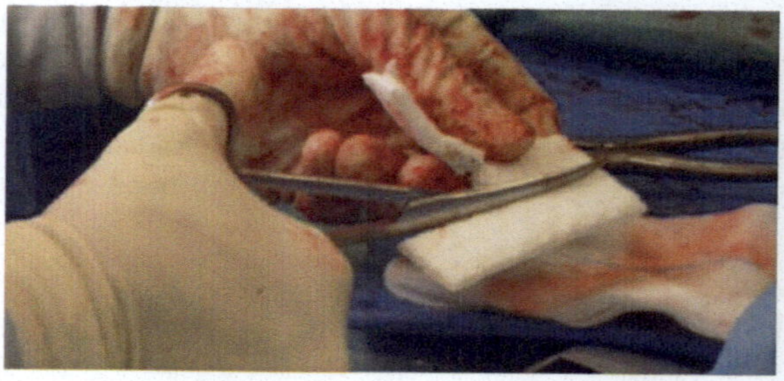

Fig. 12.3 Intraoperative image showing the malleability and flexibility of the magnesium-doped hydroxyapatite/type I collagen 3D scaffold. The device can be cut and shaped according to the surgeon's needs

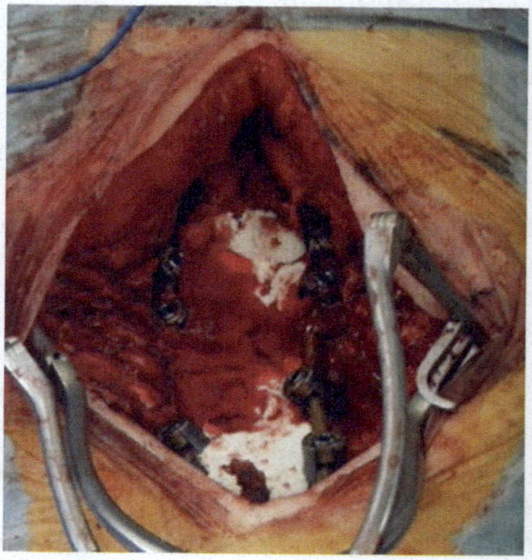

Fig. 12.4 Intraoperative imagine showing the use of the magnesium-doped hydroxyapatite/type I collagen 3D scaffold in association with local autologous bone graft for the postero-lateral fusion in the adult scoliosis surgery

ingrowth and bone formation with successful fusion. The characteristics of such new scaffold have been described by Grigolo et al. [56] that reported the case of a long posterolateral fusion for treatment of adult scoliosis with sagittal imbalance. One year later during the revision surgery, a histological examination found the whole osteointegration of the graft (Fig. 12.7).

12.8.2 Septic Bone Disease

Bone and joint infections are a major complication in orthopedic and trauma surgery [25]. The presence of implanted biomaterials or foreign bodies strongly increases the risk of local infection, due to the ability of bacteria to adhere on implant surfaces and to form protective biofilm [57–59]. The chronic bone infections are probably the most difficult condition to treat because the infected bone is partially or totally avascular, so the antibiotics are difficulty blood-delivered. Moreover, the infected

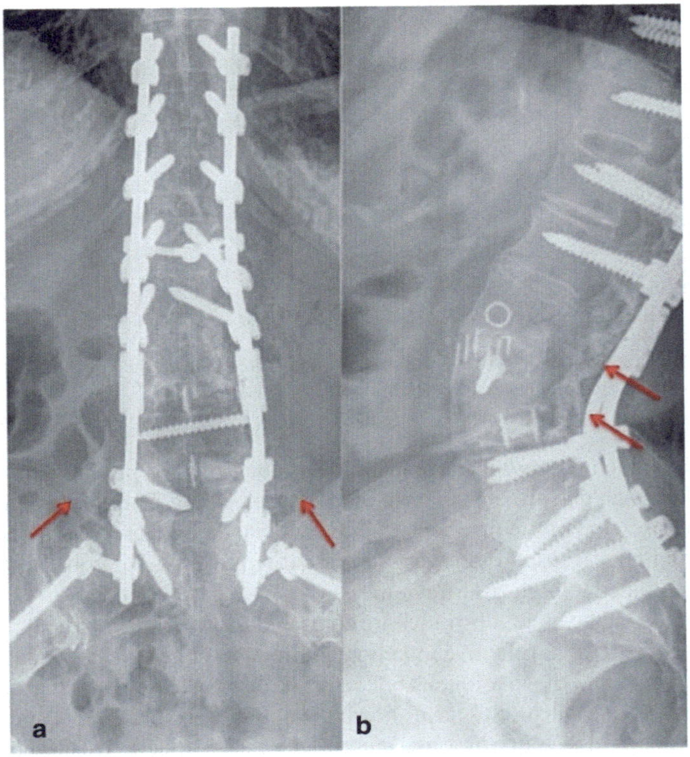

Fig. 12.5 Posterior **a** and Lateral **b** X-ray at 36 months of follow-up showing successful postero-lateral fusion using magnesium-hydroxyapatite/collagen scaffolds, as confirmed by the presence of mature bony trabeculae (arrows), in a patient who underwent T10 ileum fusion

Fig. 12.6 Intraoperative imagine of postero-lateral fusion area at 30 months (case of spine revision surgery in adult scoliosis) that shows a complete osteointegration of hydroxyapatite/type I collagen 3D scaffold

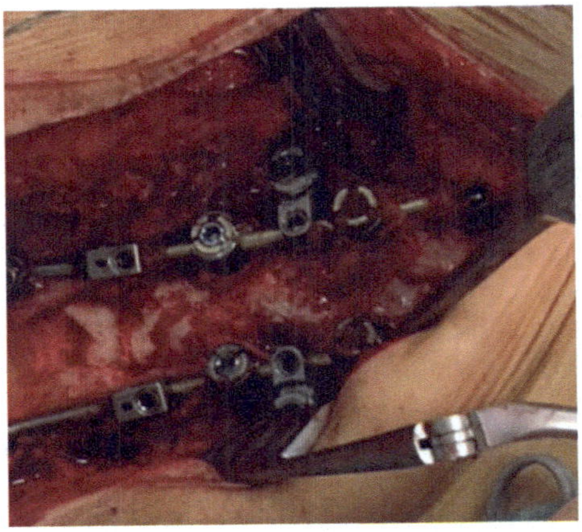

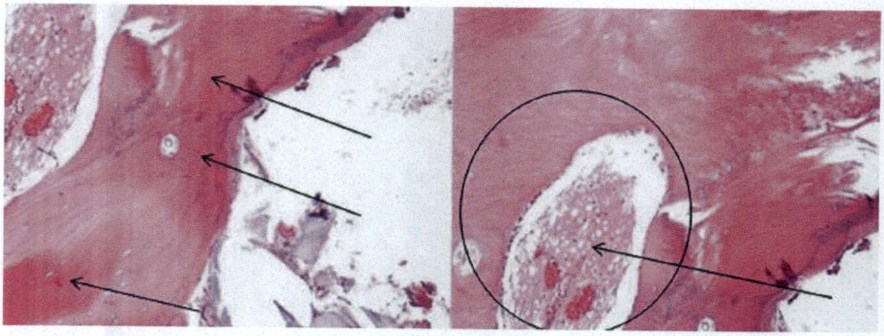

Fig. 12.7 Histological imagines showing newly formed bone tissue marked by purple-red staining (black arrows) and the complete osteointegration at 30 months post-op

bone is surrounded by fibrous tissue with low blood flow and inflammatory characteristics, thus unable to protect the bone or to induce healing. For this reason, in these cases, large excision of the infected structures is mandatory for a correct treatment. The resulting bone gaps is often very important requiring a difficult and extensive grafting associated to high risk of recurrence.

In fact, most of the bone substitutes are contraindicated in the treatment of osteitis. In this critical situations effective infection control and bone defects management is very challenging and the autologous bone graft remains the gold standard. The new generations of bone substitutes have shown the ability to be have as local carriers of antibiotics, acting at the same time as a bone filler, an osteoconductive biomaterial and a local antibacterial [10, 60, 61]. The most studied of the biocompatible carriers so far has been collagen, especially the bovine collagen. The gentamicin has been added to the collagen matrix, resulting in the so-called gentamicine containing collagen implants (GCCI) [62].

It was reported that the introduction of gentamycin in the collagen/hydroxyapatite composite resulted in an increase in the therapeutic effect of the drug due to the possibility of its slow release over a long period of time. Parent et al. determined the profile of vancomycin release from hydroxyapatite scaffolds that were treated with different concentrations of this antibiotic [63]. In this study, all the analyzed scaffolds showed good bactericidal properties against *Staphylococcus Aureus*. For this reason, the ternary system containing hydroxyapatite, collagen and vancomycin seems to be an interesting solution [64]. In a study by Egawa et al., vancomycin/hydroxyapatite/collagen composites showed higher adsorption rates and antibacterial activity in rats' model [65]. In another study [66], the authors investigated the vancomycin release and bone stimulant properties in the treatment of osteomyelitis. The porous heparinized nanohydroxyapatite/ collagen granules and vancomycin showed the ability to eliminate bacteria achieving bone regeneration. The magnesium-doped hydroxyapatite/type I collagen 3D scaffold preloaded with vancomycin or with vancomycin and meropenem shows a rate of infection control comparable with other existing methods of local antibiotic delivery but has a higher tolerability even with implanted biomaterials.

12.9 Summary

The goal of bone grafting is the restoration of functional bone, not only the filling of a gap. For this reason, is a complex process which success depends on the host and the accurate choice of the properly graft for the single case. The most actual tendency is that of obtaining and using BG with properties as close to those of autograft. The new generation magnesium-doped hydroxyapatite/type I collagen 3D scaffold is an effective and safe bone substitute and seems to be useful also in spinal fusion and orthopedic infection's field without adverse and inflammatory reactions.

Future researches are needed to clarify the mechanisms of osteointegration and local tissue responses with the aim of improving this model of biomaterial engineering.

References

1. Giannoudis PV, Dinopoulos H, Tsiridis E (2005) Bone substitutes: an update. Injury 36(Suppl 3):S20–S27
2. Langer R, Vacanti JP (1993) Tissue engineering. Science 260:920–926
3. Dimitriou R, Jones E, McGonagle D et al (2011) Bone regeneration: currnt concepts and future directions. BMC Med 9:66
4. Geiger M, Li RH, Friess W (2003) Collagen sponges for bone regeneration with rhBMP-2. Adv Drug Deliv Rev 55:1613–1629
5. Park JJ, Hershman SH, Kim YH (2013) Updates in the use of bone grafts in the lumbar spine. Bull Hosp Jt Dis 71:39–48
6. Carlson GA, Dragoo JL, Samimi B et al (2004) Bacteriostatic properties of biomatrices against common orthopaedic pathogens. Biochem Biophys Res Commun 321:472–478
7. Silber JS, Anderson DG, Daffner SD et al (2003) Donor site morbidity after anterior iliac crest bone harvest for single-level anterior cervical discectomy and fusion. Spine 28:134–139
8. Uemura T, Dong J, Wang Y et al (2003) Transplantation of cultured bone cells using combinations of scaffolds and culture techniques. Biomaterials 24:2277–2286
9. Gupta A, Kukkar N, Sharif K et al (2015) Bone graft substitutes for spine fusion: a brief review. World J Orthop 6:449–456
10. Campana V, Milano G, Pagano E et al (2014) Bone substitutes in orthopaedic surgery: from basic science to clinical practice. J Mater Sci Mater Med 25:2445–2461
11. Miyazaki M, Tsumura H, Wang JC et al (2009) An update on bone substitutes for spinal fusion. Eur Spine J 18:783–799
12. Alsaleh KA, Tougas CA, Roffey DM et al (2012) Osteoconductive bone graft extenders in posterolateral thoracolumbar spinal fusion: a systematic review. Spine 37:E993–E1000
13. Lerner T, Bullmann V, Schulte TL et al (2009) A level-1 pilot study to evaluate of ultraporous beta-tricalcium phosphate as a graft extender in the posterior correction of adolescent idiopathic scoliosis. Eur Spine J 18:170–179

14. Dai LY, Jiang LS (2008) Single-level instrumented posterolateral fusion of lumbar spine with beta-tricalcium phosphate versus autograft: a prospective, randomized study with 3-year follow-up. Spine 33:1299–1304
15. Linovitz RJ, Peppers TA (2002) Use of an advanced formulation of beta-tricalcium phosphate as a bone extender in interbody lumbar fusion. Orthopedics 25(5 Suppl):s585–s589
16. Muschik M, Ludwig R, Halbhübner S et al (2001) Beta-tricalcium phosphate as a bone substitute for dorsal spinal fusion in adolescent idiopathic scoliosis: preliminary results of a prospective clinical study. Eur Spine J 10(Suppl 2):S178–S184
17. Gao C, Deng Y, Feng P et al (2014) Current progress in bioactive ceramic scaffolds for bone repair and regeneration. Int J Mol Sci 15:4714–4732
18. Zhou H, Lee J (2011) Nanoscale hydroxyapatite particles for bone tissue engineering. Acta Biomater 7:2769–2781
19. Wang X, Grogan SP, Rieser F et al (2004) Tissue engineering of biphasic cartilage constructs using various biodegradable scaffolds: an *in vitro* study. Biomaterials 25:3681–3688
20. Clarke KI, Graves SE, Wong ATC et al (1993) Investigation into the formation and mechanical properties of a bioactive material based on collagen and calcium phosphate. J Mater Sci: Mater Med 4:107–110
21. O'Brien FJ, Harley BA, Yannas IV et al (2004) Influence of freezing rate on pore structure in freeze-dried collagen-GAG scaffolds. Biomaterials 25:1077–1086
22. Kikuchi M, Matsumoto HN, Yamada T et al (2004) Glutaraldehyde cross-linked hydroxyapatite/collagen self-organized nanocomposites. Biomaterials 25:63–69
23. Nishida J, Shimamura T (2008) Methods of reconstruction for bone defect after tumor excision: a review of alternatives. Med Sci Monit 14:RA107–RA113
24. Holt G, Murnaghan C, Reilly J et al (2007) The biology of aseptic osteolysis. Clin Orthop Relat Res 460:240–252
25. Romanò CL, Romanò D, Logoluso N et al (2011) Bone and joint infections in adults: a comprehensive classification proposal. Eur Orthop Traumatol 1:207–217
26. Larsson S, Hannink G (2011) Injectable bone-graft substitutes: current products, their characteristics and indications, and new developments. Injury 42(Suppl 2):S30–S34
27. Altman GH, Diaz F, Jakuba C et al (2003) Silk-based biomaterials. Biomaterials 24:401–416
28. White AA, Best SM, Kinloch IA (2007) Hydroxyapatite-carbon nanotube composites for biomedical applications: a review. Int J Appl Ceram Technol 4:1–13
29. Yoshikawa H, Myoui A (2005) Bone tissue engineering with porous hydroxyapatite ceramics. J Artif Organs 8:131–136
30. Liu Y, Kim YK, Dai L et al (2011) Hierarchical and non-hierarchical mineralisation of collagen. Biomaterials 32:1291–1300
31. Xie J, Baumann MJ, McCabe LR (2004) Osteoblasts respond to hydroxyapatite surfaces with immediate changes in gene expression. J Biomed Mater Res A 71:108–117
32. Wang RZ, Cui FZ, Lu HB et al (1995) Synthesis of nanophase hydroxyapatite/collagen composite. J Mater Sci Lett 14:490–492
33. Serre CM, Papillard M, Chavassieux P et al (1993) *In vitro* induction of a calcifying matrix by biomaterials constituted of collagen and/or hydroxyapatite: an ultrastructural comparison of three types of biomaterials. Biomaterials 14:97–106
34. Scabbia A, Trombelli L (2004) A comparative study on the use of a HA/collagen/chondroitin sulphate biomaterial (Biostite) and a bovine-derived HA xenograft (Bio-Oss) in the treatment of deep intra-osseous defects. J Clin Periodontol 31:348–355
35. Johnson KD, Frierson KE, Keller TS et al (1996) Porous ceramics as bone graft substitutes in long bone defects: a biomechanical, histological, and radiographic analysis. J Orthop Res 14:351–369
36. El-Amin SF, Lu HH, Khan Y et al (2003) Extracellular matrix production by human osteoblasts cultured on biodegradable polymers applicable for tissue engineering. Biomaterials 24:1213–1221
37. Celotti G, Tampieri A, Sprio S et al (2006) Crystallinity in apatites: how can a truly disordered fraction be distinguished from nanosize crystalline domains? J Mater Sci Mater Med 17:1079–1087

38. Landi E, Tampieri A, Celotti G (2005) Nucleation of biomimetic apatite in synthetic body fluids: dense and porous scaffold development. Biomaterials 26:2835–2845
39. Bigi A, Foresti E, Gregorini R et al (1992) The role of magnesium on the structure of biological apatites. Calcif Tissue Int 50:439–444
40. Chen Z, Klein T, Murray RZ et al (2016) Osteoimmunomodulation for the development of advanced bone biomaterials. Mater Today 19:304–321
41. O'Neill E, Awale G, Daneshmandi L et al (2018) The roles of ions on bone regeneration. Drug Discov Today 23:879–890
42. Song Y, Wu H, Gao Y et al (2020) Zinc silicate/nano-hydroxyapatite/collagen scaffolds promote angiogenesis and bone regeneration via the p38 MAPK pathway in activated monocytes. ACS Appl Mater Interfaces 12:16058–16075
43. Suzuki T, Katsumata S, Matsuzaki H et al (2016) A short-term zinc-deficient diet decreases bone formation through down-regulated BMP2 in rat bone. Biosci Biotechnol Biochem 80:1433–1435
44. Yu L, Rowe DW, Perera IP et al (2020) Intrafibrillar mineralized collagen-hydroxyapatite-based scaffolds for bone regeneration. ACS Appl Mater Interfaces 12:18235–18249
45. Giorgi P, Capitani D, Sprio S et al (2017) A new bioinspired collagen-hydroxyapatite bone graft substitute in adult scoliosis surgery: results at 3-year follow-up. J Appl Biomater Funct Mater 15:e262–e270
46. Barbanti Brodano G, Griffoni C, Zanotti B et al (2015) A post-market surveillance analysis of the safety of hydroxyapatite-derived products as bone graft extenders or substitutes for spine fusion. Eur Rev Med Pharmacol Sci 19:3548–3555
47. Tampieri A, Celotti G, Landi E et al (2003) Biologically inspired synthesis of bone-like composite: self-assembled collagen fibers/hydroxyapatite nanocrystals. J Biomed Mater Res A 67:618–625
48. Minardi S, Corradetti B, Taraballi F et al (2015) Evaluation of the osteoinductive potential of a bio-inspired scaffold mimicking the osteogenic niche for bone augmentation. Biomaterials 62:128–137
49. Tampieri A, Sprio S, Sandri M et al (2011) Mimicking natural bio-mineralization processes: a new tool for osteochondral scaffold development. Trends Biotechnol 29:526–535
50. Weiner S (2008) Biomineralization: a structural perspective. J Struct Biol 163:229–234
51. Kon E, Delcogliano M, Filardo G et al (2010) Orderly osteochondral regeneration in a sheep model using a novel nano-composite multilayered biomaterial. J Orthop Res 28:116–124
52. Kon E, Filardo G, Delcogliano M et al (2010) Platelet autologous growth factors decrease the osteochondral regeneration capability of a collagen-hydroxyapatite scaffold in a sheep model. BMC Musculoskelet Disord 11:220
53. France JC, Yaszemski MJ, Lauerman WC et al (1999) A randomized prospective study of posterolateral lumbar fusion. Outcomes with and without pedicle screw instrumentation. Spine 24:553–560
54. Delloye C, Cornu O, Druez V et al (2007) Bone allografts: what they can offer and what they cannot. J Bone Joint Surg Br 89:574–579
55. Sprio S, Sandri M, Panseri S et al (2012) Hybrid scaffolds for tissue regeneration: chemotaxis and physical confinement as sources of biomimesis. J Nanomater 2012:e418281. https://doi.org/10.1155/2012/418281
56. Grigolo G, Dolzani P, Giannetti C et al (2016) Use of a fully-resorbable, biomimetic composite hydroxyapatite as bone graft substitute for posterolateral spine fusion: a case report. Int J Clin Exp Med 9:22458–22462
57. Gristina AG, Shibata Y, Giridhar G et al (1994) The glycocalyx, biofilm, microbes, and resistant infection. Semin Arthroplasty 5:160–170
58. Gristina AG (1987) Biomaterial-centered infection: microbial adhesion versus tissue integration. Science 237:1588–1595
59. Gristina AG, Costerton JW (1984) Bacterial adherence and the glycocalyx and their role in musculoskeletal infection. Orthop Clin North Am 15:517–535

60. Hanssen AD (2005) Local antibiotic delivery vehicles in the treatment of musculoskeletal infection. Clin Orthop Relat Res. https://doi.org/10.1097/01.blo.0000175713.30506.77
61. Zalavras CG, Patzakis MJ, Holtom P (2004) Local antibiotic therapy in the treatment of open fractures and osteomyelitis. Clin Orthop Relat Res. https://doi.org/10.1097/01.blo.0000143571.18892.8d
62. Noah EM, Chen J, Jiao X et al (2002) Impact of sterilization on the porous design and cell behavior in collagen sponges prepared for tissue engineering. Biomaterials 23:2855–2861
63. Parent M, Magnaudeix A, Delebassée S et al (2016) Hydroxyapatite microporous bioceramics as vancomycin reservoir: antibacterial efficiency and biocompatibility investigation. J Biomater Appl 31:488–498
64. Suchý T, Šupová M, Klapková E et al (2016) The sustainable release of vancomycin and its degradation products from nanostructured collagen/hydroxyapatite composite layers. J Pharm Sci 105:1288–1294
65. Egawa S, Hirai K, Matsumoto R et al (2020) Efficacy of antibiotic-loaded hydroxyapatite/collagen composites is dependent on adsorbability for treating staphylococcus aureus osteomyelitis in rats. J Orthop Res 38:843–851
66. Coelho CC, Sousa SR, Monteiro FJ (2015) Heparinized nanohydroxyapatite/collagen granules for controlled release of vancomycin. J Biomed Mater Res A 103:3128–3138

Pietro Domenico Giorgi Dr. Giorgi is an orthopaedics and traumatology specialist and consultant spine surgeon. Dr. Giorgi also serves as the director of the spine center: traumatic and degenerative disease at ASST Grande Ospedale Metropolitano Niguarda, Emergency and Urgency Department, Orthopedic and Traumatology Unit. In addition, Dr. Giorgi is also a member of AOSpine Europe. Dr. Giorgi is skilled in traumatic vertebral fracture and spinal cord injuries as well as an expert in anterior surgical approaches and minimally invasive spinal surgeries.

Giuseppe Rosario Schirò Dr. Schirò is an orthopaedics and traumatology specialist as well as a consultant spine surgeon at ASST Grande Ospedale Metropolitano Niguarda, Emergency and Urgency Department, Orthopedic and Traumayology Unit. Dr. Schirò is skilled in traumatic vertebral fracture and spinal cord injuries.

Simona Legrenzi Dr. Legrenzi is an orthopaedics and traumatology specialist as well as a consultant spine surgeon at ASST Grande Ospedale Metropolitano Niguarda, Emergency and Urgency Department, Orthopedic and Traumayology Unit.

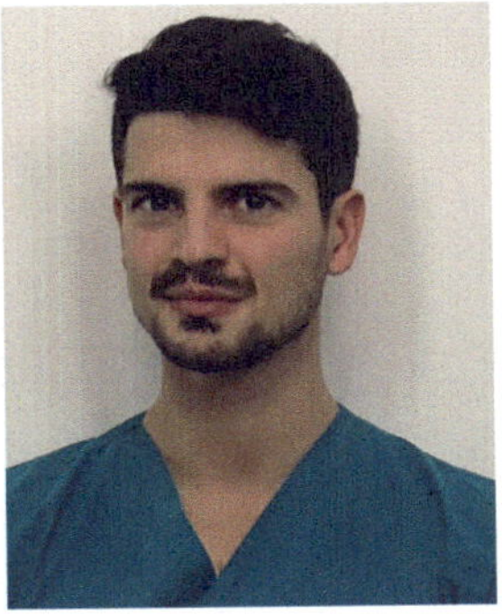

Francesco Puglia Dr. Puglia is a fellow resident at ASST Grande Ospedale Metropolitano Niguarda, Emergency and Urgency Department, Orthopedic and Traumatology Unit.

Index

© The Editor(s) (if applicable) and The Author(s), under exclusive license to Springer Nature Singapore Pte Ltd. 2022
A. H. Choi and B. Ben-Nissan (eds.), *Innovative Bioceramics in Translational Medicine II*, Springer Series in Biomaterials Science and Engineering 18,
https://doi.org/10.1007/978-981-16-7439-6

The manufacturer's authorised representative in the EU is Springer Nature Customer Service Centre GmbH, Europaplatz 3, 69115 Heidelberg, Germany. If you have any concerns regarding our products, please contact ProductSafety@springernature.com

Printed and bound by CPI Group (UK) Ltd, Croydon, CR0 4YY
23/09/2026
02221546-0006